Reiner Westermeier,
Tom Naven
Proteomics in Practice

Related title from Wiley-VCH

Reiner Westermeier

Electrophoresis in Practice
Third Edition

ISBN 3-527-30300-6

Proteomics in Practice

A Laboratory Manual of Proteome Analysis

Reiner Westermeier,
Tom Naven

Dr. Reiner Westermeier
Dr. Tom Naven
Amershan Biosciences Europe GmbH
Munzinger Str. 9
79111 Freiburg
Germany

This book was carefully produced. Nevertheless, authors and publisher do not warrant the information contained therein to be free of errors. Readers are advised to keep in mind that statements, data, illustrations, procedural details or other items may inadvertently be inaccurate.

Library of Congress Card No.: applied for

British Library Cataloguing-in-Publication Data:
A catalogue record for this book is available from the British Library.

Die Deutsche Bibliothek – CIP Cataloguing-in-Publication Data:
A catalogue record is available from Die Deutsche Bibliothek.

© Wiley-VCH Verlag-GmbH
Weinheim, 2002

All rights reserved (including those of translation in other languages).
No part of this book may be reproduced in any form – by photoprinting, microfilm, or any other means – nor transmitted or translated into a machine language without written permission from the publisher.
Registered names, trademarks, etc. used in this book, even when not specifically marked as such, are not to be considered unprotected by law.

printed in the Federal Republic of Germany
printed on acid-free paper.

Composition Kühn & Weyh, Software GmbH, Freiburg
Printing Druckhaus Darmstadt GmbH, Darmstadt
Bookbinding J. Schäffer GmbH & Co. KG, Grünstadt

ISBN 3-527-30354-5

Contents

Preface

The objective of Proteomics in Practice is to provide the reader with a comprehensive reference and manual guide for the successful analysis of proteins by 2-D electrophoresis and mass spectrometry. The idea for the book has come from the continuing success and favourable responses received from the scientific public for our on-going proteomics seminar and practical courses we have delivered in the past twelve months.

The book will include a theoretical introduction, comprehensive practical section complete with worked examples, a unique troubleshooting section designed to answer many of the frequently asked questions regarding proteome analysis and a thorough reference list to guide the interested reader to further detail.

The theoretical section will introduce the fundamentals behind the techniques currently being used in proteomics today and describe how the techniques are used for proteome analysis.

However, the practical aspects of the book will not address many of these methods, but will instead focus on the main stream methodology of 2-D electrophoresis and mass spectrometry. 2-D electrophoresis is still the most successful method of resolving a proteome with increasing reproducibility and automation. All aspects for the successful performance of 2-D electrophoresis and image analysis will be addressed in practical detail. Subsequently, the importance of mass spectrometry, sequence databases and search engines for successful protein identification will be discussed. The practical section of the book is in principle a course manual, which has been optimized over a number of years. The success of the "Electrophoresis in Practice" book range, has demonstrated that a course manual is a useful guide for daily work in the laboratory. The section will describe how to achieve good, reliable and reproducible results using a single instrumental setup, instead of presenting a wide choice of techniques and instruments. In this book some statements may be found, which do not comply with the "high end" technological achievements published. The experimental procedures are restricted to the area of robustness and routinely achievable good results.

The authors understand and wholly appreciate that the analysis of post-translational modifications such as phosphorylation and glycosylation is an integral aspect of proteomics. As such the theoretical, technical and practical issues involved will be addressed in great detail in a subsequent edition. Approaches for functional proteomics are still varying and many procedures are under development. These methods will be added in a later edition.

As the technical developments in this field are proceeding so fast, the contents of the book need to be updated every few months. The reader can have access to a web-site at WILEY-VCH: http://www3.interscience.wiley.com/XXXXXX, which will contain the updated chapters and recipes.

Reiner Westermeier
Tom Naven
January 2002

Thanks to:
Jan Axelsson, Tom Berkelman, Philippe Bogard, Josef Bülles, Maria Liminga, Tom Keough, Matrixscience.com, Staffan Renlund, Günter Thesseling.

Foreword

Proteomics is in an extraordinary growth phase. This is due to a great extent to the fact that the major undertaking of sequencing the human and other important genomes has largely been accomplished, which has opened the door for proteomics by providing a sequence-based framework for mining the human proteome and that of other organisms. It is evident that proteomics has attracted a substantial following, with an influx of investigators and of biotechnology and pharmaceutical companies that are taking an active interest in the field, as well as an influx of a new generation of scientists in training. There is undeniably a pressing need for training in proteomics and much need for textbooks that facilitate the use of related methodology. This book makes a valuable contribution by providing a clear presentation of some of the most widely utilized methods in this field.

The field of proteomics can be divided in practice into three major areas: expression proteomics, functional proteomics and proteome related bioinformatics. This book focuses primarly on methodology utilized for expression proteomics, an important component of proteomics which deals with global quantitative analysis and identification of proteins encoded in genomes and expressed to a varied extent in different tissues and cell populations. Expression proteomics relies on a mix of, on the one hand, high-tech approaches and on the other, a harvest of know-how in protein chemistry and biochemistry gained over the past half-century. Although the face of proteomics is evolving right before our eyes, it is likely that some fundamental technologies will remain in use for many years to come. This is likely to be the case for the technologies and methodologies covered in this book, namely 2-D methods and related mass spectrometry techniques for protein identification, with which the field of proteomics has been tightly associated in the past decade.

Evidently, 2-D gels have come under assault lately, due in part to the influx of new investigators to the field, most of whom have no particular leanings towards 2-D gels and consider the lack of automation and the limited sensitivity of 2-D gels as major drawbacks. While non-2-D gel based approaches such as protein microarrays and multidimensional liquid based separations, and peptide (as opposed to protein) profiling are making some inroads, a technology that is clearly superior to 2-D gels for global proteome profiling has yet to emerge. Clearly, industry is after a robust "industrial strength" proteomic platform to achieve high throughput, which 2-D gels with their limited automation at the present time do not adequately provide. However a more pressing concern of most investigators contemplating using 2-D gels is how to overcome the difficulties and challenges notoriously associated with this technique. Indeed, much "art" is needed to produce quality 2-D gels, which in the past has limited the successful use of this technique to a privileged few. Numerous "tricks" need to be learned, for which this text is quite valuable. Equally, from a mass spectrometry point of view, although spectacular progress has been made in this field with the development of instrumentation that is highly performing and much more user friendly than in the past, much needs to be mastered for the optimal utilization of mass spectrometry. There remains much challenge in protein identification, which this book is intended to facilitate, particularly for those entering the field.

A valuable contribution of this book is the manner in which the methodologies used for 2-D gels and mass spectrometry have been integrated. It is rare to find scientists with expertise in mass spectrometry that are also knowledgeable in the practical aspects of successfully producing quality 2-D gels. The combined backgrounds of the two authors is ideally suited to provide readers with a comprehensive and expert presentation of methodology utilized to successfully combine 2-D gels and mass spectrometry. In my own laboratory, I predict that this book will make a valuable contribution to the training of graduate students, post-doctoral fellows and technologists that are joining the laboratory. As the field is constantly evolving, the planned frequent updates would be valuable to keep the text current.

Sam Hanash MD, PhD
Department of Pediatrics
A 520 MSRBI
University of Michigan
Ann Arbor MI 48109
USA

Abbreviations, symbols, units

1-D electrophoresis	One-dimensional electrophoresis
2-D electrophoresis	Two-dimensional electrophoresis
A	Ampere
A,C,G,T	Adenine, cytosine, guanine, thymine
AEBSF	Aminoethyl benzylsulfonyl fluoride
API	Atmospheric pressure ionization
APS	Ammonium persulfate
AU	Absorbance units
16-BAC	Benzyldimethyl-n-hexadecylammonium chloride
BAC	Bisacryloylcystamine
Bis	N, N′-methylenebisacrylamide
BLAST	Basic local alignment search tool
bp	Base pair
BPB	Bromophenol blue
BSA	Bovine serum albumin
C	Crosslinking factor [%]
CAF	Chemically assisted fragmentation
CAM	Co-analytical modification
CAPS	3-(cyclohexylamino)-propanesulfonic acid
CBB	Coomassie brilliant blue
CCD	Charge-coupled device
CHAPS	3-(3-cholamidopropyl)dimethylammonio-1-propane sulfonate
CE	Capillary electrophoresis
CID	Collision induced dissociation
conc	Concentrated
CM	Carboxylmethyl
CMW	Collagen molecular weight marker
const.	Constant
CTAB	Cetyltrimethylammonium bromide
Da	Dalton
DB	Database

DBM	Diazobenzyloxymethyl
DDRT	Differential display reverse transcription
DEA	Diethanolamine
DEAE	Diethylaminoethyl
DGGE	Denaturing gradient gel electrophoresis
2,5-DHB	2,5-dihydroxybenzoic acid
Disc	Discontinuous
DMF	Dimethyl formamide
DMSO	Dimethylsulfoxide
DNA	Desoxyribonucleic acid
dpi	dots per inch
DTE	Dithioerythreitol
DTT	Dithiothreitol
E	Field strength in V/cm
EDTA	Ethylenediaminetetraacetic acid
ESI	Electrospray ionization
EST	Expressed sequence tag
FAB	Fast atom bombardment
FT-ICR	Fourier transform – Ion cyclotron resonance
GLP	Good laboratory practice
GMP	Good manufacturing practice
h	Hour
HCCA	α-cyano-4-hydroxycinnamic acid
HeNe	Helium neon
HEPES	N-2-hydroxyethylpiperazine-N′-2-ethananesulfonic acid
HMW	High Molecular Weight
HPLC	High Performance Liquid Chromatography
I	Current in A, mA
ICAT	Isotope coded affinity tags
IEF	Isoelectric focusing
IgG	Immunoglobulin G
IPG	Immobilized pH gradients
ITP	Isotachophoresis
kB	Kilobases
kDa	Kilodaltons
LC	Liquid chromatography
LMW	Low Molecular Weight
LOD	Limit of detection
LWS	Laboratory workflow system
M	mass
mA	Milliampere
MALDI	Matrix assisted laser desorption ionization
min	Minute
mol/L	Molecular mass per liter

mr	Relative electrophoretic mobility
mRNA	messenger RNA
MS	Mass spectrometry
MS^n	Tandem mass spectrometry where n is greater than 2
MS/MS	Tandem mass spectrometry
M_r	relative molecular mass
m/z	mass/charge ratio (x-axis in a mass spectrum)
Nonidet	Non-ionic detergent
NEPHGE	Non equilibrium pH gradient electrophoresis
NHS	*N*-hydroxy succinimide
NR	Non redundant
O.D.	Optical density
P	Power in W
PAG	Polyacrylamide gel
PAGE	Polyacrylamide gel electrophoresis
PAGIEF	Polyacrylamide gel isoelectric focusing
PBS	Phosphate buffered saline
PEG	Polyethylene glycol
pI	Isoelectric point
pK value	Dissociation constant
PMF	Peptide mass fingerprint
PMSF	Phenylmethyl-sulfonyl fluoride
PPA	Piperidino propionamide
ppm	parts per million (measure of mass accuracy)
PSD	Post source decay
PTM	Post-translational modification
PVC	Polyvinylchloride
PVDF	Polyvinylidene difluoride
QTOF	quadrupole time-of-flight
r	Molecular radius
Rf value	Relative distance of migration
Rm	Relative electrophoretic mobility
RNA	Ribonucleic acid
RP	Reversed Phase
rpm	revolutions per minute
RuBPS	Ruthenium II tris (bathophenanthroline disulfonate)
s	Second
SDS	Sodium dodecyl sulfate
S/N	Signal/noise ratio
T	Total acrylamide concentration [%]
t	Time, in h, min, s
TBS	Tris buffered saline
TCA	Trichloroacetic acid

TEMED	N,N,N′,N′-tetramethylethylenediamine
THPP	Tris(hydroxypropyl)phosphine
ToF	Time of Flight
Tricine	N,tris(hydroxymethyl)-methyl glycine
Tris	Tris(hydroxymethyl)-aminoethane
U	Volt
V	Volume in L
v	Speed of migation in m/s
v/v	Volume per volume
W	Watt
w/v	Weight per volume (mass concentration)

Glossary of terms

Term	*Definition*
Adduct peak	Results from the photochemical breakdown of the matrix into more reactive species, which can add to the polypeptide. Can also result from salt ions, Na^+ etc., that are embedded in the matrix.
Analytical 2-D electrophoresis	Proteins are loaded in amounts of 10 to 100 µg. Mostly broad pH intervals are used in the first dimension.
Average molecular weight <2093.8> average molecular weight 	The mass of a molecule of a given empirical formula calculated using the average atomic weights for each element. An average mass is obtained in MALDI-TOF-MS when a peak is not isotopically resolved (see mono-isotopic molecular weight).
Background subtraction	The process in which the background (chemical and detector noise) is subtracted, leaving the peaks above the noise at the base level.
Base peak	The most intense peak in a mass spectrum. A mass spectrum is usually normalized so that this peak has an intensity of 100%.
Calibrant	A compound used for the calibration of an instrument.
Calibration	A process where known masses are assigned to selected peaks. The purpose is to improve the mass accuracy of an MS instrument.

Term	*Definition*
Centroided mass peak	The centroided mass peak is located at the weighted centre of mass of the profile peak.
Collision induced dissociation (CID)	A process whereby an ion of interest, the precursor ion, is selected, isolated, excited and fragmented by collisions with an inert gas within the mass spectrometer.
Dalton (Da)	According to the guidelines of the SI, the use of the term Dalton for 1.6601×10^{-27} kg is no longer recommended. However it is still a current unit in biochemistry.
Daughter ion (see product or fragment ion)	An ion resulting from CID performed on a precursor ion during a product ion MS/MS spectrum.
Digestion	Cleavage of subject protein by proteolytic enzymes, including trypsin and chymotrypsin.
Dried droplet method	Sample preparation method for MALDI-TOF MS applicable to peptides, protein digests and full length proteins.
Electrospray ionisation (ESI)	An ionisation technique, which enables the formation of ions from molecules directly from samples in solution. The ions formed in this process are predominantly multiply charged. Commonly coupled with analysers capable of tandem mass spectrometry (MS/MS). It is readily coupled with HPLC or capillary electrophoresis.
Electroendosmosis	In an electric field, fixed charges on the gel matrix or on a glass surface are attracted by the electrode of opposite sign. As they are fixed, they cannot migrate. This results in a compensation by the counterflow of H_3O^+ ions towards the cathode for negative or OH^- ions towards the anode for positive charges. In gels, this effect is observed as a water flow.
Expression proteomics	The massive parallel study of highly heterogeneous protein mixtures with high throughput techniques like 2-D electrophoresis and MALDI mass spectrometry.
External calibration	A calibration is performed with a known calibration mixture. The resultant calibration constants (file) are then applied to a separate sample.
Fragment ion (product or daughter ion)	See product ion.

Term	*Definition*
Fragmentation	A physical process of dissociation of molecules into fragments in a mass spectrometer. The resultant spectrum of fragments is unique to the molecule or ion. Fragmentation data can be used to sequence peptides and resultantly provide data for protein identification.
Full length protein	An intact polypeptide chain, constituting a protein in its native or denatured state. The molecular weight of which can be determined accurately with MALDI and ESI MS.
Functional proteomics	This research is only possible with non-denatured cell extracts and requires different tools than 2-D electrophoresis and MALDI MS. A smaller subset of proteins is isolated from the highly heterogeneous protein lysate and analysed with mild techniques that do not affect protein complexes and three-dimensional structures.
Immobilized pH gradients	Polyacrylamide gels, which contain an in-built pH gradient, created by acrylamide derivatives, which carry acidic and basic buffering groups. Because an immobilized pH gradient is absolutely continuous, narrow pH intervals can be prepared, which allow unlimited resolution.
In-gel digestion	The embedded protein in the gel is cleaved using enzymes of known specificity. During the process, peptides are formed, which are extracted from the gel for subsequent analysis.
Internal calibration	Calibration where known masses in each spectrum are used to calibrate that spectrum. Greater mass accuracy than an external calibration.
Ion detector.	A detector that amplifies and converts ions into an electrical signal
Ion gate	Typically an electrical deflector that permits certain ions through to later stages of ion optics (open), or deflects unwanted ions out of the way of the later stages of the mass spectrometer. Commonly used in post-source decay (PSD) analysis for the selection of a precursor ion.
Ion source	Region of the mass spectrometer where gas phase ions are produced.
Ion transmission efficiency	Refers to the fraction of the ions produced in the source region that actually reaches the detector.
Ionisation	The process of converting a sample molecule into an ion in a mass spectrometer.

Term	Definition
Isoelectric point (pI)	The pH value where the net charge of an amphoteric substance is zero. Because the pK values of buffering groups are temperature-dependent, this is valid also for the pI. The pI of a protein that can be measured.
Isotope	Atoms of the same element having different mass numbers due to differences in the number of neutrons.
Isotope abundance	The relative amount in nature of certain atomic isotopes.
Laboratory workflow system	Database and computer network for the integrated laboratory to control the entire workflow in the proteomics factory.
Linear time-of-flight mass analyser	Simplest TOF analyser, consisting of a flight tube with an ion source at one end and a detector at the other.
Mass accuracy	The ability to assign the actual mass of an ion. This is typically expressed as an error value.
Mass analyser	Second part of the mass spectrometer, separating the ions forms in the ion source according to their m/z value. Examples of mass analysers include ion trap, quadrupole, time-of-flight and magnetic sector.
Mass range	The area of interest to be measured in an experiment. Or the capability of the analyser
Mass spectrometer	An instrument that measures the mass to-charge ratio (m/z) of ionized atoms or molecules. Comprises three parts: an ion source, a mass analyser, and an ion detector.
Mass spectrometry (MS)	A technique for analysing the molecular weight of molecules based upon the motion of a charged particle in an electric or magnetic field.
Mass spectrum	A plot of ion abundance (y-axis) against mass-to-charge ratio (x-axis).
Mass-to-charge ratio (m/z)	A quantity formed by dividing the mass of an ion (in Da units) by the number of charges carried by the ion.
Matrix	Necessary for the ionisation of sample molecules by MALDI. A small, organic compound which absorbs light at the wavelength of the laser.
Matrix-assisted laser desorption/Ionisation (MALDI)	Ionisation technique, commonly used for the ionisation of biological compounds. Sample is incorporated into the crystal structure of the matrix and irradiated with the light from a laser.
Metastable ion	An ion that decomposes into fragment ions and/or neutral species, during its passage through the mass spectrometer.

Term	Definition
Monoisotopic molecular weight	The mass of a molecule containing only the most abundant isotopes, calculated with exact atomic weights. With respect to peptide analysis, the monoisotopic peak is the ^{12}C peak, i.e. the first peak in the peptide isotopic envelope.
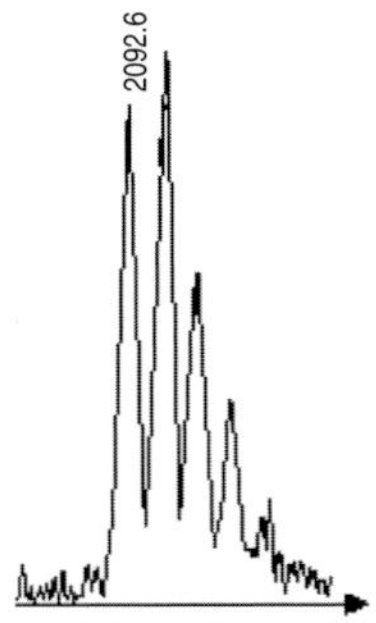 monoisotopic molecular weight	
Multiple-charged ion	Ion possessing more than a single charge. Characteristic of ESI
Normalization	All peaks are reported with peak heights relative to the highest peak height or area in the spectrum.
Neutral loss scan	A type of MS/MS experiment. Useful for the indication of individual components in a complex mixture.
N-terminal amino acid	The amino acid residue at the end of a polypeptide chain containing the free amino group.
Parent ion (precursor ion)	Refers to the peak of an ion that will be selected for fragmentation in a product ion MS/MS or PSD spectrum.
Molecular mass (M_r)	The relative molecular mass is dimensionless. In practice and in publication the dimension Da (Dalton) is used. Particularly in electrophoresis the term "Molecular weight" is frequently used.
Optical Density (O.D.)	The unit O.D. for the optical density is mostly used in biology and biochemistry and is defined as follows: 1 O.D. is the amount of substance, which has an absorption of 1 when dissolved and measured in 1 mL in a cuvette with a thickness of 1 cm.
Peptide-mass fingerprinting	Technique for searching protein databases for protein ID. Subject protein is cleaved and the resultant cleaved peptide masses are used for a database search.
Peak area	The area bounded by the peak and the base line. Can be calculated by integrating the abundances from the peak start to the peak end.
Peak height	The distance between the peak maximum and the baseline.
Peak resolution	The extent to which the peaks of two components overlap or are separated. Compare with FWHM.

Term	*Definition*
Peak width	The width of a peak at a given height.
Phosphate buffered saline (PBS):	140 mol/L NaCl, 2.7 mmol/L KCl, 6.5 mmol/L Na_2HPO_4, 1.5 mmol/L KH_2PO_4, pH 7.4.
Post-source decay (PSD)	A technique describing fragmentation of a precursor ion that occurs in the first field free region of the TOF before the reflectron.
Post-translational modification (PTM)	Modifications of proteins occuring after coding; common examples include phosphorylation and glycosylation. PTM analysis is an integral part of proteomics.
Precursor ion (parent ion)	Ion selected to undergo fragmentation within the mass spectrometer in a product ion MS/MS or PSD spectrum.
Precursor ion scan	A type of MS/MS experiment. Useful for the indication of individual components in a complex mixture.
Preparative 2-D electrophoresis	Proteins are loaded in the lower mg amounts. Mostly narrow pH intervals are used in the first dimension.
Product ion (see daughter ion)	An ion resulting from CID performed on a precursor ion during a product ion MS/MS spectrum.
Product ion scan	The principle MS/MS experiment. Involves the selection of a precursor ion to undergo fragmentation within the MS.
Protein characterization	The identification of structural aspects of a protein. Includes amino acid sequence, molecular weight, three-dimensional structure, post-translational modifications and biologic activity of a particular protein.
Protein sequencing	The determination of the order of amino acids in a subject protein or peptide.
Proteome	The complete profile of proteins expressed in a given tissue, cell or biological system at a given time.
Proteomics	Systematic analysis of the protein expression of healthy and diseased tissues.
Pulsed mass analyser	Includes time-of-flight, ion cyclotron resonance, and quadrupole ion traps. An entire mass spectrum is collected from a single pulse of ions.
Quantification	In many papers the term "quantitation" is used, which is incorrect. In gel electrophoresis only relative quantification is possible, but the adjective "relative" is omitted in most papers on this subject.
Reflectron	Improves resolution of a mass spectrometer by acting act as an ion mirror. Compensates for the distribution of kinetic energy, ions of the same mass experience in the source.

Term	*Definition*
Rehydration	The correct word is "rehydratation". Because this word is a tongue-twister and is used many times for the methodical description of the first dimension separation, the incorrect term "rehydration" is generally used.
Resolution 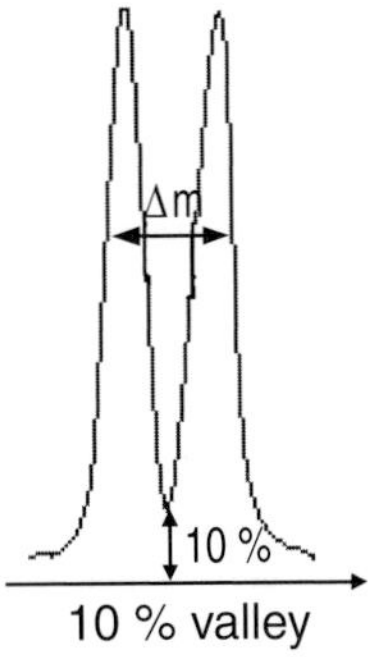	Refers to the separation of two ions where resolution, R= m/Δm. For a single peak made up of singly charged ions at mass m in a mass spectrum, the resolution maybe expressed as m/Δ where Δm is the width of the peak at a height that is a specified fraction of the maximum peak height (for instance full width at half maximum height – FWHM). A second definition for defining resolution is 10% valley. Two peaks of equal height in a mass spectrum at masses m and m-Δm are separated by a valley that at its lowest point is just 10% of the height of either peak.
Resolving power (mass) 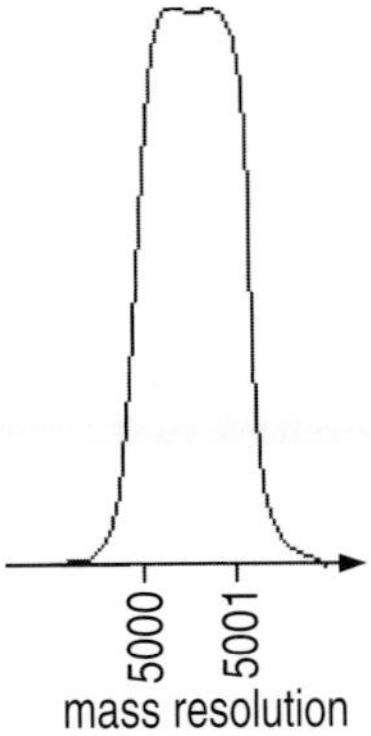	The abilitity to distinguish between ions differing slightly in mass-to-charge ratio.
Sample preparation	A crucial stage to achieve efficient, optimal ionisation.
SDS electrophoresis	Proteins are separated in a polyacrylamide gel as negatively charged protein-detergent micelles. Secondary, tertiary, quaternary structures and the individual charges of the proteins are cancelled. The migration distances of the resulting zones from the origin correlate roughly to the logarithm of the M_r of the respective protein.

Term	Definition
Seeded microcrystalline film Method	A sample preparation method. First, a thin layer of small matrix crystals is formed on the sample slide. Then a droplet containing the analyte is placed on top of this layer. This is left to dry. The deposit is washed before the sample slide is inserted into the mass spectrometer. Is a direct replacement of the dried droplet method.
Signal-to-noise ratio (S/N)	The ratio of the signal height and the noise height. An indication of the sensitivity of an instrument or analysis.
Slow crystallization Method	A sample preparation method. Here, large matrix crystals are grown. The analyte is added to a saturated matrix solution. Microcrystals are formed. The supernatant is removed and a slurry of crystals is made. The slurry is then applied to the sample slide, allowed to dry and inserted into the mass spectrometer. Can be used when the dried droplet method has failed
Tandem mass spectrometry (MS/MS)	Used to elucidate structure within the mass spectrometer. Three types of MS/MS experiment can be performed.
Thin layer method	A sample preparation method. Applying the matrix onto the substrate creates a very thin layer of matrix. A droplet of the analyte is then dried onto this layer. Contaminants can now be washed away before introducing the sample into the mass spectrometer. A variant of the dried droplet method.
Threshold fluence.	The lowest laser fluence at which analyte can be observed in MALDI, fluence for optimal resolution.
Time-ion extraction (time lag focussing or delayed extraction)	Improved resolution is obtained for a specified mass range by applying a controlled delay between ion formation and acceleration, also called delayed extraction.
Time-of-flight (TOF) analyser	Separates ions in time as they travel down a flight tube.
Tuning	The process of optimising the MALDI instrument's laser power and target position to obtain the best possible sensitivity and signal-to-noise ratio for a specific type of experiment.
Two-dimensional electrophoresis	There are more than one possibility to combine two different electrophoretic separation principles. If not further specified, 2-D electrophoresis means isoelectric focusing under denaturing conditions followed by SDS polyacrylamide gel electrophoresis.
Unit resolution	Distinguishes between ions separated by 1 m/z unit.

Part I:
Proteomics Technology

1
Introduction

Now as the complete human genome and many other genomes have been deciphered, we have to face the fact that the genomic sequence and protein function cannot be directly correlated. In a living cell, most activities are performed by proteins. When we want to study the pathways of cell metabolism and identify possible drug targets, we have to analyse the *proteome* (Wasinger *et al.*1995): the composition of all *prot*eins expressed by the gen*ome* of an organism.

Wasinger VC, Cordwell SJ, Cerpa-Poljak A, Yan JX, Gooley AA, Wilkins MR, Duncan MW, Harris R, Williams KL, Humphery-Smith I. Electrophoresis 16 (1995) 1090–1094.

Cell proteomes are very complex, they are composed of several thousand proteins. Studying the proteome of an organism requires an analytical effort beyond the capacity of a standard laboratory equipment. The fast acquisition of the human genome was only possible by the application of an industrial approach. Exactly the same happens now for proteome analysis: most of the data will certainly be delivered by "Proteomics factories".

Because of the complexity of the sample, two-dimensional polyacrylamide gel electrophoresis has been widely utilized as the standard separation and display method. The proteome is not a stable mixture of proteins. It is impossible to display it by running a single two-dimensional electrophoresis gel. Usually multiple samples are produced at different stages of a stimulation, gene deletion or overexpression, or drug treatment experiments, and separated in a number of 2-D gels (see figure 1). Sample treatment and the separation of the proteins are usually performed rapidly to avoid protein modification. Once the proteins are isolated in the gel matrix, the single proteins are much more stable and can be further identified and characterized by mass spectrometry.

Two-dimensional gel electrophoresis does not only have a very high resolution, the gels are very efficient fraction collectors The proteins can be stored in the gels until further analysis without degradation.

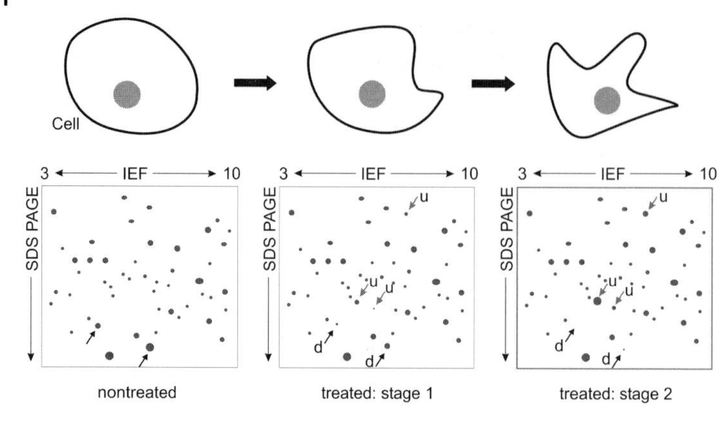

Fig. 1: Schematic representation of a series of 2-D gels showing different stages of a proteome. Up and down regulated gene products are marked with u and d respectively.

By definition *"Proteomics"* is the simultaneous analysis of complex protein mixtures like cell lysates and tissue extracts, to look for quantitative changes of expression levels. The scope of applications extends from drug discovery, to diagnostics, therapy, microbiology, biochemistry, and plant research. It has bee substantially facilitated in the past decade because of developments in mass spectrometry and the availability of genomic information. Developments in proteomics have proceeded in parallel:

2-D maps are no longer operator-dependent.

- The technology for high-resolution *two-dimensional electrophoresis* has been considerably improved, which makes the method more reliable, and reproducible. The resolution has been further increased as well. Image analysis software of these complex spot patterns has been developed to such a degree that also non-computer experts can use it and get reliable results.

Developments in mass spectrometry were a key development for proteome analysis.

- Novel ionization techniques and detectors for *mass spectrometry* have been invented, which allow the analysis of proteins and peptides with high sensitivity, accuracy and throughput. Online peptide fragmentation allows quick amino acid sequence analysis of low amounts of peptides at low running cost. Also the analysis of post-translational modification can be addressed using this technology.

Proteome analysis would be impossible without genome sequence databases.

- *Genomics:* Thanks to the development of high throughput DNA sequencing genomic databases of many different organisms have been established in a short period of time. Genomic sequence data is growing with immense speed. Unfortunately for most genes, the function is unknown. A gene code, for more than one gene product, alternative splicing of the mRNA can result in different proteins. Furthermore most proteins become modified by complex gene interactions, cellular events, and environmental influences that result in post-translational modifications. Knowledge of the DNA sequence of organisms to be analysed is very important for protein identification and characterization with mass spectrometry.

- Activities in the developing field of *bioinformatics* have been initiated to develop tools for combining and bundling the huge amount of data produced by new high throughput analysis methods. Only in this way it will be possible to draw meaningful conclusions from the huge amount of data generated.

Bioinformatics will be utilized as part of the procedures for improved sample tracing tools and good laboratory practice.

Additionally, a few developments in related fields are very helpful for proteomics analysis:

- Biochemistry and biomedical research had focused on studying the structure and function of proteins which have been proposed as key enzymes in related pathways. The technology for micro characterization of proteins has been continuously refined with respect to sensitivity and accuracy. If antibodies were available, identification of a protein was no problem. With de novo sequencing of proteins purified by HPLC or electrophoretic methods databases have been compiled. Those could be used also for cross-species protein identification using amino acid composition and short sequence information. But the techniques applied are relatively slow and allow only low throughput. In most cases tedious and expensive de novo amino acid sequencing is no longer needed. With the available genomic databases it is sufficient to acquire short sequence information for a protein by mass spectrometry to identify it, and to retrieve its complete amino acid sequence with the help of in-silico translation of the open reading frames.

A high level of skills is required for this type of research, particularly in the area of protein chemistry.
The technology for de novo sequencing with mass spectrometry continues to be further developed.

- *Labelling techniques* using fluorescent tags and stable isotopes have been made available for differential analyses of related samples. Quantitative changes in expression levels can thus be easily uncovered.

Isotope labeling requires mass spectrometry analysis.

- *Developments of complementary technology:* For the analysis of smaller protein subsets, the detection of proteins at very low expression levels, determination and assay of protein function many novel methods are under development.

The most important and promising methods will be described later in the overview chapter.

The points listed above describe an approach, which is often also defined as *"Expression Proteomics"* or *"Classical Proteomics"*. With most of the techniques employed, the proteins become denatured: their three-dimensional and quaternary structures are destroyed, informations on functions and complexing of the proteins cannot be determined. For that purpose *"Functional Proteomics"* methods are applied, which do not denature the proteins and keep complexes intact: e.g. affinity chromatography and electrospray ionization mass spectrometry.

1.1 Applications of proteomics

The greatest expectations from proteomics come from pharmaceutical research for faster new drug protein targets identification in transformed cell lines or diseased tissues. Also the validation of the detected targets, in-vitro and in-vivo toxicology studies, and checks for side effects can be performed with this approach. Clinical researchers want to compare normal versus disease samples, disease versus treated samples, find molecular markers in body fluids for diagnosis, monitor diseases and their treatments, determine and characterize post-translational modifications. In clinical chemistry it would be interesting to subtype individuals to predict response to therapy.

However, proteomics must always be the holistic approach. A "small scale" proteomics approach does not exist.

Biologists study basic cell functions and molecular organizations. Another big field is microbiology for various research areas. Proteomics is also applied for plant research for many different purposes, for instance for breeding plants of higher bacterial, heat, cold, drought, and other resistances, increasing the yield of crop and many more. For all this research a combined strategy with genomics is employed.

1.2 Separation of the protein mixtures

Stegemann H. Angew Chem 82 (1970) 640.
O'Farrell PH. J Biol Chem 250 (1975) 4007–4021.
Bjellqvist B, Ek K, Righetti PG, Gianazza E, Görg A, Westermeier R, Postel W. J Biochem Biophys Methods 6 (1982) 317–339.

As already mentioned above, high-resolution two-dimensional electrophoresis is the mainly applied separation technique. The separation according to the two completely independent physico-chemical parameters of the proteins: isoelectric point and size offers the highest resolution. Several thousand proteins can be separated, displayed and stored in one gel. The history of this technique goes back to a paper in german language by Stegemann (1970), combining isoelectric focusing (IEF) and SDS polyacrylamide gel electrophoresis. The resolution of 2-D electrophoresis was considerably increased by the introduction of denaturing conditions during sample preparation and isoelectric focusing by O'Farrell in the year 1975. With this modification the method gained a wide acceptance. But only with the application of immobilized pH gradients (Bjellqvist *et al.* 1982) for the first dimension the technique became reproducible enough for proteome analysis.

Pre-separation of cells into organelles by centrifugation is very useful to reduce the number of proteins, and for localization of a protein within the cell. Also prefractionation according to physico-chemical parameters of proteins like isoelectric points is desirable to enable high sample loads on 2-D gels, and to prevent protein-protein interactions. Because two-dimensional electrophoresis cannot easily be automated, alternatives like multidimensional chromatography and direct mass spectrometry methods are tested and further developed for the separation of these complex protein mixtures.

Two-dimensional electrophoresis will remain the major separation technique for the next decade, because its resolution and the advantage of storing the isolated proteins in the gel matrix until further analysis is unrivalled by any of the alternative techniques.

1.3 Detection

Ideally one protein expressed in a cell should be detectable. With the current state of technology this is completely impossible. With non-radioactive labelling and staining techniques (fluorescence or silver staining, LOD ca. 1 ng protein) down to 100 proteins per cell can be visualized in a 2-D gel, when 10 mg total protein – corresponding for instance to10^8 cell equivalents of lymphoma cells – are loaded on the isoelectric focusing gel (Hoving *et al.* 2000). Mass spectrometry methods are generally more sensitive than these staining methods, however in practice 10 to 20 ng of protein in a spot is required for good signals in mass spectrometry.

Hoving S, Voshol H, van Oostrum J. Electrophoresis 21 (2000) 2617–2621.

In practice Coomassie Brilliant Blue, zinc-imidazol, and fluorescence stained gels used with high throughput mass spectrometry analysis.

1.4 Image analysis

The 2-D patterns are very complex, only with informatics tools it is possible to find expressed changes in a series of gels: like up and down regulated proteins, post-translational modifications. Image analysis has still been the bottleneck in the proteomics procedure, because the spot detection parameters had to be adjusted and optimized manually. Only since a very short time new developed software can perform fully automated and hands-off evaluation. The reliability of quantitative determinations of protein amounts in spots is strongly dependent on the protein detection technique applied. Protein spots of interest are then further analysed.

The position of a spot in the 2-D map is not enough information for an exact identification of a protein. Identification can only be achieved with chemical or mass spectrometry analysis of the protein spot.

1.5 Identification of proteins

Pappin DJ, Hojrup P, BleasbyJA. Curr biol 3 (1993) 327–332.
The techniques and its background will be described in detail in the method chapters.

Peptide mass fingerprinting The easiest and fastest way to identify proteins is shown in figure 2: peptide mass fingerprinting (PMF), which was introduced by four independent groups, including Pappin *et al.* (1993). The gel plug containing the protein of interest is cut out of the gel slab, the protein is digested inside the gel plug with a proteolytic enzyme, mostly trypsin. The cleavage products, the peptides, are eluted from the plug and submitted to mass spectrometry analysis. Mostly MALDI ToF instruments are employed, because they are easier to handle than electrospray systems. The mass spectrum with the accurately measured peptide masses is matched with theoretical peptide spectra in various databases using adequate bioinformatics tools. When no match is found in peptide and protein databases, genomic databases can be searched. The DNA sequence in the open reading frames can be theoretically translated into the amino acid sequence we have to remove this because it is not very practical to search DNA with MALDI data, it is not specific enough. You can do it easily with MS/MS though. Since the cleavage sites of trypsin are known, theoretical tryptic peptide masses can be generated and compared with the experimentally determined masses. If a sufficient number of experimental peptide masses match with the theoretical peptides within a protein, then protein identification with high confidence can be achieved.

This procedure works very well for protein identification. However, the method can be compromised for a number of reasons. In these circumstances, more specific information is needed for unambiguous protein identification, specifically peptide sequence information.

Even for organisms without a genomic database, successful identifications are feasible across species, when there is enough sequence homology in a conserved area of the polypeptide.

Amino acid sequence analysis Amino acid sequence is highly discriminating information for unambiguous protein identification. During mass spectrometry analysis a peptide can be selected from the spectrum and fragmented inside the instrument, termed tandem mass spectrometry. The resultant fragment ion masses are indicative of amino acid sequence and can be used to generate a sequence ladder. Amino acid sequence derived from mass spectrometry can be used to search not only the protein databases, but also the EST databases and used for *de novo* sequencing when necessary.

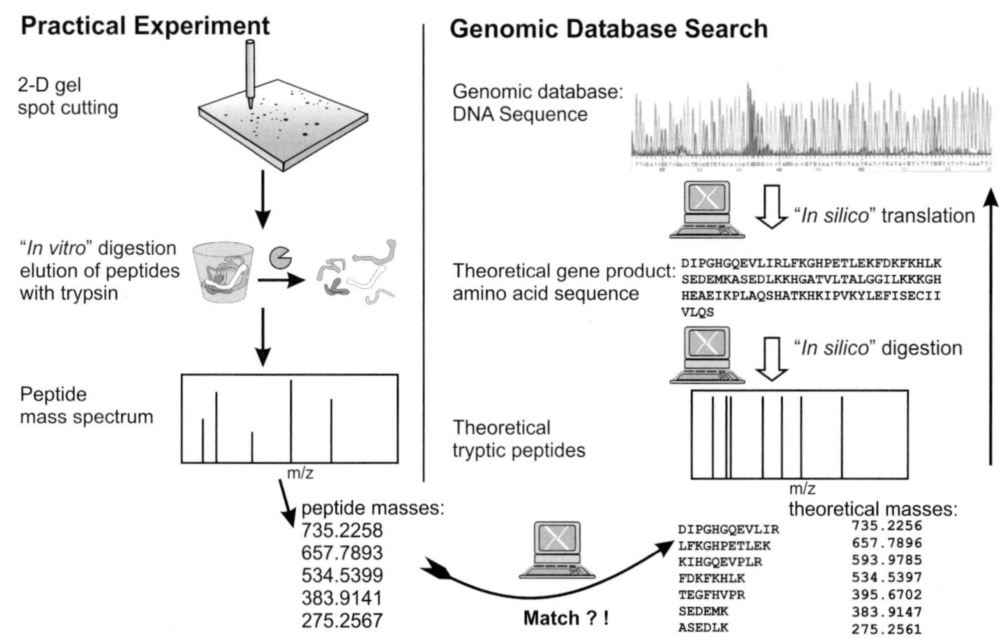

Fig. 2: Protein identification with peptide mass fingerprinting. The peptide masses of the digested protein are matched with a list of theoretical masses of peptides, which are mathematically derived from the open reading frames of the genome database of a certain organism.

1.6
Characterization of proteins

Besides amino acid sequence information also other structural data of a protein can be determined with mass spectrometry: disulfide bonds, post-translational modifications like phosphorylation, truncation, acylation and glycosylation. The identification of sites of disulfide bonds, phosphorylation and glycosylation can all be determined using tandem mass spectrometry, though quantitation of these modifications is considerably more difficult.

However, three-dimensional folding, complexes and functional informations of the proteins are cancelled by the commonly used separation and mass spectrometry methods in structural proteomics (see above).

1.7
Functional proteomics

The following strategy is pursued: At first target proteins are identified, characterized and correlated with "protein families". Once some structural informations are known, smaller subsets of proteins are analysed with milder separation and measuring techniques: for

Lamond AI, Mann M. Trends Cell Biol 7 (1997) 139–142.
Pandey A, Mann M. Nature 405 (2000) 837–846.

instance, some proteins are fished out of a cell lysate with affinity chromatography and then proteins with intact tertiary structure, or protein-protein complexes are analysed after electrospray ionization (see reviews by Lamond and Mann, 1997, and Pandey and Mann, 2000).

2
Expression proteomics

2.1
Two-dimensional Electrophoresis

Only high-resolution 2-D electrophoresis, with both dimensions run under denaturing conditions, is used in proteomics. Native 2-D separations do not play a big role in proteome expression analysis. The goal is to separate and display all gene products present. Unfortunately 2-D electrophoresis cannot meet all challenges completely. Complementary techniques are thus necessary to find the missing proteins.

Challenges

- *Spot number:* Depending on the type of organism and its metabolic state between 10,000 and 150,000 gene products are expected to be present in a cell. Because of different post-translational modifications the real number cannot be predicted from the genomic information. In order to display these proteins, the size of the gels, the sensitivity and the dynamic range of the detection method have to be adequate.

It is impossible to display all proteins in one single gel. Several gels are most probably required for one sample.

- *Isoelectric points spectrum:* According to *in-vitro* translated open reading frames from several organisms the spectrum of expected isoelectric points reaches in general from pH 3 to 13. It is expected that the alterations of pIs caused by post-translational modifications will not cause pIs outside of this range. In practice a pH gradient 3 to 13 does not exist. It is technically problematic to separate and display proteins with pI above pH 11.5.

A possible approach would be a derivatisation of the very basic protein fraction.

- *Molecular weights range:* Small peptides can be analysed by modifying the gel and the buffer of the SDS PAGE method. High molecular weight proteins with sizes above 250 kDa do not go through the 2-D electrophoresis process.

High molecular weight proteins are included when the sample is directly applied on a SDS gel. An 1-D electrophoresis sample can be run in a lane at the side of the 2-D separation.

More chemicals developments are required.

- *Hydrophobic proteins:* Some very hydrophobic proteins do not go in solution, others are lost during sample preparation and IEF.

Silver staining is still the most sensitive method, followed by fluorescence labeling and staining.

- *Sensitivity of detection:* Some of the proteins are expressed in very low copy numbers. When the most sensitive method – radiolabelling – cannot be applied, more protein must be loaded.

The loading capacity of the IEF step can be increased much easier than the capacity of the second dimension.

- *Loading capacity:* For the detection of minor components the protein load must be increased. A wide dynamic range of the SDS gel is required to prevent merging of highly abundant fractions.

Silver staining, for instance, does not give reliable quantitative data.

- *Quantification:* Because proteomics is a method of quantitative comparisons, no proteins should get lost during sample preparation and the analysis process. The detection method must give reliable quantitative information.
- *Reproducibility:* It is of highest importance that the results are reproducible. Otherwise it is impossible to find variations of proteins between different samples.

Note
A good-looking spot pattern – streak and smear free – is not a guarantee for a reproducible procedure.

2.1.1 Evolution of the 2-D methodology

In the original procedure the first dimension, isoelectric focusing, is run in thin polyacrylamide gel rods in glass or plastic tubes. The gel rods contain urea, detergent, reductant and carrier ampholytes to form the pH gradient in the electric field. Usually the sample is loaded onto the cathodal side of the gel rod, which becomes the basic end of the gradient in the electric field.

Anderson NG, Anderson NL. Anal Biochem 85 (1978) 331–354.
Anderson NG, Anderson NL. Clin Chem 28 (1982) 739–748.

This “O'Farrell” technique has been used for about two decades without major modifications. The potential of the method for a systematic approach to create a protein database has been recognized very soon. Anderson and Anderson (1978) have designed instruments and operation procedures to prepare and run multiple 2-D gels under the most reproducible conditions, in order to develop a “Human protein index” (Anderson and Anderson, 1982).

Klose J, Kobalz U. Electrophoresis 16 (1995) 1034–1059.

Gel sizes Gels with about 20 × 20 cm have become standard for an adequate spatial resolution. If one assumes, that up to 100 bands can be resolved in a 20 cm long one-dimensional gel, a theoretical separation space of 10,000 proteins could be reached in such a gel. In practice several thousands of proteins can be detected. Very few groups only manage to handle giant gel sizes like 40 × 30 cm (Klose and Kobalz, 1995).

Detection methods The highest sensitivity is reached with radioactivity / fluorography detection. This is only possible with living cells. In most cases the proteins are stained after the separation in the gel. A number of different methods are applied, with big differences in sensitivity, linearity, and dynamic range. According to calculations of Hoving *et al.* (2000) down to 100 copies per cell can be detected with sensitive silver staining in a gel.

Hoving S, Voshol H, van Oostrum J. Electrophoresis 21 (2000) 2617–2621.

Problems with the first dimension Originally the pH gradients for isoelectric focusing for 2-D electrophoresis were created by *carrier ampholytes*. Those are mixtures of a few hundred different homologues of amphoteric buffers, which are synthesized together in one reaction flask. The mixtures contain buffers with isoelectric points evenly distributed over a wide spectrum from pH 3 to 10. Their special property is a high buffering power at their isoelectric point. At the beginning all carrier ampholytes are charged, because they exist as a mixture. When an electric field is applied, they start to migrate according to their charges towards the anode or the cathode respectively, and form automatically stable pH gradients between the electrodes. When they reach their pI, they lose their charges and buffer the pH value of their pI. This works very well for native separations. But, because denaturing conditions are applied for the high-resolution 2-D technique, long migration times are needed, which lead to a destabilization of the gradient.

The soft, thin, and long gel rods demand great experimental skill. Between pH 6.7 and 7 there is very often a lack of buffering power, leading to mechanical instability of the thin gel rods, and an empty area in the spot pattern.

As long as the pH gradient for the first dimension step was generated with carrier ampholytes, the handling of the technique was rather cumbersome, and the patterns were not reproducible enough. Batch to batch variations of the carrier ampholyte mixtures were the reasons for differences in the profiles of the pH gradients. Gradient drift with prolonged isoelectric focusing time lead to losses of almost all basic and some of the acidic proteins.

2-D electrophoresis had sometimes been seen as a technique, which produces operator-dependent results. Because of technical limitations, mainly acidic proteins could be studied in the past.

Basic proteins A remedy had soon been found for the display of basic proteins: O'Farrell *et al.* (1977) have introduced a modification of the first dimension run: NEPHGE (non-equilibrium pH gradient electrophoresis). Here the sample is loaded onto the acidic end of the gel and the proteins are separated while the gradient drifts towards the cathode. The run is stopped after a defined time period. However, due to the time factor it is hard to achieve a good reproducibility. The proteins are not focused like in IEF; the resolution is limited by the number of different carrier ampholyte homologues.

O'Farrell PZ, Goodman HM, O'Farrell PH. Cell 12 (1977) 1133–1142.

The proteins are not focused, but stacked between the different carrier ampholyte homologues.

Bjellqvist B, Ek K, Righetti PG, Gianazza E, Görg A, Westermeier R, Postel W. J Biochem Biophys Methods 6 (1982) 317–339.
Righetti PG. In: Burdon RH, van Knippenberg PH. Ed. Immobilized pH gradients: theory and methodology. Elsevier, Amsterdam (1990).

Immobilized pH gradients To overcome these problems, immobilized pH gradients had been developed as an alternative to carrier ampholytes by Bjellqvist *et al.* (1982). The pH gradients in these gels are prepared by co-polymerizing acrylamide monomers with acrylamide derivatives containing carboxylic and tertiary amino groups. Because the buffering groups, forming the pH gradient, are fixed, the gradient cannot drift and is not influenced by the sample composition. Comprehensive informations on immobilized pH gradients and detailed explanations of the principle are found in the book by Righetti (1990).

Strahler JR, Hanash SM, Somerlot L, Weser J, Postel W, Görg A. Electrophoresis 8 (1987) 165–173.
Görg A, Postel W, Günther S. Electrophoresis 9 (1988) 531–546.
Görg A, Obermaier C, Boguth G, Harder A, Scheibe B, Wildgruber R, Weiss W. Electrophoresis 21 (2000) 1037–1053.

In practice, these immobilized pH gradient gels are cast in a similar way like porosity gradient gels. Thin slabs are co-polymerized with a film support, whose surface had been treated to covalently bind the gels. According to the procedure introduced by Angelika Görg, the sample proteins are separated in individual strips, which have been cut from these film-supported gel slabs (Strahler *et al.* 1987, and Görg *et al.* 1988, 2000).

Blomberg A, Blomberg L, Norbeck J, Fey SJ, Larsen PM, Roepstorff P, Degand H, Boutry M, Posch A, Görg A. Electrophoresis. 16 (1995) 1935–1945.

The use of immobilized pH gradients in the first dimension has allowed many methodical innovations for 2-D electrophoresis, which will be described later. Since ten years, pre-manufactured gel strips and dedicated instruments are available as commercial products. This was the prerequisite to make 2-D electrophoresis a reproducible and reliable method. An inter-laboratory comparison has demonstrated the high reproducibility of the method (Blomberg *et al.* 1995).

First, mainly the wide gradients pH 3–10 and 4–7 have been utilized. After the highly abundant and acidic proteins have been identified and characterized, the analysis of lowly expressed and basic proteins come into focus. Therefore new types of gel strips with narrow and basic pH gradients are developed. A narrow pH gradient strip combines two advantages: it has a higher loading capacity and allows increased spatial resolution. The running conditions, however, become more critical for these special gradients. Separations in narrow gradients require long focusing times. The stability of some proteins at their isoelectric point can become a problem, particularly in alkaline conditions. But immobilized pH gradients are the only reliable and reproducible technique to resolve and display basic proteins.

2.1.2
Sample preparation

The sample treatment is the key to adequate results. The protein composition of the cell lysate or tissue must be reflected in the pattern of the 2-D electrophoresis gel without any losses or modifications. Of course, the sample must not be contaminated with proteins and peptides not belonging to the sample. "Co-analytical modifications" (CAM) of proteins must be avoided. This is not trivial, because various protein-protein interactions may happen in such a complex mixture, and pre-purification of the sample can lead to uncontrolled losses of some of the proteins.

With higher protein loads sample preparation becomes even more critical, because also the contaminants concentrations are increased.

It should be mentioned, that the way, how samples are collected, is very important. It is relatively easy for body fluids and lysates from cultured cells, because the proteins are evenly distributed. However, when tissue material is studied, the cells to be analysed must be well defined. A highly selective procedure for tissue analysis is laser capture micro dissection (Banks *et al.* 1999).

Banks R. Dunn MJ, Forbes MA, Stanly A, Pappin DJ, Naven T, Gough M, Harnden P, Selby PJ. Electrophoresis 20 (1999) 689–700.

To avoid protein losses, the treatment of the sample must be kept to a minimum; to avoid protein modification, the sample should be kept as cold as possible; to avoid losses and modifications, the time should be kept as short as possible.

Too much salt in the sample disturbs isoelectric focusing and leads to streaky patterns. Amphoteric buffers in cell cultures, like HEPES, overbuffer the gradient in the areas of their pIs, which results in vertical narrow areas without protein spots. The chemicals used must be of the highest purity.

The proteins have to be extracted from cells or tissue material. Liquid samples have to be denatured to prevent the formation of polypeptide oligomers, aggregates and interactions. Some material contains proteolytic enzymes, which are still active under denaturing conditions.

Denaturing conditions

A frequently expressed wish of proteomics researchers is the possibility to separate the protein mixtures under native conditions, in order to conserve the three-dimensional structures of the proteins. There are a number of reasons, why high molar urea, a reductant, and nonionic or zwitterionic detergents must be present:

- Under native conditions, a great part of the proteins exists in several different conformations. This would lead to even more complex 2-D pattern, which could not be evaluated. The proteins have to be denatured in order to display them in single conformations.

In presence of more than 7 mol/L urea most polypeptides exist only in one single configuration.

- Different oxidation steps must be prevented by the addition of a reducing agent.
- All proteins, also the hydrophobic ones, have to be brought into solution. Because buffers and salt ions would disturb isoelectric focusing, a noncharged chaotrope, urea, and detergents have to be used for the solubility of all proteins.

Also the addition of carrier ampholytes supports the solubility of hydrophobic proteins.

- Protein aggregates and complexes would be too big to enter a gel matrix. They are prevented by the addition of high urea concentrations and reduction of the sample.
- Protein-protein interactions are avoided by applying denaturing conditions.
- For the analysis of the proteome, it should be possible to match the theoretically calculated isoelectric point of a polypeptide with pI position in the 2-D map. This is only possible, when the three-dimensional structure is cancelled and all buffering groups are exposed to the medium. Bjellqvist *et al.* have shown, that this is possible with immobilized pH gradients.

In the three-dimensional configuration some buffering groups are hidden, which causes a shift of the pI.
Bjellqvist B, Hughes GJ, Pasquali C, Paquet N, Ravier F, Sanchez J-C, Frutiger S, Hochstrasser D. Electrophoresis 14 (1993) 1023–1031.

The proteins are therefore extracted with a so called "lysis buffer". The major tasks of the lysis buffer and its additivies are:

- Convert all proteins into single conformations.
- Cancel different oxidations steps.
- Prevent protein aggregates.
- Prevent protein modifications.
- Get hydrophobic proteins into solution and keep them in solution.
- Deactivate proteases.
- Cleave disulfide and hydrogen bonds; uncoil the polypeptides to expose all buffering groups to the medium.

Composition of the standard "lysis buffer":

9 mol/L urea, 4 % CHAPS, 1 % DTT, 0.8 % carrier ampholytes, 0.002 % bromophenol blue.

Urea The high urea concentration is needed to convert proteins into single conformations by cancelling the secondary and tertiary structures, to get and keep hydrophobic proteins in solution, and to avoid protein-protein interactions. Only in special cases a second, stronger denaturating chaotrope has to be added to urea in order to increase the solubility of very hydrophobic proteins, like membrane proteins: thiourea. The purity of urea is very critical: Isocyanate impurities and heating must be avoided, because these would cause carbamylation of the proteins, resulting in artifactual spots.

Urea is not stable in solution, repeated freeze – thawing must be avoided.

Detergent CHAPS is a zwitterionic detergent, and preferred to non-ionic polyol mixtures like Triton X-100 and Nonidet NP-40, because of its higher purity. It increases the solubility of hydrophobic proteins.

Mass spectrometry is particularly sensitive to contaminations coming from detergents like Triton X-100.

Reducing agents The reductants DTT or DTE prevent different oxidation steps of the proteins. Both are interchangeable. 2-mercaptoethanol should not be used, because of its buffering effect above pH 8 (Righetti *et al.* 1982). Keratin contaminations, seen with mass spectrometry, have been traced back to contaminated 2-mercaptoethanol by Parness and Paul-Pletzer in (2001). Unfortunately DTT as well as DTE can become ionized above their pK of 8 and migrate towards the anode during IEF in basic pH gradients. This leads to horizontal streaks coming from some spots in the basic area, due to inadequate protection of the cysteins.

Righetti PG, Tudor G, Gianazza E. J Biochem Biophys Methods 6 (1982) 219–22.
Parness J, Paul-Pletzer K. Anal Biochem 289 (2001) 98–99.

Tributylphosphine, introduced by Herbert *et al.* (1998), is generally not recommended as a replacement because of its poor stability. As a remedy Herbert *et al.* (2001) have then proposed alkylation of proteins prior to IEF, particularly for separations in basic pH gradients. However, this is not a good solution, because complete alkylation of all proteins in a complex mixture is not easy to control. This results in additional artifactual spots. The pIs of the basic proteins become modified, which is another disadvantage.

Herbert BR, Molloy MP, Gooley AA, Walsh BJ, Bryson WG, Williams KL Electrophoresis 19 (1998) 845–851.
Herbert B, Galvani M, Hamdan M, Olivieri E, MacCarthy J, Pedersen S, Righetti PG. Electrophoresis 22 (2001) 2046–2057.

An alternative way to adequate and reproducible 2-D pattern in basic gradients is suggested by Hoving *et al.* (2001): the addition of higher amounts of DTT to the gel, adding more DTT to a cathodal paper strip, and a few more measures described later in the isoelectric focusing chapter.

Hoving S, Gerrits B, Voshol H, Müller D, Roberts RC, van Oostrum J. Proteomics 2 (2002) 127–134.

Carrier ampholytes Carrier ampholytes, which had been designed for generating pH gradients, improve the solubility of proteins considerably by substituting ionic buffers. In a mixture they are charged. They do not disturb IEF like buffer addition, because they migrate to

Note: The composition of carrier ampholytes or IPG buffers used will influence the result.

their pIs, where they become uncharged. Dedicated pH intervals, prepared for the addition to immobilized pH gradients, are called *IPG buffers*.

This should be checked in a practical optimization experiment for a different sample type.

Various IPG buffer mixtures are designed for the respective pH gradients. Sometimes in practice, the use of carrier ampholyte mixtures for wide gradients, like Pharmalyte pH 3 – 10, instead of dedicated IPG buffers, has shown better results for narrow gradients.

Bromphenol blue can be interchanged with Orange G.

Dyes The anionic dye Bromophenol blue is very useful as a control for the start and running conditions. The low amounts used do not disturb the analysis.

Protease inhibitors

Unfortunately there is no complete insurance against protease activities.

Dunn MJ. Gel electrophoresis of proteins. Bios Scientific Publishers Alden Press, Oxford (1993).

Some proteases are also active in presence of urea and detergents. Protease inhibitors can inactivate most of the proteolytic activities, however in some cell lysates not completely. *PMSF* is frequently used (8 mmol/L), but it is a toxic compound. It has to be added to the sample prior to the reductant, because it would become deactivated. *Pefabloc* (AEBSF) applied as 5 – 10 mmol/L is less toxic, but might lead to charge modifications of some proteins (Dunn, 1993). This can also happen with application of protease "cocktails". Some proteases are inhibited by the denaturing conditions, some by basic pH. Therefore Tris base – below 40 mmol/L – is sometimes be added to the lysis solution.

Proteases can be inactivated by boiling the sample in SDS buffer for a few seconds prior to the addition of urea-containing lysis buffer. Another way of protease inactivation is to precipitate the extracted proteins immediately at – 20 °C with TCA acetone or with the novel precipitation procedure described below.

Alkaline conditions

Rabilloud T, Valette C, Lawrence JJ. Electrophoresis 15 (1994) 1552–1558.

Tris base up to 40 mmol/L or 25 mmol/L spermine base (Rabilloud *et al.* 1994) is sometimes added to the lysis buffer to maximize protein extraction, precipitate nucleic acids, or to keep protease activities low. This should not be done, when basic proteins are of interest and must be displayed in the gel. Care should be taken when preparative sample loads need to be analysed: the ionic contamination can become too high.

Zhou G, Li H, DeCamp D, Chen S, Shu H, Gong Y, Flag M, Gillespiel J, Hu N, Taylor P, Buck ME, Liotta LA, Petricoin III EC, Zhao Y. Mol Cell Proteomics 1 (2002) in press.

When proteins are prelabelled with modified CyDyes for fluorescence 2-D difference gel electrophoresis, the pH of the sample must be adjusted to pH 8.5 with Tris-HCl. This has been found as the optimum pH for dye labelling (Zhou *et al.* 2002).

Frequently applied sample treatments

Cell washing: Very frequently washing of cells with PBS (phosphate buffered saline) is proposed. PBS contaminates the cell surfaces with salt, which leads to horizontal streaking. Instead, Tris-buffered sucrose should be employed (10 mmol/L Tris, 250 mmol/L sucrose pH 7): 1.21 g Tris, 85.6 g sucrose, fill up to 1 L with H_2O_{dist}, titrate to pH 7 with 4 mol/L HCl.

It is important that the washing solution contains enough osmoticum to avoid cell lysis at this stage. Sorbitol does not work not as good as sucrose.

Cell disruption: The choice of disruption methods is dependent on the type of sample. The easiest way is osmotic lysis of cultured cells, using the standard lysis solution defined above. Sometimes lyophilized cells are brought into solution with lysis solution. For analytical applications $5 - 10 \times 10^6$ cell equivalents are usually applied. Bacteria cells can be extracted by repeated freezing at –20 °C and thawing. For some organisms detergent lysis is necessary, for instance for tough cell walls of yeast and fungi. When samples are treated with the anionic detergent SDS, the solution must be diluted with lysis solution to a less than 0.1 % (w/v) SDS prior to IEF. Sonication with a probe is very helpful for solubilisation. French pressure cells and mechanical homogenizors are employed to burst tough cell walls from yeast or plant cells.

When a sonicator is used, the procedure should be performed on ice. Only short bursts should be applied in order to avoid heating of the urea.

Tissues and samples with tough cell walls are often treated with mortar and pestle at low temperatures. The handling of small sample amounts requires smaller tools. A new sample grinding kit is designed for the effective grinding of small samples for the purpose of protein extraction. The kit consists of microcentrifuge tubes each containing a small quantity of abrasive grinding resin suspended in water. Extraction solution of choice is added to the tube along with the sample to be ground. A grinding pestle is used to grind the sample. Cellular debris and grinding resin are removed by centrifugation. Intracellular organelles are also disrupted, resulting in the liberation and extraction of all proteins soluble in the extraction solution.

In general

The disruption method used is influencing the 2-D pattern. Therefore it has to be employed in a reproducible way and be thoroughly described in the analysis protocol.

Removal of contaminants

A crude extract can be contaminated with salt ions, phospholipids and nucleic acids, leading to disturbances or background streaking.

Nucleic acids are visualized with silver staining as horizontal streaks in the acidic part of the gel. Furthermore, they can precipititate with the proteins when the sample has to be applied on the basic end of

Heating must be avoided when sonicating.

the IEF gel. They can be removed with DNAse and RNAse treatment. The easiest technique is sonication, which breaks nucleic acids into little fragments. Precipitation also removes nucleic acids.

Lipids are removed with an excess of detergent (> 2%) or with precipitation. *Proteases* can also be inactivated by precipitation. *Solid material* is removed by spinning it down by centrifugation. *Salts* can be dialysed away or removed by precipitation. In the following pages the procedures are explained more detailed.

Microdialysis Desalting with microdialysis tubes shows less protein losses than with gel filtration. A mini dialysis kit has been designed for the dialysis of small samples with minimal handling and without sample loss. Each dialysis tube consists of a sample tube with a cap that incorporates a dialysis membrane. Usually the membrane with a molecular cut-off of 8,000 Da is used. The sample is pipetted into the tube, which is capped and inverted in a beaker containing the solution the sample is to be dialyzed against. The tubes are held in place with floats. Salts and molecules smaller than the molecular weight cut-off of the dialysis membrane rapidly exchange through the membrane. Following dialysis, the tube is centrifuged briefly to recover the contents. Dialysis time is 2 hours to overnight.

Görg A, Boguth G, Obermaier C, Posch A, Weiss W. Electrophoresis 16 (1995) 1079–1086.

Electrophoretic desalting There are some cases where the sample must not be dialysed. For example: Proteins in halobacteria lysate will not be soluble if the salt is removed. Another example: if the extraction solution containing bovine vitreous proteins would be desalted, they would gel. Görg *et al.* (1995) showed that the sample can be electrophoreticly desalted during the first IEF phase in the immobilized pH gradient strip: the applied voltage has to be limited to 100 V for 5 hours.

Precipitation methods

Note: Protein losses can never be completely prevented.

There are several reasons to apply a protein precipitation procedure:

- Concentration of low concentrated proteins, like in plant tissue.
- Removal of several disturbing compounds at the same time.
- Inhibition of protease activity.

The proteins are precipitated while interfering substances such as detergents, salts, lipids, phenols, and nucleic acids are left behind in solution. The proteins are then resuspended in lysis solution.

Wessel D, Flügge UI. Anal Biochem 138 (1984) 141–143.

Method by Wessel and Flügge (1984): Methanol and chloroform are added to the sample. After adding water the phases are separated and the proteins are precipitated at the chloroform–methanol–water interphase. Excess methanol is added, followed by centrifugation.

Method by Damerval et al. (1986): The plant material – or else – is frozen in liquid nitrogen and ground in a pre-frozen mortar. The powder is mixed with 10 % TCA in cold acetone (–20 °C) containing 0.07 % 2-mercaptoethanol. Precipitation occurs overnight in a freezer, followed by centrifugation and repeated washing with acetone (–20 °C) containing 0.07 % 2-mercaptoethanol. After centrifuging again, the pellet is resuspended in lysis buffer, carefully sonicated, and centrifuged at room temperature. The disadvantage is, however, that some acidic proteins get lost with this procedure, because they are not precipitated into the pellet. Figure 3 shows the comparison of results of a treated and nontreated sample.

Avoid overheating when sonicating!

Damerval C, DeVienne D, Zivy M, Thiellement H. Electrophoresis 7 (1986) 53–54.

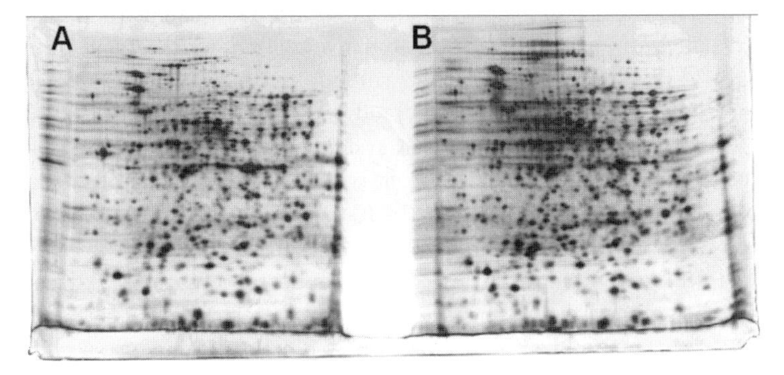

Fig. 3: 2-D electrophoresis of two different E. coli extracts in miniformat. *A:* E. coli extract was precipitated with TCA / acetone and resuspended with lysis buffer. *B:* Crude extract.

Gels: IEF in 7 cm IPG 4–7, dual SDS PAGE in a 16 × 8 cm gel of 12.5 % *T*, 1 mm thick. Silver staining. From Tom Berkelman, Amersham Biosciences San Francisco with kind permission.

The 2-mercaptoethanol can be replaced by 0.1 % DTT. Görg *et al.* (1997) have employed this procedure to extract more basic proteins from bacterial cells, yeast, and animal tissue.

Görg A, Obermaier C, Boguth G, Csordas A, Diaz J-J, Madjar J-J. Electrophoresis 18 (1997) 328–337.

Method by Mastro and Hall (1999): This procedure is applied for delipidation and precipitation of lipid-rich samples. Biological extracts with a lipid content of over 20 % of dry matter are delipitated and precipitated with a mixture of tri-n-butylphosphate, acetone and methanol at 4 °C, centrifuged, washed with the same mixture and air-dried.

Mastro R, Hall M. Anal Biochem 273 (1999) 313–315.

2-D clean-up kit: This procedure can be completed in one hour and does not result in spot gain or loss. The principle is similar to the method by Damerval *et al.* (1986), but with a detergent co-precipitant

Stasyk T, Hellman U, Souchelnytskyi S. Life Science News 9 (2001) 9–12.

proteins are more efficiently and completely removed from the solution. The wash buffer contains some organic additives that allow rapid and complete resuspension of the proteins with lysis buffer. The first experiences using this procedure are reported in a paper by Stasyk *et al.* (2001): The spot resolution is usually sharper and the number of spots higher compared to crude extracts or other precipitation methods. This method is described in detail in the practical part on page 177 ff.

Very hydrophobic proteins:
Membrane proteins do not easily go into solution. A lot of optimization work is required for different samples.

Rabilloud T. Electrophoresis 19 (1998) 758–760.
Molloy MP. Anal Biochem 280 (2000) 1–10.

Thiourea procedure: For the extraction and solubilisation of highly hydrophobic proteins like membrane proteins a combination of 7 mol/L urea plus 2 mol/L thiourea and alternative zwitterionic detergents like ASB 14 or sulfobetain in the lysis solution can be very helpful to get more proteins into solution (Rabilloud, 1998 and Molloy 2000).

This streak and the blurred spots are not sample specific.

Generally, extraction with urea and thiourea combined increases the number of spots considerably. It can be observed in the literature that many researchers have started to use this combination of chaotropes for all types of samples. However artifacts can be observed as well. In figure 4 the results of 2-D electrophoresis of rat liver extracts are shown: the lysis buffer differed only in the chaotropes used. The thiourea-containing sample shows many more spots, but also a phenomenon typical for thiourea gels: a vertical streak in the acidic area and blurred spots in the pH range below the streak. So far a remedy for this effect has not been found.

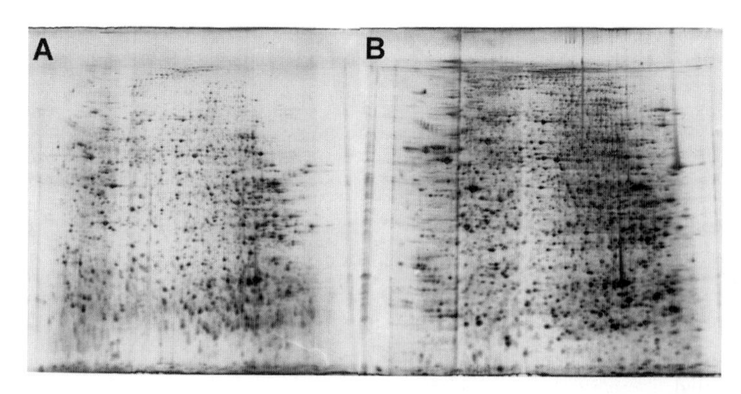

Fig. 4: 2-D electrophoresis of rat liver extracts. *A:* Lysis buffer contained 8 mol/L urea. *B:* Lysis buffer contained 7 mol/L urea and 2 mol/L thiourea. *Gels:* IEF in 18 cm IPG 3–10 nonlinear, SDS PAGE in 20 × 20 cm gels of 12.5 % T, 1 mm thick. Silver staining. From Tom Berkelman, Amersham Biosciences San Francisco with kind permission.

SDS procedure: In emergency cases even the anionic detergent SDS can be employed for sample preparation: up to 2 % SDS has been used. Before the sample is applied to denaturing isoelectric focusing the SDS sample has to be at least 20-fold diluted with urea and a non- or zwitterionic detergent containing solution. In the electric field the SDS will separate from the proteins and migrate into the anode. The major reasons for using SDS are:

The novel precipitation procedure described above can also remove the SDS from the proteins.

- Formation of oligomers can be prevented.

For instance albumin in plasma or serum.

- Organisms with tough cell walls sometimes require boiling for 5 minutes in 1 to 2 % SDS before they are diluted with lysis buffer.

For instance yeast.

- Some very hydrophobic proteins may require extraction with high percentage of SDS.

However, SDS does not always completely separate from the proteins, even under high field strength. This results in the shift of some isoelectric points to a more acidic value. SDS can be removed with the 2-D clean-up kit precipitation procedure.

New zwitterionic detergents and sulfobetains: A series of novel zwitterionic detergents (Chevallet *et al.* 1998) and non-detergent sulfobetains (Vuillard *et al.* 1995) have been tried. Some hydrophobic membrane proteins could be solubilized, which otherwise would have been lost. Up to now none of these additives has performed so well in general that it has replaced the standard cocktail. A perfect solution has not yet been found.

Chevallet M, Santoni V,. Poinas A, Rouquie D, Fuchs A, Kieffer S, Rossignol M, Lunardi J, Gerin J, Rabilloud T. Electrophoresis 19 (1998) 1901–1909.
Vuillard L, Marret N, Rabilloud T. Electrophoresis 16 (1995) 295–297.

More informations on sample preparation for 2-D electrophoresis can be found in the 2-D electrophoresis handbook by Berkelman and Stenstedt (1998), which can be downloaded from the website: http://proteomics.amershambiosciences.com, and in an overview by Rabilloud and Chevallet (2000).

Berkelman T, Stenstedt T. Handbook: 2-D Electrophoresis. Amersham Biosciences 80-6429-60 (1998).
Rabilloud T, Chevallet M. In Rabilloud T, Ed. Proteome research: Two-dimensional gel electrophoresis and identification methods. Springer, Berlin Heidelberg New York (2000) 9–29.

Dual detergent 2-D electrophoresis

Very hydrophobic proteins can become lost during isoelectric focusing. When membrane proteins are separated first in an acidic gel at pH 2.1 in presence of the cationic detergent 16-BAC, followed by an SDS electrophoresis (Langen *et al.* 2000), many more of these hydro-

Langen H, Takács B, Evers S, Berndt P, Lahm H-W, Wipf B, Gray C, Fountoulakis M. Electrophoresis 21 (2000) 411–429.

phobic proteins can be found in the gel. The resolution, however is not as good as in IEF / SDS PAGE, but this alternative 2-D electrophoresis can display proteins which are otherwise lost. Attempts to extract samples with 16-BAC and run them on an isoelectric focusing gel have not lead to an increase of hydrophobic proteins in the 2-D map (Peter James, personal communication).

High molecular weight proteins

It seems that this subject may not fit into the sample preparation part, because nothing can be done here to improve the performance of 2-D electrophoresis for displaying high molecular weight proteins with sizes above 150 kDa. It is known, that HMW proteins get lost during the isoelectric focusing step. But they are included, when they are just run in a one-dimensional SDS gel.

In general the number of HMW sample is not so high, that they could not be resolved in a one-dimensional SDS PAGE separation. If HMW proteins are of interest, there is the possibility to prepare a portion of the sample with SDS sample buffer and run it as a 1-D sample, either on the 2-D gel together with the IPG strip or on a separate SDS gel.

Overloading effects

Therefore estimation of the protein content in the sample is very important.

When a single protein is applied in very high quantity on an IPG strip, an overloading effect can occur: When the electric field focuses the protein at the isoelectric point, the gel will build a high ridge at this point. When the amount of a protein is too high, the focusing force will exudates a part of the protein through the gel surface. The protein will diffuse over the surface (see figure 5). The effect will be visible as strong horizontal streaks over the 2-D gel.

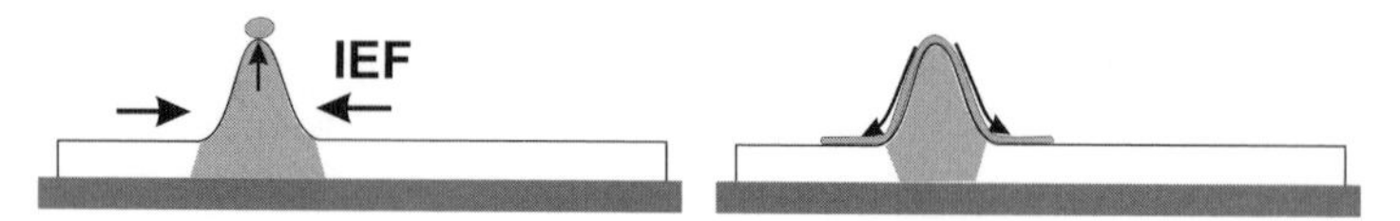

Fig. 5: Effect from overloading a protein in an IPG strip.

Quantification

Ramagli LS. In Link AJ. (Ed) 2-D Proteome Analysis Protocols. Methods in Molecular Biology 112. Humana Press, Totowa, NJ (1999) 99–103.

The usually applied additives, like urea, detergents, and reductants, interfere with the standard protein assays (e.g. Bradford, BCA, Lowry). False values often lead to over- or under-estimation of the protein load applied on gels. Ramagli (1999) had modified an existing method to a relatively reliable procedure by acidifying the sample with 0.1 mol/L HCl.

A novel protein assay has been introduced lately: The procedure works by quantitatively precipitating proteins while leaving interfering substances behind. The assay is based on the specific binding of copper ions to protein. Precipitated proteins are resuspended in a copper-containing solution and unbound copper is measured with a colorimetric agent. The color density is inversely related to the protein concentration. The assay has a linear response to protein in the range of 0–50 μg.

The precipitation method is the same like in the novel 2-D clean-up kit.

For a reliable comparison of 2-D gels the protein load should be similar. Therefore the protein amount needs to be checked. Variations of protein loads must be in certain limits to allow for corrections by the normalisation function of image analysis (see pages 94 and 245).

The knowledge of the protein content of a sample can be very useful to avoid under- or overloading. However, when the protein composition is unbalanced, this information does not help very much. A typical example is serum or plasma, where albumin makes up ca. 60 % and the gammaglobulines ca. 20 % of the total protein. These highly abundant proteins can easily cause overloading effects (see above). Thus for such sample types the total protein loading capacity of an IPG strips is considerably reduced.

Analytical versus preparative sample load Sample loads are often described as *analytical* or *preparative*. In practice it is not easy to differentiate clearly between these terms: A strong protein spot in an analytical gel can contain enough material for further analysis whereas a weak spot in a preparative gel can be insufficient. Roughly, when 50 to 150 μg of total protein is applied on a gel, and silver or sensitive fluorescent staining methods need to be employed, it is an analytical gel. Preparative gels are loaded with 1 mg total protein and more. These gels are usually stained with Coomassie Brilliant Blue, medium sensitive fluorescent dyes, or zinc-imidazol.

Special cases

Analysis of human body fluids Body fluids are an important source for detection and monitoring of disease markers. The Anderson group had started many years ago to collect data for the Human protein index with 2-D electrophoresis, for instance analyzing plasma and urinary proteins (Anderson *et al.* 1977, 1979). A plasma protein map produced with immobilized pH gradients has been published by Hughes *et al.* (1992).

Anderson L, Anderson NG. Proc Nat Acad Sci USA. 74 (1977) 5421–5425.
Anderson NG, Anderson NL, Tollaksen SL. Clin Chem. 25 (1979) 1199–1210.
Hughes GJ, Frutiger S, Paquet N, Ravier F, Pasquali C, Sanchez JC, James R, Tissot JD, Bjellqvist B, Hochstrasser DF. Electrophoresis 13 (1992) 707–714.

Harrington MG, Merrill C. J Chromatogr 429 (1988) 345–358.
Yun M, Mu W, Hood L, Harrington MG. Electrophoresis 13 (1992) 1002–1013.
Zerr I, Bodemer M, Otto M, Poser S, Windl O, Kretzschmar HA, Gefeller O, Weber T. Lancet 348 (1996) 846–849.
Burkhard PR, Rodrigo N, May D, Sztajzel R, Sanchez J-C, Hochstrasser DF, Shiffer E, Reverdin A, Lacroix JS. Electrophoresis 22 (2001) 1826–1833.

Besides plasma and serum, cerebrospinal fluid is frequently used as a sample for detecting and monitoring disease markers. Here is a selection of papers on 2-D electrophoresis of cerebrospinal fluid: Harrington and Merrill (1988), Yun *et al.* (1992), Zerr *et al.* (1996), Burkhard *et al.* (2001).

Unfortunately most of the body fluids are highly loaded with abundant proteins or salt ions, which interfere with the first dimension. Microdialysis and/or low-voltage start conditions for isoelectric focusing are required to avoid distorted patterns.

Albumin For serum and plasma usually the high abundance of albumin and globulins limit the loading capacity for the rest of the proteins. If the albumin content in samples like plasma, serum, and cerebrospinal fluid could be reduced, the sensitivity of detection for other proteins could be considerably improved. When albumin is removed with blue dextrane or a Sepharose Blue column, unfortunately many other proteins, which stick to albumin are removed at the same time. Currently there is no procedure available to get rid of albumin without losses of other proteins.

Plant proteins

Hurkman WJ, Tanaka CK. Plant Physiology 81 (1986) 802–806.

Sample preparation from plant tissues is particularly troublesome, because interfering substances such as polysaccharides, nucleic acids, lipids and phenolic compounds are present in high concentrations, whereas the proteins exist in low abundance. Usually the proteins have to be precipitated as described above on page 21. More complicated procedures are required for the analysis of plant membrane proteins (Hurkman and Tanaka, 1986). The protein pellet has to be extensively washed before it can be resuspended in lysis buffer for IEF.

Integration of sample preparation into a laboratory workflow system

Parness J, Paul-Pletzer K. Anal Biochem 289 (2001) 98–99.

It is very important to link the sample preparation procedure to the result data even down to mass spectrometry analysis. A good example case for this need is described in the – already mentioned – paper by Parness and Paul-Pletzer (2001), where the origin of the keratin of the protein spots has been found in the reductant.

Peculiar results in mass spectrometry analysis must initiate an automated search in the entire laboratory workflow. Therefore all data, including the source and the LOT number of each reagent used, have to be entered together with the sample code into the database of the laboratory workflow system.

Further developments of sample preparation procedures

Whereas the methodology for the separation and further analysis of the proteins has become much easier, reliable and reproducible, there is still a substantial wish list for improvements of sample preparation:

- A very important goal in for creating reliable data is a certain standardization of sample preparation and a systematic reduction of protein complexity prior to 2-D electrophoresis.
- A wider choice of detergents and other additives, which support the solubility of hydrophobic proteins, and would reduce the losses of proteins.
- Efficient reductants, which do not become charged in basic environment, and lead to improved pattern in basic pH gradients.
- Multiple automated sample preparation will reduce the manual influence and allow higher throughput. Because most of the problems in practice are created by contaminants of all different kind, the 2-D clean-up precipitation and resolubilisation method could be applied to many sample types.
- A pre-fractionation of proteins according to their isoelectric points into defined packages, which are applied on narrow pH interval gels could substantially reduce the load of sample not displayed in these gels, and in this way reduce losses of proteins.

2.1.3 First Dimension: Isoelectric focusing

Theoretical background

In order to understand a few effects occurring during IEF, it is useful to look at some basic facts in isoelectric focusing and the pH gradient construction.

Isoelectric focusing Isoelectric focusing is performed in a pH gradient. Proteins are amphoteric molecules with acidic and basic buffering groups. Those become protonated or deprotonated depending on the pH environment. In basic environment the acidic groups become negatively charged, in acidic environment the basic groups become positively charged. The net charge of a protein is the sum of all negative or positive charges of the amino acid side chains. The net charges can be plotted over the pH scale. Each protein has an individual net charge curve. The intersection of the net charge curve with the x-axis is the isoelectric point – the pH value where the net charge is zero. This is shown for two model proteins in figure 6.

When a protein is placed at a certain pH value of the gradient, and an electric field is applied, it will start to migrate towards the electrode of the opposite sign of its net charge. Because it migrates inside a gradient, it will arrive at a pH value of its isoelectric point (see figure 6). At the pI it has no net charge anymore and stops migrating. Should it diffuse away – above or below its pI – it will become charged again and migrate back to its pI. This is called the "focusing effect", which results in very high resolution.

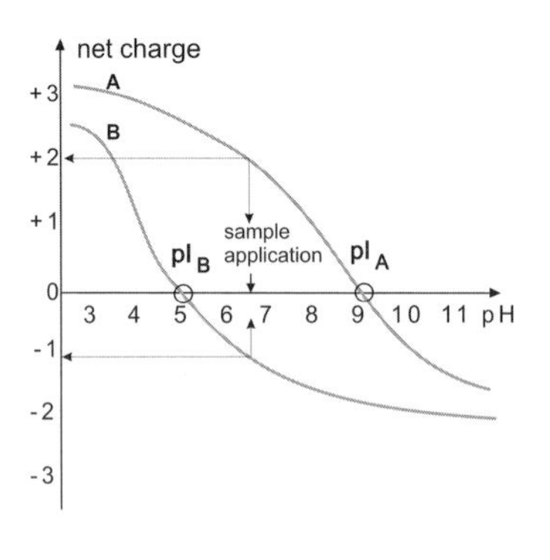

Fig. 6: The principle of isoelectric focusing. *Left:* Net charge curves of two model proteins A and B. At the point of application, A will have two positive, B will have one negative charge(s). *Right:* Migration of A and B to their pIs in the pH gradient of an isoelectric focusing gel.

In principle

Isoelectric focusing is a very high resolution separation method, and the pI of a protein can be measured.

Rosengren A, Bjellqvist B, Gasparic V. In: Radola BJ, Graesslin D. Ed. Electrofocusing and isotachophoresis. W. de Gruyter, Berlin (1977) 165–171.

Titration curve analysis The net charge curves are also called "titration curves". The titration curves of proteins can be displayed by a simple method (Rosengren *et al.* 1976): A square gel – containing carrier ampholytes, but no samples – is submitted to an electric field until the carrier ampholytes form a pH gradient 3 – 10. Then the gel is rotated by 90 degrees, the sample proteins are pipetted into a groove in the gel across the pH gradient. Now an electric field is applied perpendicular to the pH gradient: the carrier ampholytes are uncharged at their pIs and will not move. But, as shown in figure 7, the proteins will migrate towards the cathode or the anode according to their charge sign and mobility, and will thus form titration curves (net charge curves).

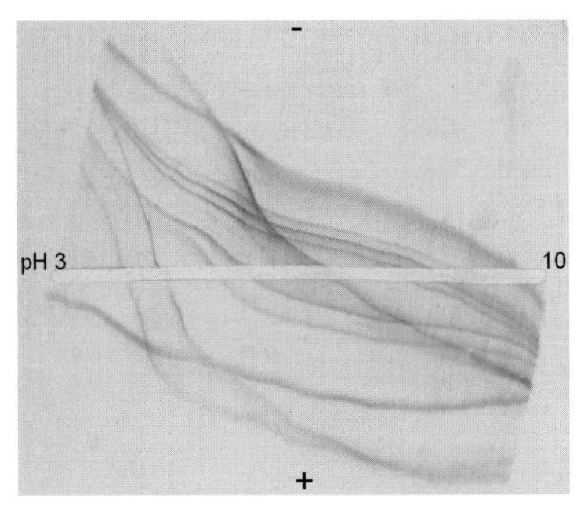

Fig. 7: Titration curves of a mixture of pI marker proteins under native conditions. The steeper the curve, the faster will be the migration of a protein in IEF. Note, that some proteins have a flatter curve above their pI.

The proteins show very individually shaped curves also under denaturing conditions. Practical experience has shown, that more proteins have a steeper curve below their pI. This is the reason, why mostly sample application close to the anode results in better pattern.

pH gradients In practice there are two ways to establish a pH gradient in a gel:

- pH gradients which are formed in the electric field by amphoteric buffers, the carrier ampholytes.
- Immobilized pH gradients in which the buffering groups are part of the gel medium.

Carrier ampholytes generated pH gradients This was the first developed technique for isoelectric focusing. Svensson (1961) has designed the theoretical basis for preparing "natural" pH gradients, while the practical realization is the work of Vesterberg (1969): the synthesis of a heterogeneous mixture of isomers of aliphatic oligoamino-oligocarboxylic acids. The result is a spectrum of low molecular weight ampholytes with closely neighbouring isoelectric points, and with high buffering power at their pI.

Svensson H. Acta Chem Scand 15 (1961) 325–341.
Vesterberg, O. Acta Chem. Scand 23 (1969) 2653–2666.

The general chemical formula is the following:

$$
\begin{array}{ccccc}
-CH_2- & N & -(CH_2)_x- & N & -CH_2- \\
 & | & & | & \\
 & (CH_2)_x & & (CH_2)_x & \\
 & | & & | & \\
 & NR_2 & & COOH &
\end{array}
$$

R = H or $-(CH_2)_x-COOH$, x = 2 or 3

These carrier ampholytes possess the following properties:

- High buffering capacity and solubility at the pI,
- Good and regular electric conductivity at the pI,
- Absence of biological effects,
- Low molecular weight.

Alternative synthesis chemistries have been developed later. The chemical structures of other amphoteric buffers are different. Their function is the same, but the properties of the gradients can be different.

Naturally occurring ampholytes such as amino acids and peptides do not have their highest buffering capacity at their isoelectric point. They can therefore not be used.

By controlling the synthesis and the use of a suitable mixture the composition can be monitored so that regular and linear gradient result.

The pH gradient is established by the electric field. At the beginning, the gel with carrier ampholyte has an uniform average pH value. Almost all the carrier ampholytes are charged: those with the higher pI positively, those with the lower pI negatively.

The anodal end becomes acidic and the cathodal side basic.

When an electric field is applied, the negatively charged carrier ampholytes migrate towards the anode, the positively charged ones to the cathode, with velocities depending on their net charges. The carrier ampholytes align themselves in between the two electrodes according to their pI and will determine the pH of their environment. A – relatively – stable, gradually increasing pH gradient from, for instance pH 3 to 10, results.The carrier ampholytes lose part of their charge so the conductivity of the gel decreases.

To maintain a gradient as stable as possible, electrode solutions are applied between the gel and the electrodes, an acid is used at the anode and a base at the cathode. Should, for example, an acidic carrier ampholyte reach the anode, its basic buffering group would acquire a positive charge from the medium and it would be attracted back by the cathode.

Carrier ampholytes as solvents for proteins Carrier ampholytes have another very important function: they help to solubilise proteins, which stay in solution only in presence of buffering compounds (see page 17). Therefore they are also needed for the immobilized pH gradient technique.

Particularly under denaturing conditions the matrix is highly viscous leading to slow migration. Additionally, unfolded polypeptides migrate slower in the gel than native proteins.

Isoelectric focusing is a relatively slow separation method, because close to their isoelectric points proteins have only a low net charge, and therefore low mobility. High spatial resolution requires long separation distances, which causes long migration time under high electric field strength.

Because the carrier ampholyte are in free solution, the gradient will become instable during long IEF time: the gradient drifts. The pattern becomes time-dependent and most of the basic proteins are drifting out of the gel together with the basic part of the gradient. Another problem: proteins of the sample behave like additional carrier ampholytes and modify the profile of the gradient. Thus the gradient becomes also sample-dependent.

Carrier ampholyte generated gradients are also influenced by the sample load and salt contents in the sample.

Immobilized pH gradients

These problems are solved with the application of immobilized pH gradients generated by buffering acrylamide derivatives, which are copolymerised with the gel matrix. The acrylamide derivatives containing the buffering groups are called *Immobilines.* An Immobiline is a weak acid or weak base defined by its pK value.

The general structure of an Immobiline is the following:

```
CH2=CH–C–N–R
        |  |
        O  H
```

R contains either a carboxylic or an amino group.

The exact chemical structures are found in the book by Righetti (1990). The acidic substances have dissociation constants in the range from pK 0.8 to pK 4.6, the basic from pK 6.2 to pK 12. In order to buffer at precise pH values, at least two different Immobilines are necessary, an acid and a base. Figure 8 shows a diagram of a poly-

Righetti PG. In: Burdon RH, van Knippenberg PH. Ed. Immobilized pH gradients: theory and methodology. Elsevier, Amsterdam (1990).

The wider the pH gradient, the more different Immobiline homologues are needed.

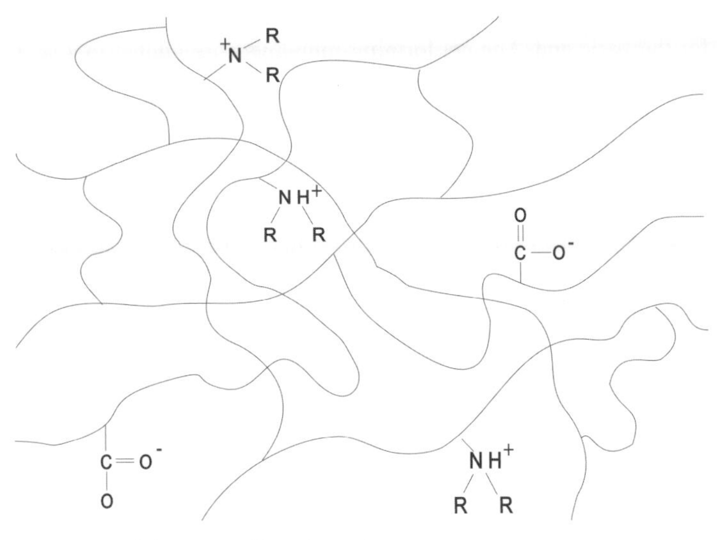

Fig. 8: Immobilized pH buffer. Schematic drawing of a polyacrylamide network with co-polymerized buffering groups.

acrylamide gel with polymerized Immobilines, the pH value is defined by the ratio of the Immobilines in the mixture.

This concept generates an absolutely continuous pH gradient.

A pH gradient is obtained by continuously varying the ratio of Immobilines. The principle is that of an acid / base titration, and the pH value at each stage is defined by the Henderson-Hasselbalch equation:

$$\mathrm{pH} = \mathrm{pK_B} + \log \frac{C_B - C_A}{C_A}$$

when the buffering Immobiline is a base.

C_A and C_B are the molar concentrations of the acid, and basic Immobiline, respectively.

If the buffering Immobiline is an acid, the equation becomes:

$$\mathrm{pH} = \mathrm{pK_A} + \log \frac{C_B}{C_A - C_B}$$

Celentano F, Gianazza E, Dossi G, Righetti PG. Chemometr Intel Lab Systems 1 (1987) 349–358.
Altland K. Electrophoresis 11 (1990) 140–147.

Celentano *et al.* (1986) and Altland (1990) have developed PC programs for the calculation and simulation of immobilized pH gradients for the optimisation of buffering power and ionic strength distribution. Practice has shown, that a pre-calculated pH gradient recipe does not automatically result in a usable IEF gel. Every recipe has to be tested and optimised with practical experiments.

Preparation of immobilized pH gradients

Immobiline stock solutions with concentrations of 0.2 mol/L are used. Two solutions with acrylamide monomers, crosslinker, and catalysts are required to prepare an immobilized pH gradient: one contains the Immobiline cocktail for the acidic end, the other the cocktail for the basic end. The gel forming monomers are diluted to 4 % *T* and 3 % *C* (for definition see page 59). Usually the acidic solution is made denser by adding glycerol. The gel is prepared by linear mixing of the two different monomer solutions with a gradient maker (see figure 9), like for pore gradients. In principle a concentration gradient is poured. 0.5 mm thick Immobiline gels, polymerized on a support film have proved most convenient. During polymerisation, the buffering carboxylic and amino groups covalently bind to the gel matrix. Because of the low electric conductivity of immobilized pH gradients, all contaminating compounds, like acrylamide monomers and the catalysts, have to be removed by washing the gels several times with deionised water.

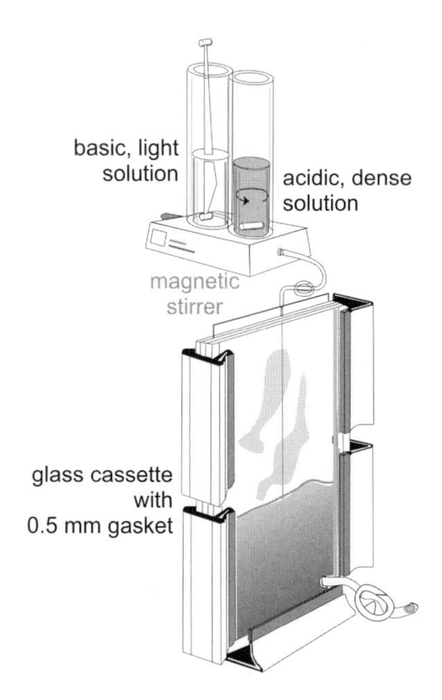

Fig. 9: Preparing an immobilized pH gradient. Pouring a gradient into a gel cassette.

Immobilized pH gradient strips

Gels containing an immobilized pH gradient are the only electrophoresis gels, which do not show edge effects during the run when they are cut into strips. This is done after the gels have been dried down on the film. Instructions and pH gradient recipes for preparing immobilized pH gradient strips in the laboratory can be found in several books: Görg *et al.* (1994, 2000) and Westermeier (2001). This is a multistep procedure, which needs some expertise and skill.

Görg A. In: Celis J, Ed. Cell Biology: A Laboratory Handbook. Academic Press Inc., San Diego, CA. (1994) 231–242.
Görg A, Weiss W. In Rabilloud T, Ed. Proteome research: Two-dimensional gel electrophoresis and identification methods. Springer, Berlin Heidelberg New York (2000) 107–126.
Westermeier R. Electrophoresis in Practice. WILEY-VCH, Weinheim (2001).

Much more reproducible results are obtained, when ready-made strips are used. Commercially produced strips are prepared according to GMP industry standards, and they are quality controlled. A wide choice of different gradients is available on the market. Additional gradients are being developed for achieving higher resolution.

It is very annoying, when the valuable sample is not separated well, just because of a little mistake occurring in the laboratory.

Dried gel strips can be stored at –20 to –80 °C from months to years before reconstitution.

Advantages of immobilized pH gradients As already mentioned above, isoelectric focusing in pre-manufactured IPG strips is a highly reproducible method compared to gels with carrier ampholytes generated pH gradients. There are several reasons for this fact:

Readymade IPG strips are usually produced according to GMP routines.

- Industrial production reduces variations caused by human interference.

In contrast to carrier ampholytes, which are mixtures of several hundred homologues.

- The chemistry of the buffering acrylamide derivatives is better controllable.

Thin gel rods require high experimental skill.

- The film-supported gel strips are easy to handle.

Very important for reproducible results.

- The fixed gradients are not modified by the sample composition, and they do not drift with IEF time.

In contrast to NEPHGE, where the basic proteins are not focused at their pI.

- Stable basic pH gradients allow reproducible separation and display of basic proteins.

The concept of immobilized pH gradients offers a number of additional beneficial features:

This procedure prevents proteins from precipitating at the application point.

- Different ways of sample applications are feasible; the dried strips can be directly rehydrated with the sample solutions.

This allows also the detection of low expressed proteins.

- Higher protein loads are achievable.

Some of them would inhibit gel polymerisation.

- Various additives, like detergents and reductants, can be added to the rehydration solution.

Righetti PG, Gelfi C. J Biochem Biophys Methods. 9 (1984) 103–119.

- Less proteins are lost during equilibration in SDS buffer, because the fixed charged groups of the gradient hold the proteins back like a weak ion exchanger (Righetti and Gelfi, 1984).
- Probably the most powerful feature is the possibility to reach almost unlimited spatial resolution with very narrow pH intervals. This is needed for protein identification and characterization.
- Gradients can be engineered according to special needs; they are absolutely continuous.

Gel sizes The commercial strips are 3 mm wide and have to be reconstituted to the original thickness of 0.5 mm with rehydration solution prior to IEF. Thicker and wider strips would have a higher protein loading capacity. This would, however, not be an advantage, because the SDS polyacrylamide gel of the second dimension would show overloading effects of proteins and a high detergent background. This could be compensated by using SDS gels thicker than the conventional 1 to 1.5 mm. Staining of such thick gel slabs would be very time consuming and not allow a high-throughput approaches.

Strips are available in several different lengths from 7 cm to 24 cm. For the proteomics approach usually 18 and 24 cm strips are used, because the highest possible resolution is required for proteome analysis. Miniformat strips with 7 cm length are ideal for optimization of the sample preparation method.

With miniformat gels results are obtained after a few hours, with large formats after days.

When resolution does not have the highest priority, like for protein identification with western blotting, small gel formats from 7 to 13 cm are sufficient.

Gradient types One of the advantages of immobilized pH gradients is the possibility to produce absolutely linear gradients from pH 3 to 10. However, there are samples with an uneven distribution of the proteins over the pH gradient from 3 to 10. The nonlinear gradient pH 3 to 10 is flat in the acidic range to accommodate more different and higher concentrations of acidic proteins. So far the widest gradient was designed and applied by Görg *et al.* (1999): pH 3 to 12. For increased resolution and higher protein loads acidic (pH 4 to 7, pH 3 to 7) and basic (pH 6 to 11, pH 6 to 9) gradients are employed. Narrow intervals with only one or two pH units allow very high protein loads and excellent spatial resolution. The narrowest intervals, so called "ultra zoom gels" with 0.5 pH units, have been applied to 2-D electrophoresis by Hoving *et al.* (2000).

Görg A, Obermaier C, Boguth G, Weiss W. Electrophoresis 20 (1999) 712–717.
Hoving S, Voshol H, van Oostrum J. Electrophoresis 21 (2000) 2617–2621.

Rehydration of IPG strips
Composition of the standard "rehydration solution":

8 mol/L urea, 0.5 % CHAPS, 0.2 % DTT, 0.5 % carrier ampholytes, 10 % (v/v) glycerol, 0.002 % bromophenol blue.

- Glycerol reduces electroendosmotic effects, prevents drying of the gel, and urea crystallization.
- The concentrations of the additives are lower than in the lysis buffer, in order to reduce unwanted side effects like
  - crystallization of urea.
  - instability of the patterns due to conflicts of the buffering groups, caused by overloading with carrier ampholytes.
  - formation of micelles between the different detergents in the second dimension, which can cause dark and dirty background with some of the staining techniques (Görg *et al.* 1987a).

for electroendosmosis see page 59 or glossary.

See also page 18 for choice of IPG buffers or carrier ampholytes.

Görg A, Postel W, Weser J, Günther S, Strahler JR, Hanash SM, Somerlot L. Electrophoresis 8 (1987a) 45–51.

- The bromophenol blue allows a good control of the liquid distribution during rehydration.

If this U-shaped film is not added, the pore size of the strips will be too small for high-molecular weight proteins to enter the gel.

Rehydration cassette In the original rehydration procedure, which is still in use in some laboratories, the strips are placed into a glass cassette, in order to control the gel thickness to prevent over swelling (see figure 10). An additional U-shaped gasket cut from 0.2 mm thick film is added to achieve 0.5 mm thick gels on the 0.2 mm thick film support.

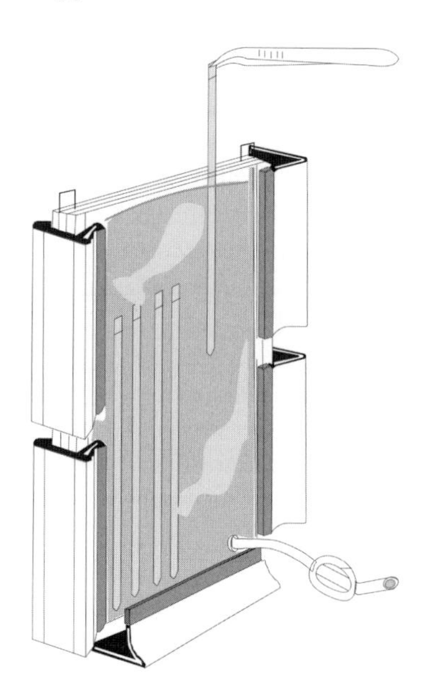

Fig. 10: Rehydration cassette with 0.5 mm gasket. Note the U-shaped film added to achieve 0.5 mm thick gels on film supports.

Some laboratories claim, that they obtain the best results, when they rehydrate the strips with this method.

With this procedure many strips can be rehydrated under identical conditions. The disadvantages of the cassette method:

- High volume of rehydration solution needed.
- Cassettes sometimes leak because of the urea and the detergent in the liquid.
- Rehydration loading of different samples is not possible.

Sanchez J-C, Rouge V, Pisteur M, Ravier F, Tonella L, Moosmayer M, Wilkins MR, Hochstrasser DF. Electrophoresis 18 (1997) 324–327.

Reswelling tray The disadvantages of the cassette technique are avoided when using a reswelling tray (see figure 11), as suggested by Sanchez *et al.* (1997). The liquid volume offered to the strips must be exactly controlled:

- If the liquid volume is too big, the strip will preferably reswell with the low-molecular weight compounds and leave reagents with higher viscosity and higher molecular weights outside. There would not be any control over the reagent concentrations inside the strip. Over swelling of a strip results in liquid exudation during IEF. This would cause protein transport to the surface, resulting in background smearing.
- If the liquid volume is too small, the resulting pore size will be too small to allow high-molecular weight proteins to enter the gel.

According to a theoretical calculation the volumes for the 3 mm wide and 0.5 mm thin strips would be 360 μL for a 24 cm and 270 μL for an 18 cm strip. In practice the optimal reswelling volumes are:

This has been found out by experiments.

7 cm strip	125 μL
18 cm strip	340 μL
24 cm strip	450 μL

Reswelling is performed at room temperature, because the urea would crystallize in a cold room. Figure 11 shows a reswelling tray for up to twelve strips of different strip length up to 24 cm.

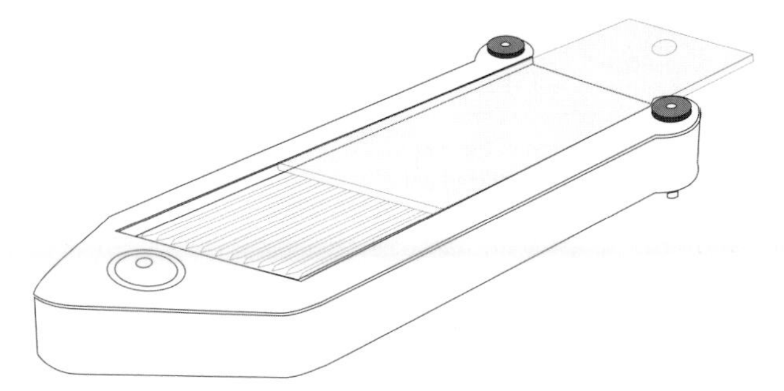

Fig. 11: IPG Drystrip reswelling tray for rehydration of IPG strips in individual grooves.

Sample application on IPG strips

As shown in figure 12, there are two ways to apply the sample to these immobilized pH gradient strips.

This concept was first proposed by: Rabilloud T, Valette C, Lawrence JJ. Electrophoresis 15 (1994) 1552–1558; and modified by: Sanchez J-C, Rouge V, Pisteur M, Ravier F, Tonella L, Moosmayer M, Wilkins MR, Hochstrasser DF. Electrophoresis 18 (1997) 324–327.

- *Rehydration loading:* The sample in lysis buffer is diluted with rehydration solution to the wanted amount of protein. Rehydration of a strip occurs in an individual strip holder or in a reswelling tray in an individual groove. The dry gel matrix takes up the fluid together with the proteins. Because the pores are very small in the beginning, the smaller molecules (water, urea, detergent, reductant) go into the gel matrix faster than the proteins. Many of the proteins diffuse into the fully rehydrated gel later. This is the reason for the much longer rehydration time required for rehydration loading. IEF can be performed with the gel surface up or down.

Note: This has been the original procedure for IEF in IPG strips.

- *Cup loading:* The strip is pre-rehydrated with rehydration solution. The sample is applied into a loading cup at a defined pH of the gradient, either on the acidic or the basic end. The optimal application point is critical, and has to be determined for each sample type and the gradient used. The proteins are transported into the strip electrophoretically. IEF is performed in the strips with the gel surface up.

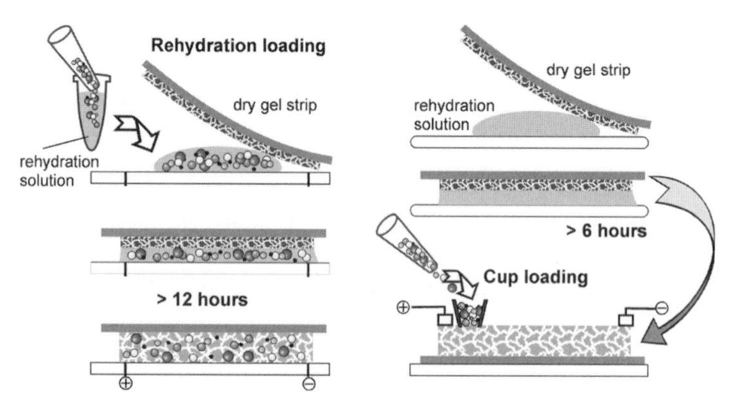

Fig. 12: Schematic representation of the two ways to apply the sample into immobilized pH gradient strips.

Silicon oil is not recommended, because it can contain dissolved oxygen and it is not easy to remove from the instruments and the benches.

Cover fluid The gel strips are covered with *paraffin oil* during rehydration and IEF to prevent drying of the strips, crystallisation of the urea, and oxygen and carbon dioxide uptake. It has never been observed, that hydrophobic proteins change over from the rehydration solution into the oil phase.

Rehydration time

**) Large protein molecules need a long time to diffuse into the strip.*

without sample	> 6 h
including sample	> 12 h or overnight *)

See page 28, titration curves.

Rehydration loading versus cup loading The way of sample loading will have an effect on the 2-D electrophoresis pattern. This is easy to imagine: Each protein has an individually shaped titration curve, the mobility of the protein is usually different, depending on whether it is negatively or positively charged.

- Rehydration loading: Each protein exists in both forms: negatively and positively charged, the proteins approach their pI from both sides with different mobilities.
- Cup loading: All proteins are charged in the same way, all proteins approach their pIs only from the same side.

This facts can explain why some samples show better results with cup loading compared to rehydration loading*).

*See below under "Pro, cup loading".

In the following table the advantages and disadvantages of the two methods are listed:

Tab. 1: Comparison of rehydration and cup loading.

Rehydration loading	*Cup loading*
Pro	**Pro**
The proteins are distributed evenly over the gradient, precipitation at the sample application point cannot happen.	IEF in very acidic narrow gradients and in basic pH gradients works much better with cup loading.
Rehydration, sample application and IEF can be combined to one single step, resulting in less manipulation steps.	The sample gets immediately transported into the gel and becomes separated, thus the chances of protein interactions are reduced, for instance by proteases.
When the sample is diluted, a higher protein load can be applied.	The proteins can be loaded onto both ends of the strip in order to include proteins with pIs close to the pH of their application points.
Voltage can be applied during rehydration, which improves the entry of high molecular weight proteins.	Because of mostly unknown reasons*) some sample types display more spots after cup loading, which are obviously lost with rehydration loading.
Con	**Con**
When basic gradients like pH 6 – 11 and pH 6 – 9 are employed, rehydration loading leads to severe protein losses, because some proteins do not like basic environment.	Proteins with isoelectric points close to the pH of the application point have low mobility and solubility, they tend to precipitate on the surface and build a streak in the second dimension gel.
The protein mixture is kept for many hours at room temperature, thus the danger of exposure to protease activities can increase.	Also when proteins applied on both ends, proteins precipitate at the application points. Quantification of proteins becomes a problem.

Rehydration loading	Cup loading
During the rehydration process the protein concentration outside the gel strip increases, which can lead to aggregation of proteins; those cannot enter the gel.	During the first phase of IEF the proteins become concentrated at the point of entry and can form aggregates and precipitate.
With preparative sample loads, proteins with low solubility tend to precipitate inside the gel during the run, which lead can to blurred spot pattern.	Cup loading on both ends does not always work fine, for example, when the mobilities of negatively and positively charged proteins are too different.

For more details see sample entry effects on page 43.

Gels with high protein loads should always be run with the surface up, also when rehydration loading is performed, see page 48.

In several papers the statement can be found, that higher protein loads can be applied with rehydration loading than with cup loading. This should be checked for each sample type, for some samples it might be vice versa.

see page 52.

A strategy for selection of the optimal procedure is described later.

IEF conditions

In general, separation conditions have to be adjusted to the nature and composition of the sample. Horizontal streaking in the 2-D pattern, for instance, can have many different reasons: overloading effects, too short focusing time, too long focusing time (some proteins became unstable at their pIs), oxidation of cysteins, too much salt, nucleic acids, phospholipids etc.

Görg A, Postel W, Friedrich C, Kuick R, Strahler JR, Hanash SM. Electrophoresis 12 (1991) 653–658.

Electric conditions and temperature control are closely related.

Temperature It is very important to run IEF at a defined temperature, even when the proteins are denatured. It had been demonstrated that spot positions of certain proteins can vary along the x-axis of the gel dependent on the temperature (Görg *et al.* 1991). Running the strips at 20 °C is optimal, because it is above the temperature where crystallisation of urea can be critical, and it is below overheating temperature, which can cause carbamylation of proteins during IEF. Even, when the applied electric power, the product of current and voltage, is very low in IPG strips, it is not sufficient to run IEF just in a thermostated room with 20 °C. Active temperature control is necessary to dissipate local heat developed at the ion fronts in the strips ("hot spots", see figure 13).

Hot spots In an IPG strip, there are always salt and buffer ions present, which have not been washed out. They are bound to the buffering groups of the immobilized pH gradient. A second group come from ionic sample contaminants.

In the electric field they start to migrate towards the electrodes with the opposite signs. During the run they are forming visible ion fronts between the region with low ion concentration and the regions with high ion concentrations (fig. 13), which are moving towards the electrodes. Immobilized pH gradients have very low conductivity, because the buffering groups of the gradient are fixed and cannot diffuse. Because of the differences in conductivity, the electric field strength in these regions is varying considerably. Punctual heat development can be observed at these fronts, so called "hot spots".

In IEF, all non-amphoteric compounds are transported out of the gradient towards the electrodes.

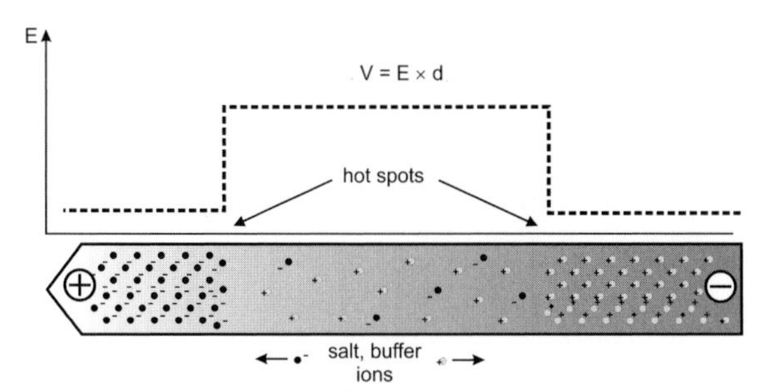

Fig. 13: Schematic representation of the distribution of salt and buffer ions in an IPG strip during IEF and the resulting electric field strength. The ion fronts are migrating slowly towards the electrodes. V = voltage, E = electric field strength, d = electrode distance.

Electric conditions As already mentioned above, immobilized pH gradients have very low conductivity. The current is usually limited at 50 – 70 μA per strip. Higher current settings are not recommended, because the strips could start to overheat and burn in certain areas of the gradient. The strips should never be prefocused, because the higher conductivity produced by salt and buffer ions in the beginning is advantageous for sample entry and for the start conditions of proteins.

The electric conditions are controlled with the voltage setting. First low voltage is applied in order to avoid sample aggregation and precipitation in the loading cup or overheating in some strip areas. The voltage is slowly raised to reach electric field strength as high as possible in the focusing phase. The voltage changes are either programmed in steps or with ramping (see figure 14).

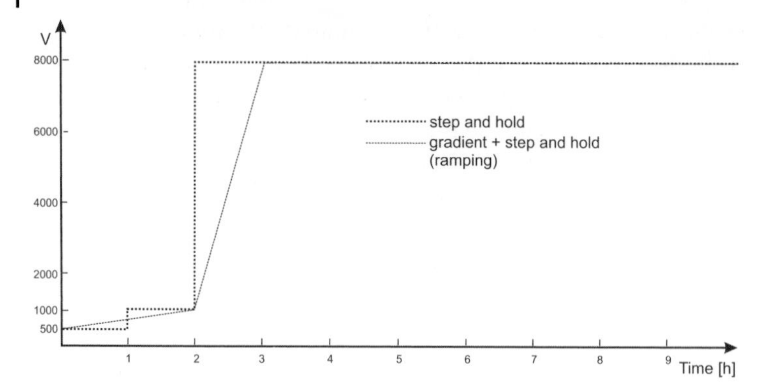

Fig. 14: Voltage settings for IEF in IPG strips. For rehydration loaded samples mostly the "step and hold" mode is applied, for cup loaded samples the "gradient" mode for the start phase.

> Note
> ***The set current can limit the achievable voltage, when the sample contains too much salt, or when buffers are included in the rehydration or lysis solution.***

The electric conditions change sometimes dramatically during the first phase as shown in figure 15:

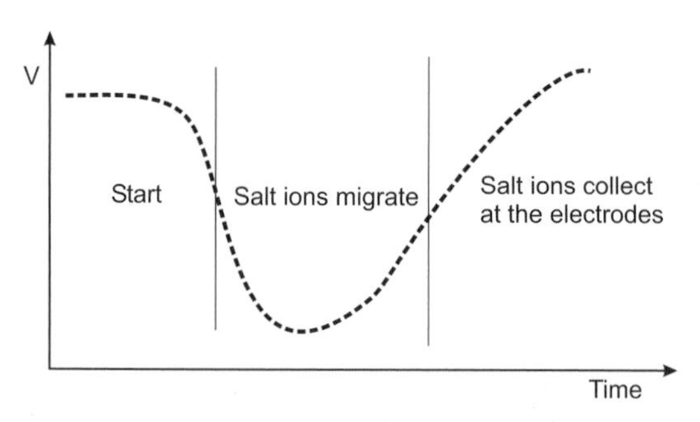

Fig. 15: Development of the actual voltage course during the first IEF phase in an IPG strip.

The separation can be speed up with applying higher electric field strength. For these high voltages the instrumentation has to be designed to comply with a high safety level.

Volthours The complete voltage load is defined in volthour integrals (Vh): the amount of voltage applied over a certain time. For example: 5000 Vh (5 kVh) can be 5000 V in one hour or 1000 V over five hours.

- The Vh definition corrects the running conditions for different conductivities in different strips caused by different protein compositions and salt loads.
- In many cases increasing the voltage gradients (ramping) rather than in steps improves the result considerably. In order to apply comparable voltage loads, the Vh value is a good measurement.

Under focusing When the applied volthours are insufficient, not all spots are round. Horizontal streaks are produced instead. Higher volthours loads are needed for samples containing more high molecular weight proteins, more hydrophobic proteins and preparative runs.

Over focusing When the proteins are focused too long, some other negative effects can happen:

These effects are result of too long time, not of too high voltage.

- Cysteins become oxidized, the pI of the protein changes.
- Some proteins become instable at their isoelectric point. The modified proteins or its fragments have different pI and start to migrate again. The result: some horizontal streaks coming out from some spots.
- Basic narrow gradients are particularly sensitive to over focusing. Additionally to possible protein breakdowns, the basic extreme (> pH 10) of the gel can become instable because of the very high pH.

The best results are obtained with a focusing phase as short as possible at a voltage as high as possible.

Should the optimal Volthours have been achieved during the night or another time of no attendance, no voltage should be applied on the strip. Prior to removal the strips from the strip holders the highest possible voltage is applied for 15 minutes, in order to refocus the slightly diffused bands.

Sample entry effects During the first phase of IEF with the cup loading procedure interesting effects are observed:

Josef Bülles: personal communication.

- If the voltage setting in the beginning is low (100 V, 18 cm IPG strip), the proteins enter the gel, but a part of them precipitates or gets stuck within the first few cm of the strip, when the voltage is raised to quickly after this phase. A vertical

When low voltage is applied for the start, it should be continued to use low voltage during the first few hours.

streak, which is displaced a few centimetres from the application point, is visible in the stained second dimension gel. Obviously the precipitated proteins are locked into the strip.

- If the voltage setting for the start is high (500 V, 18 cm IPG strip), a part of the proteins accumulate and precipitate on the surface of the application point and show a vertical streak in the second dimension. Sometimes the precipitated proteins at the application point are washed off during equilibration: no streak is visible at all.

In the second case the 2-D separation shows a clearer pattern than in the first.

> ***Both matters are critical: the voltage level during sample entrance and the beginning of protein migration, and the duration of the focusing phase at high voltage.***

Instrumentation

The Multiphor The IPG procedures had been optimised on the Multiphor chamber (see figure 16). With this equipment, rehydration can only done in the reswelling tray or a rehydration cassette outside the instrument. The separation is performed in a tray for up to 12 strips.

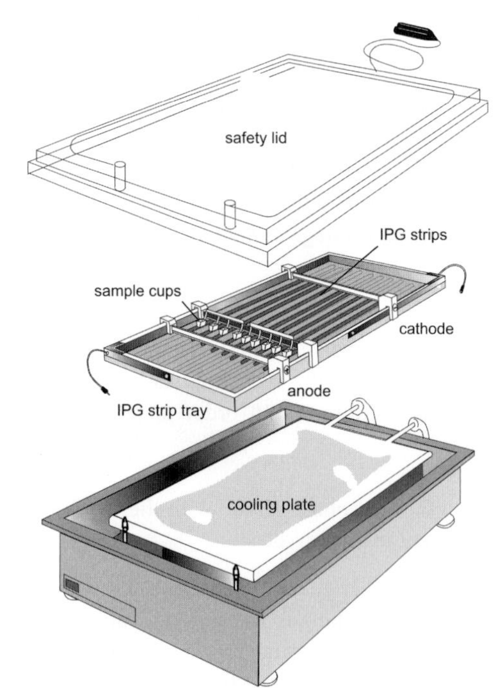

Fig. 16: Multiphor chamber with tray for IEF in IPG strips. An external programmable power supply for high voltages and a thermostatic circulator is required.

Due to safety regulations the maximum voltage allowed is 3,500 V, because in the modular set-up power supply and cooling are connected to the chamber via cables and tubings. The handling of the IPG strips and the entire IEF procedure were improved by developing a dedicated instrument for IEF in IPG strips for 2-D electrophoresis.

The IPGphor In the instrument shown in figure 17, rehydration and IEF can be combined. In this case the IPG strips are run facing down. With individual trays, called strip holders, any carry-overs of samples or additives are prevented. The high thermal conductivity of the ceramics material is very important to remove the locally developed heat from some areas of the IPG strips efficiently. The programmable power supply, which generates up to 8000 V, and a Peltier cooling system are integrated into the electrophoresis chamber.

When everything is integrated in one instrument, much higher voltages can be applied than on modular systems with separate power supplies and thermostatic circulators.

The strip holders contain platinum contacts at fixed distances. The electric field is applied at the gel strips through these contacts, which are positioned on the respective contact area of the cooling plate. The anodal contact area is large to accommodate all different lengths of strip holders.

Up to 14 regular strip holders fit into the instrument.

At the bottom of the apparatus is a serial port for possible software updates and to connect a computer or serial printer to the instrument for a report on the electric conditions after every five minutes of the run. With this interface the instrument can be integrated into the laboratory workflow system and allow procedures according to GLP (good laboratory practice).

For optimisation of running conditions the report of the actual electric parameters is very useful.

It is not recommended to run strips with different lengths and different pH gradients at the same time. Also, the temperature setting of 20 °C should not be modified (see above).

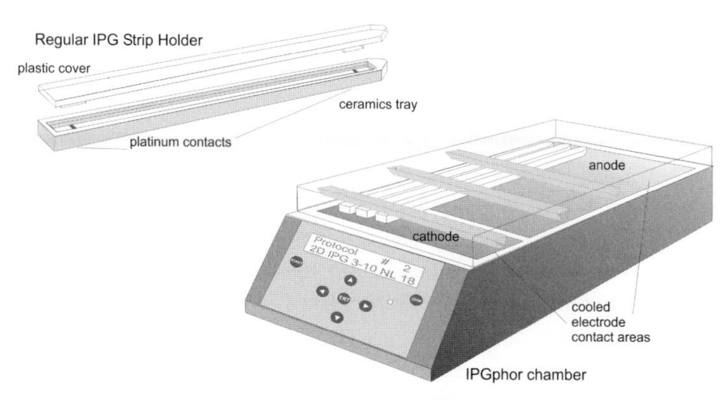

Fig. 17: IPGphor regular strip holder and IPGphor chamber for IEF in IPG strips.

Regular Strip Holders In figure 18 the procedure how to apply sample and IPG strips is shown. The trays, made from specially treated aluminium ceramics are placed on the thermostated electrode contact areas of the power supply. Rehydration as well as the IEF separation are carried out at +20 °C.

Ceramics material is the optimum, plastic does not have a good thermal conductivity. Other materials have electric conductivity, bind proteins, or show EEO effects.

The advantages are:

In most laboratories this method has become the default procedure.

- The number of manipulations and hands-on steps is reduced.
- Up to an entire day can be saved.
- Rehydration can be performed under low voltage (30 – 60 V) for 10 hours, which facilitates the entry of high molecular weight proteins larger than 150 kDa into the strips (Görg *et al.* 1998).

Görg A, Boguth G, Obermaier C, Harder A, Weiss W. Life Science News 1 (1998) 4–6.

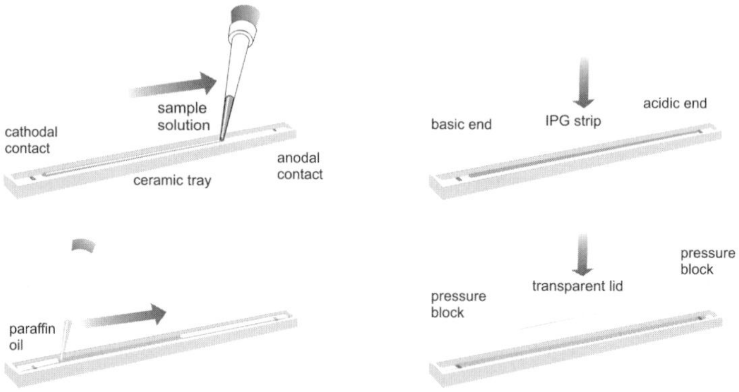

Fig. 18: Rehydration loading into IPG strips in individual strip holders of the IPGphor.

The distance between the platinum contact and the end of the tray is shorter at the anodal end.

The pressure blocks on the coverlids hold the strips down, in order to maintain the contact also when electrolysis gas is produced during IEF. The bars of the safety lid of the chamber press the coverlids down. The most reproducible results are obtained when the acidic (pointed) end of the strip is always placed in direct contact with the anodal end of the tray.

Clean strip holders should be handled with gloves to avoid contamination.

Cleaning of the strip holders The strip holders must be carefully cleaned after each IEF run. The solutions must never dry in the strip holder. Cleaning is very effective, if the strip holders are first soaked a few hours in a solution of 2–5% of the specially supplied detergent in hot water.

Usually the strip holder slot is vigorously brushed with a toothbrush using a few drops of undiluted IPGphor Strip Holder Cleaning Solution. Then it is rinsed with deionised water.

Sometimes protein deposits are left on the bottom of the strip holder after IEF. This happens when highly abundant proteins have been squeezed out of the gel surface at their isoelectric point (see figure 19). It is not always easy to remove these proteins, particularly when they are sticky like serum albumin. In this case the strip holders should be boiled in 1 % (w/w) SDS with 1 % (w/v) DTT for 30 minutes before the slot is cleaned with the toothbrush.

SDS solution in absence of buffer has a neutral pH.

Important
Strip holders may be baked, boiled or autoclaved. But, because of the specially treated surface they must not be exposed to strong acids or basis, including alkaline detergents.

Note
The strip holder must be completely dry before use.

For most analytical applications this procedure works very well and delivers qualitative good and highly reproducible results. However, there are a few situations, where running the strips gel side up works better. In some cases cup loading delivers the best results (see also comparison on page 39).

For cup loading the gels are always run gel side up.

Preparative runs It has been observed, that when the sample loads are increased from analytical (ca. 100 µg) to preparative (> 1 mg) amounts, the quality of the 2-D pattern decreases, when the strips are run with the surface down. The increasing amount of the highly abundant proteins cause this phenomenon. In the end phase of IEF every focused protein forms a little ridge. These little ridges can be easily seen on the surface of the strips after IEF. When very high abundant proteins are focused, they form much higher ridges than the other proteins. In the case of regular strip holders the weight of the strips rests on these ridges, and the proteins are partly squeezed out. These proteins diffuse along the interface between the gel surface and the strip holder bottom (see figure 19), and create a smear. When the strips are run with the surface up, no pressure is applied on the ridges. Thus, the results are better, when preparative runs are performed with the surface up in a so-called cup loading strip holder.

Another advantage of running preparative samples with the surface up is, that the electrodes are easier accessible for the insertion of filter paper pads between the electrodes and the strip. In this way salt and buffer ions as well as proteins with isoelectric points outside the pH range of the strip's gradient can be collected in the paper pads.

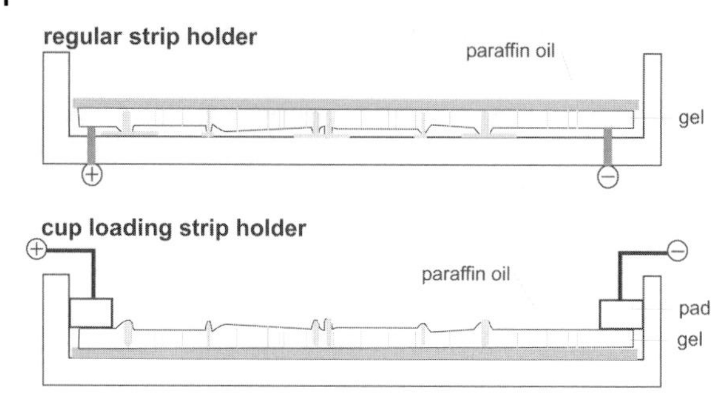

Fig. 19: Schematic representation of the running conditions for IPG strips with high protein loads. Comparison of runs with the surface down and with the surface up.

Narrow pH gradients High sample load also means high load of contaminants. This is particularly important for runs in narrow pH intervals. The advantage of narrow pH gradients is the high spatial resolution, but also the high loading capacity. Most of the protein load, however, will accumulate at the electrodes, because most proteins have isoelectric points outside the pH range of the strip. If these proteins cannot migrate out of the gel, they will precipitate there. Because they are charged, they can cause electroendosmotic effects: the resulting local water transport pushes non-precipitated proteins towards the gel center, which are visible as horizontal streaks at the lateral sides of the 2-D gel.

If the sample has to be electrophoretically desalted (see page 20) the filter pads have to be exchanged several times against new ones.

The name "cup loading strip holder" is sometimes misleading.

Cup loading strip holders (fig. 20) These strip holders cannot only be used for cup loading, but also strips, which have been loaded with proteins by rehydration.

- Very alkaline and very acidic gradients give better results with *cup loading*. They are applied onto the area of the milder pH rather than with rehydration loading.
- High sample loads (> 1 mg) often require runs with the gel surface facing up. The samples are either applied with *rehydration loading* in the external reswelling tray (Fig. 11) or in regular strip holders.

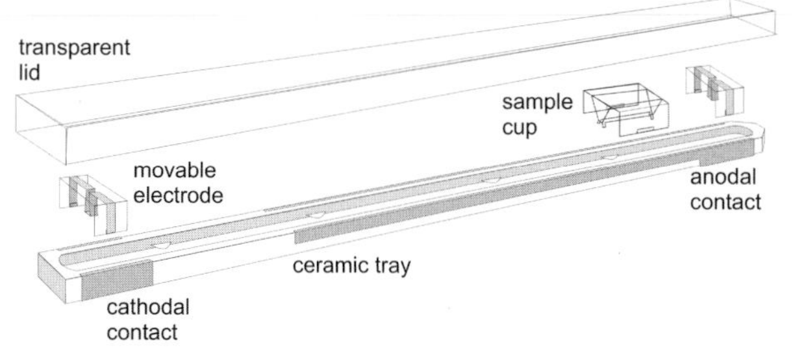

Fig. 20: Cup loading strip holder.

Oil leakage leads to drying of the strip surface and sometimes to crystallization of urea on the surface.

4 mL paraffin oil are pipetted into the strip holder before filter paper pads and the loading cup are put into position. Filter paper pads are soaked in deionised water and inserted between the electrodes and the IPG strips to trap salt ions, and those proteins, which possess pIs above and below the pH gradient. Sometimes high protein loads, salt and Tris ions cause oil leakage out of the strip holder. This is prevented with applying only 300 V for 3 hours and slowly increasing the voltage by ramping over several hours. Example: Figure 21 shows a proposal for the voltage setting for cup loading of 1 to 2 mg protein on an 18 cm long strip pH 3–10 non-linear, with final 50 kVh.

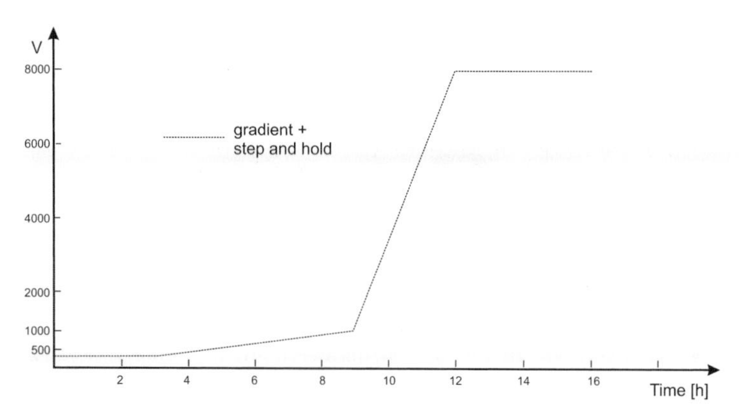

Fig. 21: Voltage setting for a preparative cup loading procedure on an 18 cm IPG strip pH 3–10 non-linear.

The gel strips are never rehydrated in the cup loading strip holder, because it is too wide. Either they are rehydrated in the reswelling tray (fig.11) or in the regular strip holder (fig.18). Of course, rehydration loading can be applied with those as well. Sometimes rehydration under low voltage improves the result. The strips can be transferred in the morning after the initial rehydration / run from the regular strip holders to cup loading strip holders.

Cleaning Cleaning of the cup loading strip holder is less critical than for the regular strip holder, because the gel surface and the proteins do not have direct contact with the surface. It is – like the regular strip holder – brushed with a toothbrush using a few drops of undiluted IPGphor Strip Holder Cleaning Solution. Then it is rinsed with deionised water. The loading cups can be reused for most of the samples after thorough cleaning with laboratory detergent.

Basic pH gradients It has already been mentioned above, that rehydration loading in basic gradients must be avoided. Here are a few hints how to improve the results for basic gels:

- Samples are applied by cup loading onto the anodal end of the gradient of the pre-rehydrated gel.
- The lowest He load possible must be applied: short time at high voltage.
- A filter paper pad soaked in at least 0.4 % DTT solution should be placed close to the cathodal pad (Görg *et al.* 2000).

Görg A, Obermaier C, Boguth G, Harder A, Scheibe B, Wildgruber R, Weiss W. Electrophoresis 21 (2000) 1037–1053.

- The rehydration solution should contain additives, which suppress electroendosmosis effects: 10 to 16 % isopropanol and sometimes 0.2 % methylcellulose (Görg *et al.* 1997).

Görg A, Obermaier C, Boguth G, Csordas A, Diaz J-J, Madjar J-J. Electrophoresis 18 (1997) 328–337.

Basic pH gradient with high sample loads For high protein loads in basic gels, Hoving *et al.* (2002) suggest some additional measures to achieve good 2-D patterns in the gradients pH 6–11, 6–9, and narrow ranges:

Hoving S, Gerrits B, Voshol H, Müller D, Roberts RC, van Oostrum J. Proteomics. 2 (2002) 127–134.

- Rehydration of the IPG strips overnight in 7 mol/L urea, 2 mol/L thiourea, 4 % CHAPS, 2.5 % DTT, 10 % isopropanol, 5 % glycerol, 2 % IPG buffer pH 6-11.
- Cup loading at anodal side of 18 cm strips.
- At the cathodal side a "CleanGel" paper wick soaked in rehydration solution plus 3.5 % DTT is added. The protocol was optimized in the Multiphor electrophoresis chamber. For the cup loading strip holder in the IPGphor a 25 × 8 × 1 mm IEF electrode strip can be used like the paper bridge in figure 20, but without the intermediate electrode pad.

Paper bridge loading For application of very high sample volumes a modification of the cup loading principle has been proposed: paper bridge loading. Solution containing up to 5 mg protein can be loaded on an 18 cm long narrow pH range IPG strip (Sabounchi-Schütt *et al.* 2000). A large sample volume requires that a large paper pad be applied at the other end of the IPG strip to absorb excess water. Figure 22 shows the arrangement used when sample is applied to a paper bridge positioned between the anode and an 18 cm long IPG strip. Paper bridge and electrode pad are cut from the 1 mm thick CleanGel electrode strips to a size of 8 × 45 mm and 8 × 35 mm to fit at the anodic and cathodic end of the cup loading strip holder. The electrode pad is soaked with distilled water and blotted with a tissue paper to become damp, not wet. Sample solution is applied to the paper bridge (450 µL for anodic sample application and 350 µL for cathodic sample application). The rehydrated IPG strip is first positioned in the bottom of the stripholder. Then the paper bridges and electrode pads are positioned as indicated in figure 22. With anodic application the anode is positioned as far out as possible in the electrode holder, while the cathode is positioned close to the end of the IPG strip to ensure good contact between electrode pad and IPG strip. A 6 mm long soft plastic tubing is positioned as indicated. When the cover is placed over the strip holder, it will press down the tubing and ensure good contact between the anodic paper bridge and IPG strip.

Sabounchi-Schütt F, Aström J, Olsson I, Eklund A, Grunewald J, Bjellqvist B. Electrophoresis 21 (2000) 3649–3656.

Note
The application point (anodic or cathodic) is of importance to obtain good results.

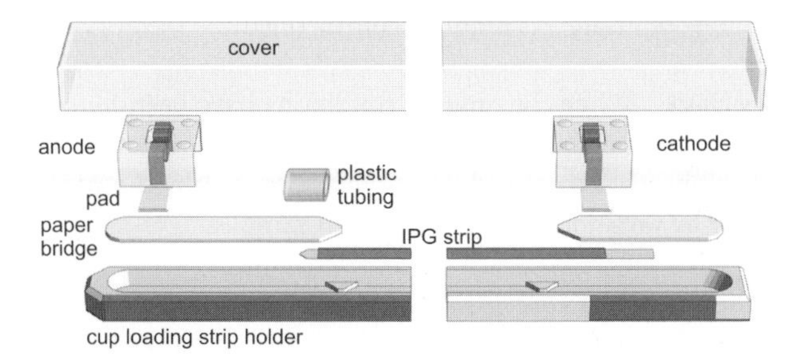

Fig. 22: Paper bridge loading of large sample volumes.

Strategy for optimal IEF running conditions

In the following an example is given how to plan optimisation work and to check the reliability of the analysis. Details on running conditions are given in the protocol Step 2: Isoelectric Focusing, pages 190 ff.

Tab. 2: Strategy for IEF

Miniformate gel		
1	Wide gradient 3–10 and/or 4–7 in 7 cm gels	Optimisation of sample preparation, and pre-check for optimal sample application method
Large gels		
2	Wide gradient pH 3–10 linear or non linear	Application of ca. 100 μg total protein by rehydration loading at 50 V for 12 hours.
3	Wide gradient pH 3–10 linear or non linear	Rehydration loading in reswelling tray and IEF in cup loading strip holder.
4	Wide gradient pH 3–10 linear or non linear	Cup loading on pre-rehydrated strip: one application at the anodal end and one on the cathodal end.
Decision point: Selection of the best procedure.		
5	Semi-wide gradients pH 4–7, 6–11, etc.	Way of loading is dependent on the results in the wide gradients.
6	Narrow gradients, like one pH unit.	Way of loading is dependent on the results in the wide gradients.
7	Basic gradients	Only: Cup loading on pre-rehydrated strip: one application at the anodal end.

Methodology check If the result is not satisfying – streaky and smeary pattern, only few or no spots – it is important to find out whether the problem is caused by inadequate sample preparation or something goes wrong during the separation. The trouble shooting guide in the appendix of this book is very useful. But a complete run with a test sample – E.coli lyophilisate – is highly recommended.

Replicate gels are anyhow highly recommended for statistically reliable results.

Reproducibility check When the optimal conditions have been found, samples have to be run in at least doublets or triplets, in order to check, whether observed pattern differences are caused by the noise of the system or by variations between different samples.

–20 to –40 °C are not enough: some proteins become modified.

After IEF Either the strips containing the focused proteins are equilibrated in SDS buffer and run on the second dimension right away, or they are stored at –60 to –80 °C in a deep-freezer.

Staining of an IPG strip

Acid violet 17 Sometimes it can be useful to check, whether the separation in the first dimension worked well before all the work with the second dimension run is started. Because of the high urea and detergent concentrations in the gel silver and Coomassie Brilliant blue staining produce a dark background. The most sensitive technique is Acid violet 17 staining according to Patestos *et al.* (1988). The bands are visible after 40 minutes, during this time the proteins in the other strips can be kept focused with the application of a medium high voltage, like 2000 V.

Patestos NP, Fauth M, Radola BJ. Electrophoresis 9 (1988) 488–496. Of course, the stained strip does no longer release the proteins into the second dimension gel.

Imidazol zinc reversible staining As mentioned in the preface, this book is derived from a 2-D electrophoresis course manual. When the time for a course is very limited, several time consuming steps can be run in parallel. The second dimension, SDS PAGE, can be started shortly after the first dimension with prerun IEF strips. Usually those are stored in a freezer at – 80 °C. Such a freezer is not always available at the location of the course. In such a case it is very useful to have prerun strips which can be stored in a refrigerator and transported at room temperature.

With Imidazol zinc staining according to Fernandez-Patron *et al.* (1998) also IPG strips can be stained and conserved until use. The proteins are not modified, but locked into the gel strip by the imidazol-zinc complexes in the rest of the gel. The strips are stored in the gel preserving solution sealed in plastic bags or tubes in a refrigerator. They are at least stable for half a year. A few days at room temperature for transport do not matter. Short before use the imidazol-zinc complexes are dissolved by placing the strips for 5 minutes into a mobilization buffer containing EDTA.

Fernandez-Patron C, Castellanos-Serra L, Hardy E, Guerra M, Estevez E, Mehl E, Frank RW. Electrophoresis 19 (1998) 2398–2406.

It must be mentioned, that in this procedure there is a loss of about 50 % of the proteins – independent from the molecular weight. Therefore this procedure can only be recommended for courses. The IPG strips should be loaded with double the protein amount for this use.

The detection of protein bands with this procedure is not as good as with acid violet 17.

Measurement of the pI

The pH gradient in an IPG strip cannot be determined with a surface electrode, because the conductivity is too low. The calculation of the pH gradient based on the pK values and the concentrations of the Immobilines is rather complex, because the presence of urea and the running temperature has to be taken in account. The addition of protein standards is generally not recommended, because absolute purity of these proteins would be required, and adding additional proteins to a complex mixture can easily lead to errors in interpretation of the pattern.

Two different ways of pI measurement are proposed for 2-D electrophoresis:

Amersham Biosciences Data file: Immobiline DryStrip visualisation of pH gradients. (2000) 18-1140-60.

1. pH gradient graphs Graphs of operational pH gradient profiles are published in a data file by Amersham Biosciences (2000) as a basis for pI estimation. More information and larger graphs can be found at this website: http://proteomics.amershambiosciences.com. These gradients are the calculated target pH gradients at 20 °C in presence of 8 mol/L urea. The pH values are plotted over the gel length given in per cent. Two examples are shown in figure 23.

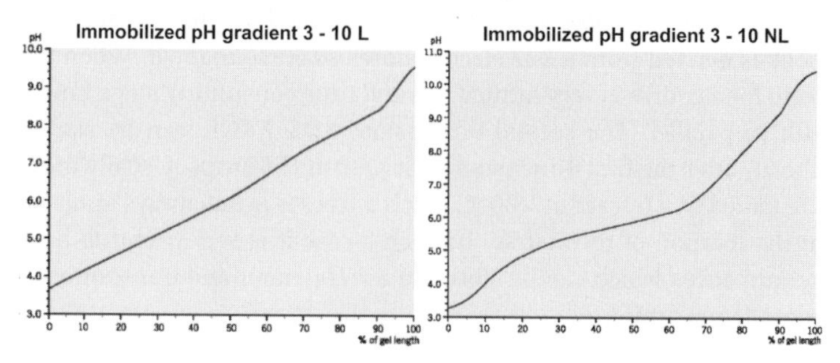

Fig. 23: pH gradients in IPG strips 3–10 and 3–10 NL. Downloaded from the following website: http://proteomics.amershambiosciences.com

The pI of a protein is estimated by relating the position of the protein in the SDS gel to its original position in the IPG strip, plotting the band position versus pH, and read out the pI. The strips do not contain pH plateaus at the ends: the graph must be aligned from one strip end to the other. Typical standard deviations for different batches of commercial IPG strips are given in the data file, mentioned above, and website.

Note

Different strip lengths are used, and non-backed gels can swell or shrink.

This procedure is usually carried out with the image analysis software as a "1-D calibration".

Bjellqvist B, Hughes GJ, Pasquali C, Paquet N, Ravier F, Sanchez J-C, Frutiger S, Hochstrasser D. Electrophoresis 14 (1993) 1023–1031.
Bjellqvist B, Basse B, Olsen E, Celis JE. Electrophoresis 15 (1994) 529–539.

2. Interpolation between identified sample proteins with known pI Prominent spots showing up in each 2-D map of a sample type can be analysed for identification and amino acid sequence information. The theoretical pIs can then be used as keystones for interpolating the pIs of the other proteins. Bjellqvist *et al.* (1993) have successfully correlated the calculated pIs from the amino acid sequences of proteins with the protein position in immobilized pH gradients and put the procedure into practice for human cells (Bjellqvist *et al.* 1994).

The second procedure is performed with the image analysis software as "2-D calibration".

This subject will be further explained in the chapter on image analysis of 2-D electrophoresis gels.

Prefractionation of the sample by electrophoresis and isoelectric focusing
It has already been mentioned at the end of the sample preparation chapter, that it would be desirable to preseparate complex protein mixtures according to their isoelectric points. Unfortunately, prefractionation can cause losses of proteins. Electrophoretic separation techniques are the most interesting candidates for this step, because they can be carried out with the least surface effects.

Preparative electrophoretic techniques are often limited by the cooling possibilities, because the Joule heat has to be efficiently dissipated. IEF techniques produce less heat, because the current flow is low. The separated fractions should preferably be available in gel-free liquid form. The two techniques described here have a good practical potential without too much protein losses:

Free flow electrophoresis and IEF In the free-flow approach, originally developed by Hannig (1982), a continuous stream of buffer flows in a 0.5 to 1.0 mm wide layer in a cooled glass cuvette. The buffer is supplied to the cuvette over the entire width. At one end the sample is injected at a defined spot and at the other end, the fractions are collected in an array of tubes. The electrical field is applied perpendicular to the buffer flow.

Hannig K. Electrophoresis 3 (1982) 235–243.

The varying electrophoretic mobilities perpendicular to the flow lead to differently heavy but constant deviations of the components so that they reach the end of the separation chamber at different though stable positions (see figure 24).

Wagner H, Kuhn R, Hofstetter S. In: Wagner H, Blasius E. Ed. Praxis der elektrophoretischen Trennmethoden. Springer-Verlag, Heidelberg (1989) 223–261.

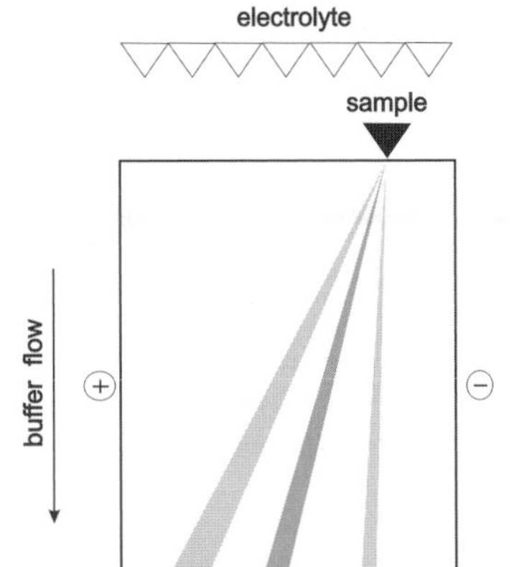

Fig. 24: Schematic drawing of continuous free flow electrophoresis. According to Wagner *et al.* (1989).

Different separation methods can be applied:

Discontinuous field electrophoresis The sample solution is supplied in a wide zone through the central openings, the buffer solutions on the right and on the left side have a twenty times higher conductivity than the sample solution. The sample ions are more or less strongly deflected towards the anode or the cathode, depending on their charges. When they reach the borderline between sample and buffer stream, their electrophoretic mobility is considerably reduced, resulting in a concentration of the fraction at the borderline.

Isoelectric focusing Either carrier ampholytes or multi-components buffers, which are mixtures of amphoteric and non-amphoteric buffers are added. The latter mixture cannot develop linear pH gradients, but they reduce the costs. With IEF a higher number of fractions can be obtained than with the discontinuous field electrophoresis.

Buffering isoelectric membranes

Wenger P, de Zuanni M, Javet P, Righetti PG. J Biochem Biophys Methods 14 (1987) 29–43.

The principle of this technique is based on the concept of immobilized pH gradients. The separation occurs in the liquid phase in a multicompartment apparatus, which is divided by buffering isoelectric membranes with defined pIs (Wenger *et al.* 1987). The electrodes are located in the two outer segments. The membranes are prepared by polymerizing a polyacrylamide gel layer with basic and acidic Immobilines around a microfibre filter or polyester grid. The amounts of Immobilines needed for a certain pH value are precalculated in the same way like for the immobilized pH gradients. Figure 25 shows the principle of fractionation with isoelectric membranes.

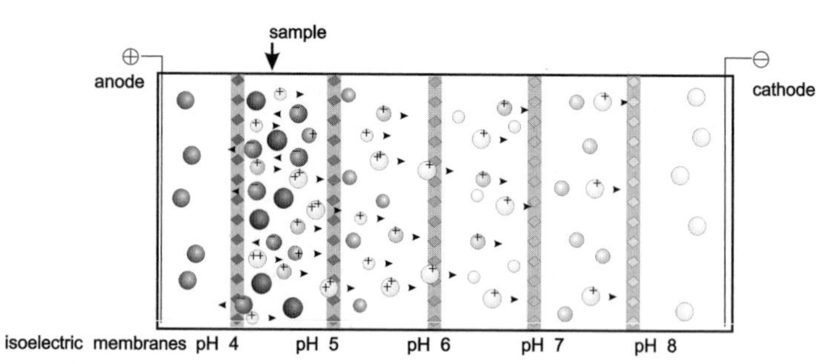

Fig. 25: Prefractionation of a protein mixture in the liquid phase with isoelectric membranes.

In a mixture the sample proteins are charged. In the electric field, a protein with a high pI is protonated and migrates through the compartments towards the cathode until it reaches a membrane with a higher pI. It cannot pass through this membrane, because it will become deprotonated. If it will get a negative net charge there, it will migrate back towards the anode until it reaches the previous membrane with a lower pI. There it will become protonated again, and so on. It is thus trapped in the compartment.

Herbert and Righetti (2000) could show, that sample prefractionation *via* multicompartment electrolysers with isoelectric membranes greatly enhances the load ability, resolution and detection sensitivity of 2-D maps in proteome analysis.

Herbert B, Righetti PG. Electrophoresis 21 (2000) 3639–3648.

2.1.4 Second dimension: SDS-PAGE

The methodology for the second dimension did not change as dramatically as for isoelectric focusing. But also for SDS-PAGE some developments in chemistry and instrumentation contributed to improved handling and reproducibility of the spot positions.

Theoretical background

Sodium dodecyl sulphate (SDS) electrophoresis With SDS-PAGE (SDS-polyacrylamide gel electrophoresis) the polypeptides are separated according to their molecular weights (Mr). SDS and proteins form complexes with a necklace-like structure composed of protein-decorated micelles connected by short flexible polypeptide segments (Ibel *et al.* 1990). As a result of the necklace structure large amounts of SDS are incorporated in the SDS-protein complex in a ratio of approximately 1.4 g SDS/g protein. SDS masks the charge of the proteins themselves and the formed anionic complexes have a roughly constant net negative charge per unit mass. Usually a reducing agent such as DTT is added to the sample to cleave the disulfide bridges between cysteins.

Ibel K, May RP, Kirschner K, Szadkowski H, Mascher E, Lundahl P. Eur J Biochem 190 (1990) 311–318.

All proteins – also those with basic pIs – will migrate towards the anode. The electrophoretic mobility of proteins treated with SDS and DTT depends largely on the molecular weight of the protein. At a certain polyacrylamide percentage there is an approximately linear relationship between the logarithm of the molecular weight and the relative migration distance of the SDS-polypeptide complexes of a certain molecular weight range. The molecular weights of the sample proteins can be estimated with the help of co-migrated standards with known molecular weights.

Mostly gels with T-values of 12 to 13 are used for the optimal separation in the range between 20 and 100 kDa.

Note
SDS PAGE cannot deliver the exact molecular mass of a protein, it allows only an estimation.

Exact masses can only be determined with mass spectrometry.

Lämmli UK. Nature 227 (1970) 680–685.

Buffer and gel The standard buffer system for second-dimension SDS-PAGE is based on the discontinuous Tris-chloride / Tris-glycine system described by Laemmli (1970). In the classical procedure according to O'Farrell the unmodified Laemmli system was employed, including a stacking gel – with Tris-chloride pH 6.8 – polymerised on top of the resolving gel, which contains Tris-chloride pH 8.8. For 2-D electrophoresis the stacking gel is not needed, because

- The proteins are already pre-separated by IEF and will therefore not aggregate while they enter the resolving gel.
- The proteins migrate from a gel into another gel, and not from a liquid phase into a gel.

Therefore the running buffer must not contain any chloride ions.

This is still a discontinuous buffer system. The spots are well resolved, because there is a stacking effect happening between the highly mobile chloride in the gel and the lowly mobile glycine in the running buffer.

Omitting the stacking gel solves a few technical problems and makes the procedure easier:

- The edge between stacking and resolving gel contain incompletely polymerised acrylamide monomers and oligomers, which stick to proteins and partly modify proteins.
- Some proteins get caught between stacking and resolving gel.
- One source of problems with reproducibility is abolished.
- Ready-made gels cannot contain a stacking gel with a buffer different from the resolving gel because of diffusion.
- The additional step of polymerizing a stacking gel short before the run is abolished.

Buffer composition and volume The running buffer and the gels contain 0.1 % SDS. Sometimes the gels are cast without SDS, because the SDS migrating into the gel from the cathodal buffer is sufficient.

During electrophoresis the negatively charged chloride, SDS and glycine ions migrate towards the anode, the positively charged Tris ions migrate towards the cathode. The buffer reservoirs of the electrophoresis chamber must be large enough to prevent depletion of the buffer ions. Or the concentrations of the running buffers must be increased. Here are two examples:

The buffer concentrations in the polyacrylamide buffer strips used in flatbed systems has to be five times higher than in a liquid tank buffer.

For novel high-throughput instruments running large vertical gels the volumes of the upper – cathodal – tanks has been reduced to simplify the handling of the parts. Therefore the buffer in the upper tank must be one and a half to two times more concentrated than the buffer in a conventional apparatus.

Gel composition Polyacrylamide gels are polymerised from acrylamide monomers and a cross-linking reagent – usually N,N′-methylenebisacrylamide. The reaction is started with ammonium persulfate as catalyst; TEMED provides the tertiary amino groups to release the radicals. The pore size can be exactly and reproducibly controlled by the total acrylamide concentration *T* and the degree of cross-linking *C*:

$$T = \frac{(a + b) \times 100}{V} \, [\%], \quad C = \frac{b \times 100}{a + b} \, [\%]$$

a is the mass of acrylamide in g,
b the mass of methylenebisacrylamide in g,
V the volume in mL.

When *C* remains constant and *T* increases, the pore size decreases. When *T* remains constant and *C* increases, the pore size follows a parabolic function: at high and low values of *C* the pores are large, the minimum being at $C = 4\%$. Gels with $C > 5\ \%$ are brittle and relatively hydrophobic. They are only used in special cases.

In order to get reproducible gels, the concentrations of the catalysts have to be balanced in favor of a higher TEMED and a lower ammonium persulfate concentration. At basic pH, ammonium persulfate can react with the Tris; this effect is minimized by adding more TEMED, and by reducing the ammonium persulfate content.

Electroendosmosis effects This phenomenon occurs, when fixed charges are present in an electric field: for instance carboxylic groups. These groups become ionized in basic and neutral buffers: in the electric field they will be attracted by the anode. As they are fixed in the matrix, they cannot migrate. This results in compensation by the counter flow of H_3O^+ ions towards the cathode: electroendosmosis. In gels, this effect is observed as a water flow towards the cathode, which carries solubilized substances along.

When immobilized pH gradient gels are placed into neutral or basic buffers, they become deprotonated: The carboxylic groups become negatively charged, the amino groups are neutral. Thus the IPG strips acquire negative net charges during equilibration with the SDS buffer. It has been observed already during the first approaches utilizing immobilized pH gradients in 2-D electrophoresis, that this can cause electroendosmotic effects (Westermeier *et al.* 1983). Those result in protein losses due to a water flow towards the cathode, see figure 26. Several measures have to be taken to minimize this effect during the protein transfer from the IPG strips to the SDS gel.

Westermeier R, Postel W, Weser J, Görg A. J Biochem Biophys Methods 8 (1983) 321–330.

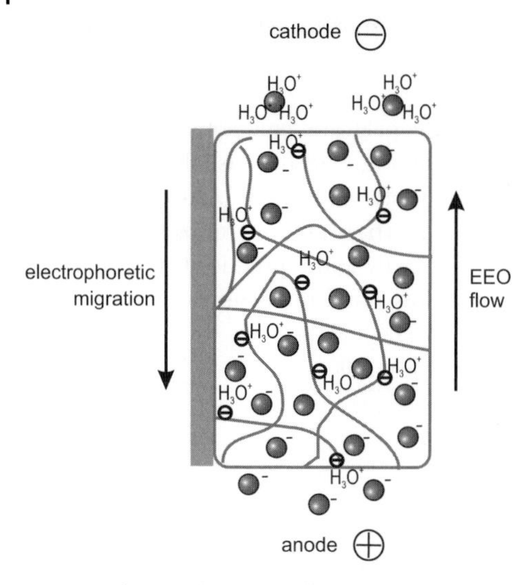

Fig. 26: Electroendosmosis effect caused by negatively charged IPG strips. The water transport can carry proteins away into the cathodal direction.

Görg A, Postel W, Günther S, Weser J. Electrophoresis 6 (1985) 599–604.

The water transport is minimized by adding urea and glycerol to the equilibration solution and reducing the electric field during the phase of protein transfer into the second dimension gel (Görg *et al.* 1985).

SDS Gels

Homogeneous gels Usually a homogeneous gel with 12.5 % *T* and 3 % *C* is used; the proteins of major interest in the size range from 14 to 100 kDa are optimally resolved. Only in special cases the matrix concentration is modified to increase the resolution in certain molecular size ranges.

The thumb rule is:

- Lower *T* value: better resolution for high molecular weight proteins.
- Higher *T* value: better resolution for low molecular weight proteins.

Berkelman T, Stenstedt T. Handbook: 2-D electrophoresis. Amersham Biosciences 80-6429-60 (1998).
Langen H, Röder D, Juranville J-F, Fountoulakis M. Electrophoresis 18 (1997) 2085–2090.

A table with optimal gel concentrations for different protein size ranges can be found in the 2-D Handbook by Berkelman and Stenstedt (1998). However, as demonstrated by Langen *et al.* (1997), it is not easy to predict the correct acrylamide concentration according to a mathematical function: In an example shown in this paper a group of proteins with ca. 50 kDa, which were co-migrating in 9–16 % *T* gradient gels were best resolved in a gel with 7.5 % *T*. This shows, that a try and error procedure must be applied for optimization, when certain protein groups need to be completely resolved.

Gradient gels Generally, in gradient gels the overall separation interval is wider, also the linear relation interval between the logarithm of M_r and the migration distance is wider than for homogeneous gels. Also the spots are sharper because the pore sizes are continuously decreasing. This effect can be very useful for separating proteins, which are highly glycosylated. Usually, when homogeneous gels are run under optimal conditions, resolution and spot definition is high enough: a gradient gel is not necessary. The preparation of gradient gels is much more work, it is more difficult to obtain reproducible gel properties than for homogeneous gels.

Peptide gels The resolution of peptides below 14 kDa is not sufficient in conventional Tris-chloride/Tris-glycine systems. The peptides smaller than 10 kDa co-migrate with the SDS front. A number of modifications of the standard gel and buffer system have been proposed, for instance the addition of 8 mol/L urea to the gel by Hashimoto *et al.* (1983), and, additionally, increasing the Tris concentration in the gel to 1 mol/L Anderson *et al.* (1983). The most efficient technique has been developed by Schägger and von Jagow (1987): An additional spacer gel is introduced between stacking and resolving gel, the gel buffer concentration is increased to 1 mol/L Tris-chloride, and the pH lowered to 8.4, tricine is used as terminating ion instead of glycine. In this way the destacking of peptides and SDS is much more efficient than in the standard Tris-glycine system. This method yields linear resolution from 100 to 1 kDa.

Hashimoto F, Horigome T, Kanbayashi M, Yoshida K, Sugano H. Anal Biochem 129 (1983) 192–199.
Anderson BL, Berry RW, Telser A. Anal Biochem 132 (1983) 365–275.
Schägger H, von Jagow G. Anal Biochem 166 (1987) 368–379.

However, the Schägger system requires long running time because of the high buffer concentration in the gel, which can become overheated with high electric power. By adding 30 % v/v ethylenglycol to the monomer solution, the buffer concentration can be reduced to 0.7 mol/L (Westermeier, 2001). The resolution is still very good and cracking of glass plates and smiling effects are avoided.

Westermeier R. Electrophoresis in Practice. WILEY-VCH, Weinheim (2001) 211–213.

Reversible gels Practical experiences in mass spectrometry analysis of peptide mixtures show, that better signals are obtained from samples digested in liquid phase rather than in gel plugs. Besides methylenebisacrylamide a number of other cross-linking reagents exist, listed and compared by Righetti (1983). Some of them possess a cleavage site, which allows to solubilise the gel matrix after electrophoresis (see table 3).

Righetti PG.: Isoelectric focusing: theory, methodology and applications. Elsevier Biomedical Press, Amsterdam (1983).

Tab. 3: Alternative crosslinkers for polyacrylamide gels (examples)

Substance	*Cleavage site*	*Cleavage agent*	*Comment*
N,N′-methylenebis-acrylamide (Bis)	none	not possible	standard cross-linker
N,N′-(1,2-Dihydroxyethylene)-bisacrylamide (DHEBA)	1,2 -diol	Periodic acid	
N,N′-Diallyl-ditartar-diamine (DATD)	Ester bonds	Hydrolysis with a base	Problems with inefficient polymerisation
N,N′-Bisacryloylcystamine (BAC)	disulfide bond	thiol reagents	DTT must be completely scavenged by iodoacetamide

Buffer systems

With the discontinuous Laemmli buffer system, reproducible and well resolved patterns are obtained even with high protein loads. It should be noted, that different laboratories prepare the buffers in slightly different ways, this can cause differences of the running conditions. When the gel buffer is titrated with a pH meter, the measuring electrode should be well calibrated. Some laboratories measure the added hydrochloric acid by the volume.

The problem is not only the limitation in shelf life, but also the lack of reproducibility because of the ongoing hydrolysis of the matrix.

Because of the very alkaline pH 8.8 in the gel of the Laemmli system, the polyacrylamide matrix becomes hydrolysed during storage. After two months the sieving property of the matrix is completely gone. In the normal laboratory practice this does not matter, because the gels are consumed within one week after preparation. The situation is different, when ready-made gels are used.

Long shelf life SDS gels For long shelf life, the pH value of the gel buffer has to be reduced to or below pH 7. This requires the modification of the buffer composition.

Because of the tricine in the cathodal buffer, these gels show a very good separation of small peptides.

Tris-acetate/Tris-tricine Tris-acetate buffer with a pH of 6.7 has proven to have a good storage stability and separation capacity for flatbed gels. Glycine has to be replaced by tricine as the terminating ion. Since tricine is more expensive, it is only used in the cathodal buffer. The anodal buffer contains Tris-acetate. The running buffer can be applied as polyacrylamide electrode strips, instead of connecting the gels to liquid buffers in tanks, see also page 70. The molecular weight distribution of proteins obtained with this buffer system is slightly different from that achieved with the Laemmli system.

Bis-Tris/Bis-Tris-tricine This is similar to the above system. Tris is replaced by Bis-Tris to obtain improved buffer capacity at neutral pH. Also with this buffer system the spot patterns are different from Laemmli buffer patterns.

PPA-chloride/Tris-glycine Here the Tris in the gel is replaced by PPA (piperidino propionamide) and titrated with hydrochloric acid to pH 7. This gel buffer can be combined with the standard running buffer Tris-glycine in the cathodal buffer. For the anodal buffer PPA would be required. Because this compound is rather expensive, the anodal buffer is usually composed of diethanolamine-acetate (DEA).

The M_r distribution is very similar to Laemmli gels.

Tris-borate for improved separation of glycoproteins Glycoproteins migrate too slowly in SDS electrophoresis, since the sugar moiety does not bind SDS. When a Tris-borate-EDTA buffer is used, the sugar moieties are also negatively charged, so the speed of migration increases (Poduslo, 1981). The glycine in the running buffer is simply replaced by boric acid. The use of gradient gels is also beneficial for better MW estimations.

Poduslo JF. Anal Biochem 114 (1981) 131–139.

SDS polyacrylamide gels

Flatbed gels Because the IPG strip is just placed on the surface of the gel, a stacking gel with low acrylamide concentration is necessary for optimal protein transfer and separation (Gyenes and Gyenes, 1987). Resolving and stacking gel is usually polymerised in one piece. These gels can only be cooled from one side; rarely gels thicker than 0.5 mm are used. These gels are polymerised onto a support film.

Gyenes T, Gyenes E. Anal Biochem 165 (1987) 155–160.

Lab-made gels are usually run with liquid buffers, connected to the gel with paper wicks. The electrodes are then placed into the buffer tanks. The Laemmli buffers are used. Instructions and recipes for preparing gels for flatbed systems in the laboratory can be found in several books: Görg *et al.* (1994, 2000) and Westermeier (2001).

Görg A. In: Celis J, Ed. Cell Biology: A Laboratory Handbook. Academic Press Inc., San Diego, CA. (1994) 231–242.
Görg A, Weiss W. In Rabilloud T, Ed. Proteome research: Two-dimensional gel electrophoresis and identification methods. Springer Berlin Heidelberg New York (2000) 107–126.
Westermeier R. Electrophoresis in Practice. WILEY-VCH, Weinheim (2001).

For ready-made gels polyacrylamide buffer strips are available, which contain the necessary buffer ions for one run (see also page 70). Electrode wires are placed on top of these electrode strips. The buffer strip concept reduces chemical and radioactive liquid waste considerably (Kleine *et al.* 1992).

Kleine B, Löffler G, Kaufmann H, Scheipers P, Schickle HP, Westermeier R, Bessler WG. Electrophoresis 13 (1992) 73–75.

The IPG strip is placed onto the SDS gel sideways.

Vertical gels As already mentioned, no stacking gel is needed. The standard thickness of SDS slab gels run in vertical equipment are 1 and 1.5 mm. Thinner gels cannot be used, because the IPG strip including the film support is 0.7 mm thick, and it swells during equilibration with SDS buffer.

Tab. 4: Comparison of gel thickness

Advantages of 1.0 mm gels	*Advantages of 1.5 mm gels*
Faster separations possible, leading to sharper spots	Higher loading capacity
Detection methods faster and more sensitive	Higher mechanical stability

Gel casting First of all, it is very important to clean glass plates and equipment very thoroughly. Mass spectrometry analysis is very sensitive, and every contamination will show up in the mass spectrogram and might lead to wrong results. Keratin stemming from hair and skin is the most frequently found protein contamination; sometimes even in reagents purchased from a supplier. It has become general practice in most proteomics laboratories to filter each monomer and buffer solution through a membrane filter before use.

Also for the SDS PAGE step only high quality chemicals should be used. The author of this chapter could write a Who-done-it book on pitfalls caused by dirty or old reagents, which were happening in various electrophoresis laboratories. Particularly for such a sophisticated and multistep procedure like 2-D electrophoresis the search for trouble causing sources is very cumbersome and time consuming.

Meticulously cleaned instruments and glass plates as well as high quality reagents are of high importance.

Gels for miniformate and medium size instruments can be cast as single gels (see figure 27) as well as in multicasters, which are built similar to those used for large gels (see figure 29).

For a high throughput method anyhow multiple gels are needed.

Large format gels, which are usually used in proteomics, can only be prepared in multicasters, because the glass plates of the cassettes would bend out and form a bulbous gel with a higher thickness in the center than along the edges.

For the handling of multiple gels it would be very impractical to use glass plates with separate spacers and clamps. Therefore the cassettes contain fixed spacers and a hinge on one side, as shown in figure 28.

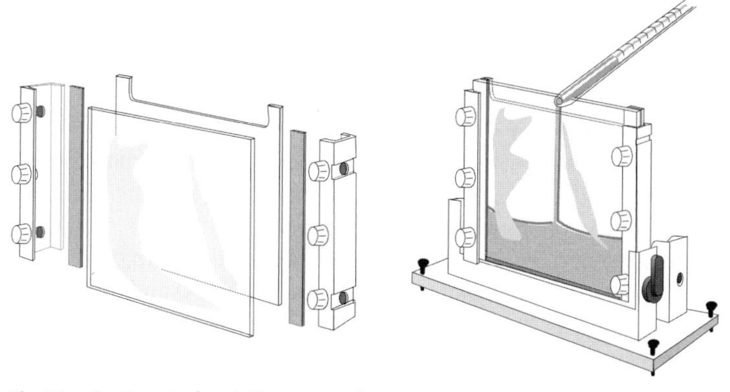

Fig. 27: Casting single miniformate and medium size SDS polyacrylamide gels for 2-D electrophoresis.

Fig. 28: Glass cassette for multiple SDS gel casters and instruments. Spacers are fixed; glass plates are connected with a hinge.

Without the plastic sheets, the cassettes would stick together because of the polymerised gel layers.

Depending on how many gels to be cast at a time, different types of multiple gel casters have been designed. They have in common, that the monomer solution flows into the cassettes from the bottom. Because the liquid will not only flow between the spacers, but also between the cassettes, plastic sheets have to be inserted between the cassettes.

As shown in figure 29, the casters are filled from the top, but the liquid will flow first to the bottom where the stream will be split between the different cassettes. In the multicaster type *A*, which is designed for preparing up to fourteen gels at a time, a coloured displacement solution containing a high percentage of glycerol is filled into a reservoir. When enough monomer solution is filled in, the tube is removed from the reservoir and the dense solution will flow down the channel and displace monomer solution in order to keep the channel free from polymerised gel.

Catalysts composition When multiple gel casters are used, the composition of catalysts in the polymerisation solution is more critical than for single gel casting. If the polymerisation runs too fast, very much heat is produced in the center of the multicasters, leading to thermal convection, which results in curved gel edges. In order to get reproducible gels with a straight edge, more TEMED and less ammonium persulfate should be added than the amounts proposed in some instructions and papers. This prevents overheating.

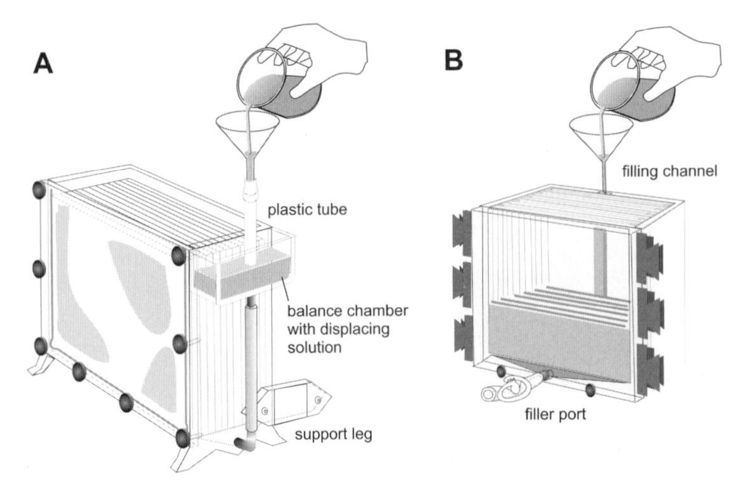

Fig. 29: Multicasters for large SDS gels. Type **A** is designed for up to 14 gels, type **B** for up to 6 gels at a time. Casting of homogeneous gels is shown.

When homogeneous gels are cast in the caster type *B*, no displacing solution is needed, because the dead volume is very low.

Before the polymerisation starts, each single gel has to be overlaid with water-saturated butanol or 70 % (v/v) isopropanol to achieve a straight upper gel edge.

> Note
> ***It is very important to obtain a straight gel surface, because the IPG strip is placed on the gel edge to edge: in this direction the film support it is absolutely rigid and cannot be bent to follow a curved edge.***

The quality and the age of chemicals, particularly the catalysts, greatly influence the polymerisation efficiency of the monomer solution.

Note: The quality of the edge is influenced by the quality of the chemicals.

Gradient gels For the preparation of gradient gels a gradient maker has to be connected to the gel caster. For the caster type *A* from figure 27 the gradient maker is connected with a tubing to the glass tube sitting in the reservoir for the displacement solution. The caster type *B* has to be set up in a different way: The prism-shaped rubber plug has to be removed from the box; the tubing of the gradient maker is connected to the filling port at the bottom of the cover plate. The gradient solution have to be delivered to the box slowly to avoid mixing – 5 to 10 minutes. The speed can be adjusted by using a laboratory peristaltic pump, or by selecting a certain level for the liquid beaker and reducing the diameter of the tubing with a pinchcock clamp (see figure 30).

Using a pump means a lot of work. Everything has to be set up very quickly, because the casting and the overlaying procedure must be finished before the gel solution starts to polymerize. The casting procedure must be as exact as possible to prepare reproducible gels.

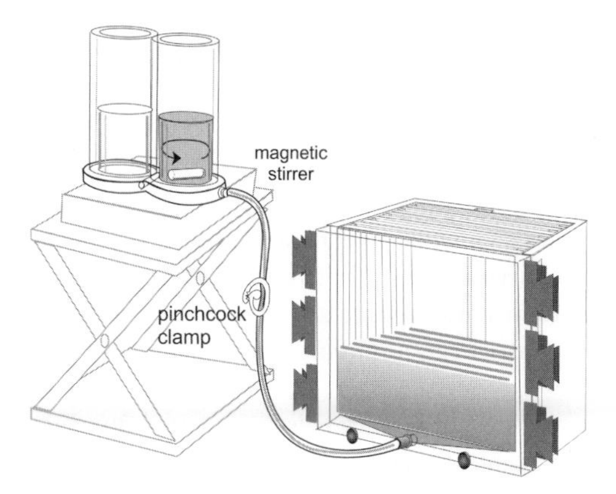

Fig. 30: Casting multiple gradient gels.

Gels for spot cutting After the 2-D patterns have been compared with image analysis software, proteins of interest have to be identified and characterized. For these further analyses the gel plugs containing these proteins have to be cut out of the gel slab. For high accuracy, reliability, and automation of this procedure robotic spot cutters are employed (see page 98 f.). The most accurate and reliable way is to use the x/y coordinates of the respective spots from the image analysis results in the automatic spot cutter. It is, however, important, that the gel does not change its shape by shrinking or swelling. Therefore gel slabs have to be fixed to a rigid film or glass plate backing.

It is not easy to produce these vertical gels on a film support in the laboratory, because the polyester films are flexible. To set up a casting system for film-supported gels would mean a big investment in equipment.

Gels fixed to a glass plate Because it is not easy to bind a gel to a glass plate after staining, gels are already covalently bound to the film or glass plate during polymerisation. Laboratory made gels are usually polymerised onto a glass plate, which has been treated with Bind-Silane. Now, mechanical stability is no longer an issue. Because staining solutions can diffuse into the gel only from one side, the gel should be as thin as possible for faster and efficient staining. These gels are usually 1.0 mm thick.

Non-fluorescent support film was not available at the time this book has been written.

When fluorescence labeling or fluorescence staining is applied, non-fluorescent glass plates are needed. Because this type of glass is very expensive, the gel layer is removed from the glass plate after all analysis has been done. When the glass plates are reused, the gel layer is removed with a plastic scraper. Remaining gel pieces disappear after vigorous treatment with a dishwashing brush.

It is very annoying, when the valuable sample is not separated well, just because of a little mistake occurring while preparing gels.

Ready-made gels for vertical systems Ready-made gels are more expensive than laboratory-cast gels. However, one should not forget, that gel casting is a lot of work – particularly for cleaning the equipment – and working time costs money as well. Commercially produced gels are prepared according to GMP industry standards, and they are quality controlled.

There are two different concepts:

- *Gels in glass or plastic cassettes.* The handling does not differ from that of laboratory-cast gels. The glass cassettes are either sent back to the producer or they are disposed of. The gels contain usually the standard Laemmli buffer and have to be used instantly.
- *Gels on film-support.* The 1 mm thick gels are inserted into specially designed re-usable cassettes (figure 31). Conventional glass cassettes, as shown in figure 28, cannot be used, because air pockets between the glass plate and the backing develop mechanical pressure on the gel, leading to an irregular front.

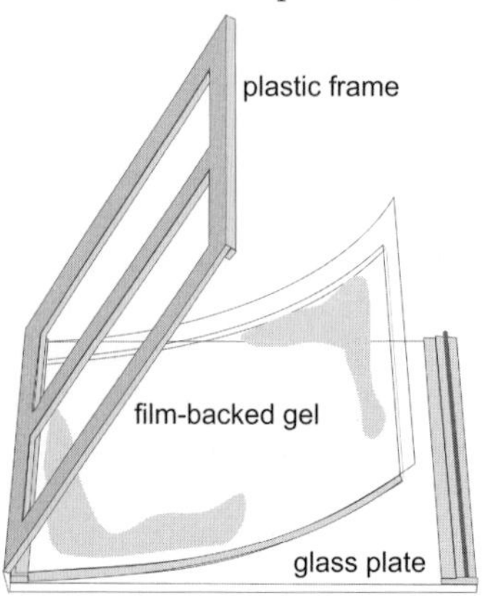

Fig. 31: Support cassette and ready-made film-backed vertical gel. The gel is shorter at the upper side to accommodate the IPG strip.

These 12.5 % homogeneous polyacrylamide gels contain PPA-chloride pH 7.0 instead of the standard Tris-chloride buffer pH 8.8, in order to achieve a long shelf life. Because polymerisation at pH 7 is more efficient than at pH 8.8, the gels have a very high mechanical stability, and the patterns are highly reproducible. The gels are packed airtight in flexible aluminium bags.

The sieving properties of a 12.5 % T PPA gel are comparable to those of 14 % T polymerised with a Laemmli buffer.

Kits with buffer concentrates and agarose sealing solutions are available. The kit for the film-backed PPA gels contains also a bottle with the gel buffer. The best way to insert the gel into the cassette is to apply a streak of 1 mL of the gel buffer onto the glass plate along the spacer of the closing side, and placing the gel on the glass plate first touching the buffer streak with the respective edge. The gel position can be easily adjusted, that the edge at this side is in contact along the entire spacer and the lower gel edge is flush with the edge of the glass plate. When the gel is lowered on the rest of the glass plate carefully, no air bubbles are caught between gel surface and glass plate. The liquid near the closing side and some air bubbles – mostly close to the hinged side – are squeezed out with a roller.

Of course, also these cassettes have to be thoroughly cleaned before and after each use to avoid contaminations.

A liquid layer between gel surface and glass plate has to be avoided. It would cause blurred spots.

A narrow gap of less than 1 mm will remain between one spacer and the gel edge: this will be sealed with hot agarose solution after the IPG strip has been inserted.

For western blotting analysis or for high sensitivity fluorescent staining the support film has to be removed. It can be cut off with a simple instrument using a steel wire (see figure 32). The large format gels are easy to handle also without the film-backing because they are well polymerised.

For those who have only used film-backed gels: **Note:** *Handling of non-backed slab gels requires some practice and skill.*

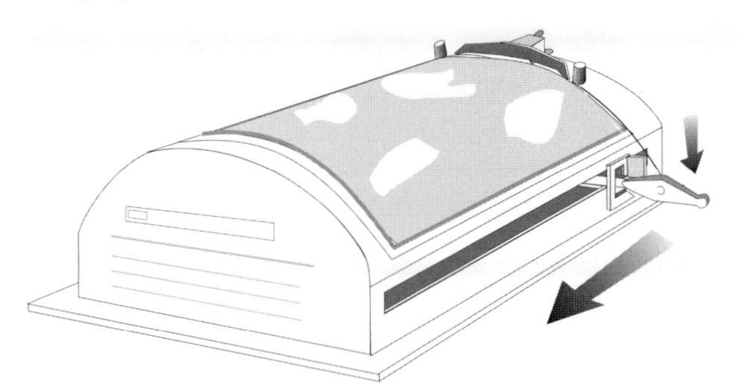

Fig. 32: Film remover for complete separation of gels from film-backing.

Instrument and gel setup for the second dimension

Figure 33 shows a Multiphor flatbed chamber used as SDS PAGE instrument and a mini vertical electrophoresis system for miniformat gels. These two instrument types are not for high throughput and thus mostly employed for optimization work.

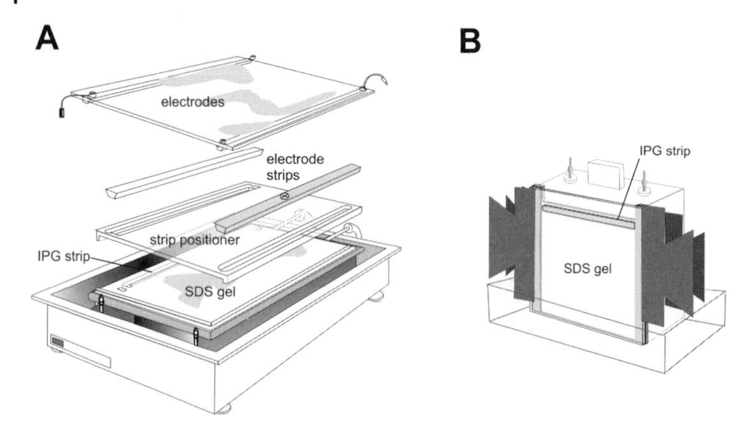

Fig. 33: Flatbed and vertical mini electrophoresis apparatus for SDS PAGE.

In the flatbed apparatus the standard gel size is 25 × 19 cm. The electrode distances can be adjusted to shorter separation distances. For example, SDS electrophoresis of three miniformat IPG strips with 7 cm length can be run together in one gel under identical conditions (see figure 34).

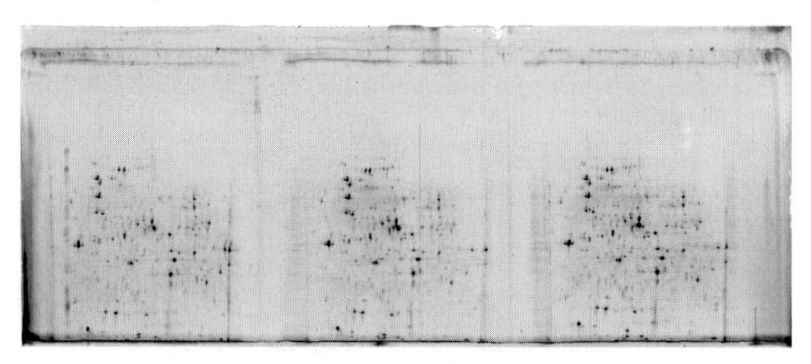

Fig. 34: Three miniformat 2-D separations run in one flatbed gel. IEF in three 7 cm IPG strips 4–7, SDS PAGE in a 25 × 11 cm gel with a stacking zone of 5 % *T* and a homogeneous resolving area of 12.5 % *T*, 0.5 mm thick. Silver staining. From Tom Berkelman, Amersham Biosciences San Francisco with kind permission.

Ready-made gels for vertical miniformat chambers are available from several suppliers. An alternative is to run also several short IPG strips together on a larger vertical gel. Some vertical systems can also be equipped with shorter glass plates. As shown in figure 3 on page 21, two miniformat IPG strips with 7 cm length can be run together in a short vertical cassette in a midsize apparatus.

Tab. 5: Comparison of the properties of flatbed and vertical systems

Flatbed Systems	***Vertical Systems***	
One gel per instrument	Multiple gel runs possible.	
Can also be used for IEF in IPGs	Dedicated for SDS electrophoresis	
Gel thickness is limited, because cooling is only possible from one side.	Higher protein loading capacity, because thicker gels can be used, which are cooled from both sides	*Up to 3 mm thick gels can be used in vertical systems.*
Very versatile for different gel sizes and methods	Gel sizes fixed by glass plate sizes	*The flatbed system has its strength in optimization work.*
Thin layer gels can be used, easy application of the IPG strips.	Thin gels cannot be used, because the IPG strip would not fit between the glass plates.	*Very thin gels show higher sensitivity of detection, they are easier and quicker to stain.*
Buffer strips (polyacrylamide or filter paper) can be used instead of large volumes of liquid buffers.	Blotting is easier because of higher gel thickness.	*Reduced chemical and radioactive liquid waste.*
The IPG strip has to be removed from the SDS gel after 40 minutes because of EEO effects.	The IPG strip does not have to be removed.	*EEO occurs in both systems, therefore urea, glycerol in the equilibration buffer and low initial voltages are needed.*

Görg *et al.* (1995) have performed a systematic comparison of results, obtained with a flatbed and a vertical system for the same sample: The spot patterns are very similar. It is demonstrated, that sharper spots are obtained in thinner gels, which can be employed in the flatbed system, but not in a vertical setup.

Görg A, Boguth G, Obermaier C, Posch A, Weiss W. Electrophoresis 16 (1995) 1079–1086.

Design of vertical instruments

The following functions are needed in the instrument for the second dimension:

- High throughput
- Large gel cassettes
- Efficient cooling for fast runs with straight front
- Leakage free
- Reproducible results independent from gel position
- Different gel thickness should be possible
- Easy handling
- Buffer consumption not too high

Anderson NG, Anderson NL. Anal Biochem 85 (1978) 341–354.

The first high-throughput chamber was developed by Anderson and Anderson (1978): the DALT box, where the gel cassettes are tilted by 90 degrees and inserted between insulating rubber flaps (see figure 35). In contrast to conventional vertical systems, the migration direction of the proteins is from left to right instead of from the top to the bottom. Handling of the instrument is easy. As only one single tank is used, there is no problem of buffer leakage from an upper to a lower tank.

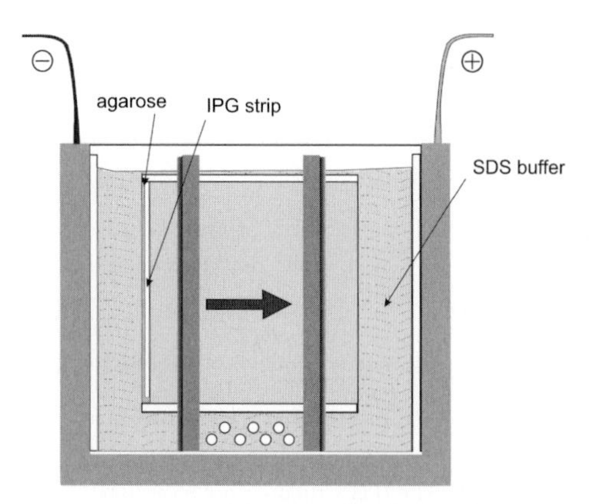

Fig. 35: Drawing of the DALT chamber for multiple SDS PAGE runs. The migration direction is from the left to the right. Insulating rubber flaps prevent current flow between the cassettes.

The gels are usually run overnight, because the cooling efficiency is limited. The entire box is filled with Tris-glycine buffer. The buffer should not be used more than once, because it will become enriched with chloride ions from the gel buffer. When the Tris-glycine buffer contains chloride ions, the initial stacking effect is disturbed – resulting in a loss of resolution – and the running time is extended. The buffer volume of 20 L is rather high.

A limitation is the design of the buffer tank: the cathodal and anodal buffer reservoir is identical. This makes peptide separations with Tris-tricine buffer very expensive. And – the stable PPA buffer system cannot be applied, because it would require two completely separate tanks for the anodal and the cathodal buffer.

When the buffer tanks are completely separate, the anodal buffer can be used repeatedly, because chloride ions do not disturb there.

For vertical instruments with completely separate tanks, several construction concepts are possible. In the following figure 36 three different designs are shown.

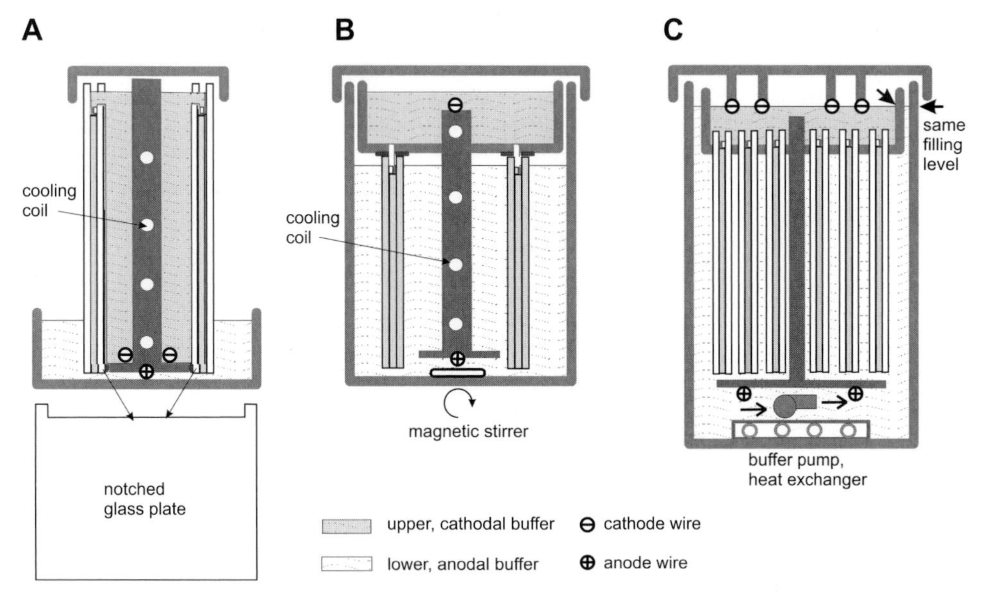

Fig. 36: Three different designs of vertical chambers. All of them can be cooled. *A*. Buffer-backed chamber. *B*. Chamber with tight upper buffer tank. *C*. Chamber for fast multiple runs.

Vertical electrophoresis chambers

- In the buffer-back chambers (type *A*) the gels are cooled via the cathodal buffer. Notched glass or aluminum oxide ceramics plates are used to enable the contact between the cathodal buffer and the gel. Maximal two gels can be run; therefore the principle is mostly applied on miniformat apparatus.
- The medium sized chamber with tight upper buffer tank (type *B*) can accommodate up to four gels at the same time, when notched glass plates are inserted between each two gel pairs. The gels are cooled via the anodal buffer. In this concept the edges of the glass plates must be absolutely intact to prevent leakage of the upper tank.
- Multiple runs with large gels can be performed in chambers of type *C*. It is, however, important to fill upper and lower buffer to the same level to avoid buffer mixing. The heat removal in this chamber type is very efficient, because the cooling anodal buffer is vigorously pumped around in the lower tank. This design is also applied on an integrated system for running up to fourteen gels (see figure 37). The buffer concentration for the upper tank must usually be doubled to prevent depletion of the cathode buffer.

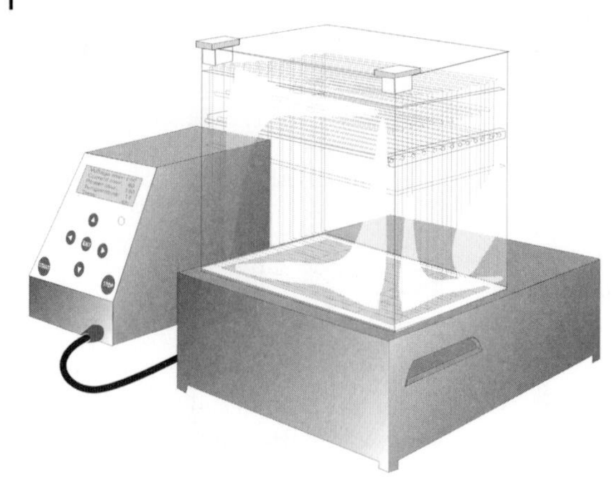

Fig. 37: Integrated apparatus Ettan DALT™ twelve for multiple 2-D PAGE. The programmable power supply controls also the Peltier cooling system and the pump for circulating the lower buffer.

At the backside of the separation unit sits a draining valve for convenient removal of used buffer. At the back panel of the control unit of the Ettan DALT™ there is a serial port for possible software updates and to connect a computer or serial printer to the instrument for a report on the electric conditions after every five minutes of the run. With this interface the instrument can be integrated into the laboratory workflow system and allow procedures according to GLP (good laboratory practice).

The Tris moves from the anodal to the cathodal buffer. Glycine moves from the cathodal to anodal buffer, it is not needed in the anodal buffer.

The following hint goes for all vertical chambers with separate tanks: When the lower – anodal – buffer is mixed with the used upper – cathodal – buffer after each run, the Tris concentration is high enough for repeated use as anodal buffer. This saves work and reagent costs. Of course, the cathodal buffer must be new for each electrophoresis.

Equilibration of the IPG strips

Prior to the run in the second dimension the strips have to be equilibrated with SDS buffer to transform the focused proteins into SDS-protein complexes, which are completely unfolded and carry negative charges only.

Equilibration stock solution

2 % SDS, 50 mmol/L Tris HCl pH 8.8, 0.01% Bromophenol Blue, 6 mol/L urea, 30 % glycerol

Equilibration is performed twice on a shaker:

15 min 10 mL equilibration stock solution plus 1 % DTT
15 min 10 mL equilibration stock solution plus 2.5 % iodoacetamide

SDS With the amount of 2 % SDS there is also sufficient SDS for preparative protein loads.

Tris HCl buffer pH 8.8 In former protocols it was proposed to add the stacking gel buffer pH 6.8, because the use of a stacking gel was standard procedure. However, it is better to add the more basic resolving gel buffer with pH of 8.8 to improve the alkylation of the cysteins.

Note
With iodoacetamide the solution can easily become acidic. This would disturb the formation of the SDS-protein complexes.

Bromophenol Blue The tracking dye allows a control of the running conditions, the shape of the migration front, and the separation time.

Urea and glycerol These additives have been introduced to keep electroendosmotic effects (see above) as low as possible. Urea is also supporting the solubility of hydrophobic proteins.

Dithiothreitol After the proteins have been focused they have to be treated with the reductant again. When very high protein loads are analysed, the concentration of DTT needs to be increased.

Iodoacetamide The alkylation agent has several functions:

- Complete alkylation of the cysteins, to avoid partial modification by acrylamide for increased spot sharpness and improved protein identification with mass spectrometry.
- Elimination of point streaking as described by Görg *et al.* (1987b).

Görg A, Postel W, Weser J, Günther S, Strahler JR, Hanash SM, Somerlot L. Electrophoresis 8 (1987b) 122–124.

- Avoidance of the artifactual horizontal lines across the SDS gel in the size range of 40 to 50 kDa.

The iodoacetamide functions as scavenger of the excess reductant.

Equilibration time Two times fifteen minutes seem to be a rather long time, and some might fear considerable losses of proteins due to diffusion. As already mentioned on page 34, immobilized pH gradients keep proteins back like a weak ion exchanger. Thus, only a little amount of proteins – those from the surface – are washed out.

The long equilibration time is necessary for the complete formation of SDS-protein complexes, because the negatively charged SDS is repelled by the negative charges on the carboxylic groups of the strips. It has been observed, that too short equilibration leads to vertical streaks and losses of high molecular weight proteins.

The second dimension can be run in vertical as well as in flatbed systems. Ready-made gels on film support and buffers are available for both systems. Depending on the system, the IPG strips are applied on the second dimension in different ways: on the gel surface in a flatbed, on the gel edge in the vertical systems, see figure 38.

Transfer of the IPG strip onto the SDS polyacrylamide gel

In order to make examination and evaluation of 2-D maps easier, the first and second dimension should always be set together in a standardized way. The majority of the scientific world has agreed, that it makes sense to orientate the gel in the way of a Cartesian coordinate system: the low values are located at the left bottom and the high values at the right top.

The IEF gel should always be placed on the SDS gel in a standard orientation: low pH to the left, high pH to the right.

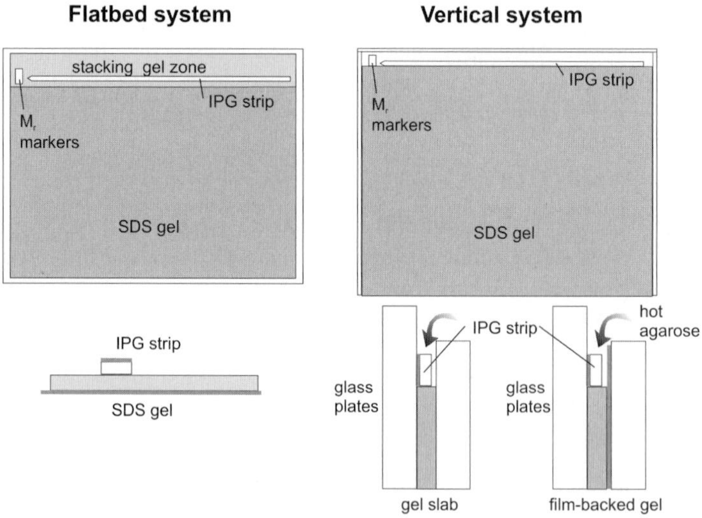

Fig. 38: Application of the IPG strip on the different types of second dimension gels.

Flatbed system It is highly recommended to use the strip positioner plate (fig. 33) for

- Improved alignment of the electrode strips.
- Preventing the electrode strips to slide away.
- Reproducible positioning of the electrode strips and the IPG strip on the SDS gel.
- Achieving a straight front, because the gel surface is covered.

The gels are placed directly on a cooling plate, which is connected to a thermostatic circulator. Usually the running temperature is 15 °C. After equilibration of the IPG strip, the excess buffer is removed by blotting it slightly with clean filter paper. The strip is placed – gel surface down – parallel to and 5 mm apart from the cathodal electrode strip. The acidic end of the strip points to the left side.

No agarose sealing solution is needed.

M_r marker proteins are applied with IEF sample application pieces as shown in figure 38.

After 40 minutes of low voltage – for sample entry – the IPG strip is removed from the gel; the cathodal electrode strip is moved from its original position to the contact area of the IPG strip. This measure prevents drying out and burning of the SDS gel along this contact line. The separation can now be continued with higher voltage settings.

In the flatbed system EEO has stronger effects, because the water is transported out of the SDS gel surface, causing drying of the SDS gel.

Vertical system The IPG strip is placed onto the edge of the SDS gel sideways, with its support film touching one glass plate. Hot agarose embedding solution, pipetted onto the strip has several functions:

Ca. 1 mL of agarose embedding solution is necessary per gel.

- achieve a gel continuity,
- prevent air bubbles between the gels,
- avoid floating of the strip in buffer solution,
- eventually seal the gaps between cassettes spacers and the ready-made gels.

Note
The agarose solution should not be hotter than 60 °C; it could cause carbamylation of some proteins because of the presence of urea.

Composition of the agarose embedding solution

0.5 % Agarose, SDS cathode buffer (1 × conc), 0.01 % Bromophenol Blue

Molecular weight marker proteins Best results are obtained when the molecular weight marker protein solution is mixed with an equal volume of a hot 1% agarose solution prior to application to the IEF sample application piece. The resultant 0.5% agarose will gel and prevent the marker proteins from diffusing laterally prior to the application of electric current.

The alternative is to apply the markers to an IEF sample application piece directly, in a volume of 15 to 20 μL. For less volume, cut the sample application piece proportionally. The markers should contain 200 to 1000 ng of each component for Coomassie staining and about 10 to 50 ng of each component for silver staining.

Running conditions for vertical gels

Electric conditions Also vertical gels must be run with low voltage for the first 40 minutes to reduce the electroendosmosis effects (see page 59 f.).

In discontinuous buffer systems, first the conductivity is high, because the gel contains chloride ions, which have high mobility. When more of the glycine or tricine ions – with low mobilities – are migrating into the gel, the conductivity decreases, and the field strength increases. In order to avoid the production of too much Joule heat at the end of the run, the gels are usually run at low current settings. A possibility to speed up the separation is to adjust the current setting from time to time.

Ideally the power supply offers the feature to run the separation at constant power. Then the current is high at the beginning and low at the end without overheating the gels and the buffers. In this way a separation can be run faster than under conventional conditions.

Temperature Multiple gels produce a substantial amount of Joule heat, which has to be removed. If this is not efficient enough the gels show the "smiling effect": faster migration in the center than on the lateral sides. It was the general opinion, that SDS gels should be cooled to 10 to 15 °C to remove the heat efficiently enough to obtain a straight front. Systematic studies in chambers with more powerful buffer circulation have shown, that better results and faster runs are obtained when the gels are run at 25 °C. Of course there are also certain limits how much temperature can be removed from the gels and the buffers.

The high settings are applied after the 40 minutes protein transfer.

Benefits of quick runs Fast separations of large gels – within a few hours instead of overnight – result in less diffused spots compared to overnight runs. This is particularly obvious in the low molecular weight area. Less diffused spots have three advantages:

- Increased spatial resolution in homogeneous gels.
- Increased detection sensitivity.

Less need for gradient gels.

In comparative experiments over 50 % more spots could be detected in these gels.

- Improved enzyme kinetics during tryptic in-gel digestion.

Because of the higher protein concentration in the gel plug.

Molecular weight determination For the determination of molecular weights in 2-D electrophoresis gels two different procedures are used:

1. Co-running molecular weight standard proteins

Molecular weight marker proteins can be applied as a separate track to the 2-D gel. The molecular weights of the proteins in the 2-D map are then interpolated with the molecular weight curve obtained from the positions of the marker proteins. This procedure is usually carried out with the image analysis software as "1-D calibration".

The problem with this method is its limited accuracy. The markers have to be applied at the lateral sides of the gels. Often the space is very limited and the bands are curved because of the edge effects in SDS PAGE. Furthermore, the estimation of molecular weights with SDS PAGE is not accurate at all.

2. Interpolation between identified sample proteins with known M_r *values*

Prominent spots showing up in each 2-D map of a sample type can be analysed for identification and amino acid sequence information. The theoretical M_r values can then be used as keystones for interpolating the M_r values of the other proteins. The second procedure is performed with the image analysis software as "2-D calibration". This method is much more exact than the method described above.

2.1.5 Detection of protein spots

There is a demand list of properties required for the ideal spot detection technique in 2-D gels in proteomics: it should

- be sensitive enough for low copy number proteins,
- allow quantitative analysis,
- have a wide linearity,
- have a wide dynamic range,
- be compatible with mass spectrometry,
- be non toxic,
- be environment friendly,
- be affordable.

Unfortunately there is no method that affords all these features together. The following table gives an overview over currently used detection principles and their features. Several different protocols exist for most of these methods. Those of major importance are quoted in the table.

2.1.5.1 Comparison of detection methods

Tab. 6: Detection methods for protein spots used in proteomics

	Method	*Advantages*	*Disadvantages*
Neuhoff V, Arold N, Taube D and Ehrhardt W. Electrophoresis 9 (1988) 255–262.	Coomassie Brilliant Blue staining (colloidal)	Steady state method, good quantification, inexpensive, mass spectrometry compatible	Low sensitivity: LOD only ca. 100 ng of BSA, slow, dye particles can cause problems in image analysis
Jochen Heukeshoven, personal communication.	Coomassie Brilliant Blue staining (alcohol free, hot, monodispers)	Steady state method, fast, good quantification, inexpensive, very environment friendly, mass spectrometry compatible	Low sensitivity: LOD only ca. 200 ng of BSA, background destaining necessary
Fernandez-Patron C, Castellanos-Serra L, Hardy E, Guerra M, Estevez E, Mehl E, Frank RW. Electrophoresis 19 (1998) 2398–2406.	Zinc imidazol reverse staining	Medium sensitivity: LOD ca. 10 ng of BSA, fast, very good compatible with mass spectrometry	Bad for quantification, negative staining not easy for documentation
Heukeshoven J, Dernick R. Electrophoresis 6 (1985) 103–112.	Silver staining (silver nitrate)	High sensitivity: LOD ca. 0.5 ng, can be made mass spectrometry compatible	Poor dynamic range, limited quantification possibilities, multistep procedure
Rabilloud T. Electrophoresis 13 (1992) 429–439.	Silver staining (silver diamine)	High sensitivity: LOD ca. 0.5 ng, stains basic proteins better than the protocol above	Poor dynamic range, limited quantification possibilities, multistep procedure, high silver nitrate consumption
Rabilloud T, Strub J-M, Luche S, van Dorsselaer A, Lunardi J. Proteomics 1 (2001) 699–704.	Fluorescent staining with RuBPS	Medium to good sensitivity: LOD ca. 5 ng of BSA, very good for quantification, wide dynamic range, mass spectrometry compatible	Overnight procedure, less sensitive than mass spectrometry compatible silver staining, fluorescence scanner necessary, dye particles can cause problems in image analysis

Method	Advantages	Disadvantages	
Fluorescent labelling for difference 2-D gel electrophoresis	Medium to good sensitivity: LOD ca. 5 ng, direct comparison of up to three samples in one gel, good for quantification, wide dynamic range, mass spectrometry compatible	Labelling protocols must be optimized for different samples, fluorescence scanner necessary, relatively expensive	*Ünlü M, Morgan ME, Minden JS. Electrophoresis 18 (1997) 2071–2077.*
Radioactive labelling	High sensitivity: LOD below pg, good quantification, very wide dynamic range with phosporimager	Limited to living cells, gels have to be dried, phosphor imager necessary, radioactivity needed	*Johnston RF, Pickett SC, Barker DL. Electrophoresis 11 (1990) 355–360.*
Stable isotope labelling	High sensitivity: LOD below pg, wide dynamic range, good quantification	Methods still under development also for 2-D electrophoresis, expensive, mass spectrometer needed	*Smolka M, Zhou H, Aebersold R. Mol Cell Proteomics 1 (2002) 19–29.*
Western blotting	Ideal for highly sensitive and selective detection of certain proteins, general protein staining is also possible, proteins are well accessible on the blotting membrane	Additional electrophoresis step, uneven transfer of proteins, limited mass spectrometry compatibility because of membrane material, specific detection works only when antibodies are available	*Dunn MJ. In. Link AJ. Ed. 2-D Proteome Analysis Protocols. Methods in Molecular Biology 112. Humana Press, Totowa, NJ (1999) 319–329.*

In practice the mostly applied techniques in proteomics laboratories are Coomassie Brilliant Blue, silver and fluorescence staining.

2.1.5.2 Staining with visible dyes

It is very important to know, that different staining techniques stain proteins differently. There are proteins, which do not stain at all with Coomassie Brilliant Blue, but with silver and vice versa. Also different silver staining protocols deliver different patterns of the same sample.

Zinc imidazol negative staining When image analysis of the pattern is not an issue, and mass spectrometry analysis of some spots is the major goal, this procedure can be recommended.

Coomassie Blue staining The “classical” alcohol/acetic acid Coomassie Blue staining recipes should not be used for 2-D electrophoresis, because during destaining with the alcohol-containing solution the protein spots are partly destained as well. Some proteins – for

The alcohols used are: methanol, ethanol, and isopropanol.

instance collagen – lose the dye earlier than the background of the gel is destained. Because no steady state is reached, quantification is not reliable and not reproducible.

Colloidal Coomassie Blue staining Colloidal Coomassie Blue staining according to Neuhoff *et al.* contains also alcohol, but in presence of ammonium sulfate. Ammonium sulfate increases the strength of hydrophobic interactions between proteins and dye. The methanol allows a much faster staining process. Coomassie G-250 is used. Repeated staining overnight and fixing during the day with 20 % ammonium sulfate in water for several times gives a sensitivity approaching that of silver staining. But this procedure takes a very long time and needs many steps; it is not ideal for high throughput.

This is the most environment friendly staining procedure, and it is well compatible with mass spectrometry analysis.

Hot Coomassie Blue staining Much quicker results are achieved with a direct fixing/staining procedure using alcohol-free Coomassie R-350 in 10 % acetic acid at elevated temperature. The easiest way is to heat the solution to 80 – 90 °C on a heating stirrer and to pour this solution over the gel, which lies in a stainless steel tray placed on a rocking platform. Staining takes 10 minutes. The gel has to be destained with 10 % acetic acid at room temperature for several hours. Staining as well as destaining solutions can be used repeatedly. The dye can be removed from the destainer by adding paper towels to the destainer or filtering it through activated carbon pellets.

No loss of sensitivity has been observed, when glass- or film-backed gels are stained with Coomassie Blue.

Silver staining The silver nitrate protocol is mostly preferred to the silver diamine protocol, because it needs only 10 % of the amount of silver nitrate, and it is less dangerous to get a silver mirror on the gel surface. When glass- or film-backed gels are stained, the staining steps have to be prolonged, and, unfortunately there is a loss of sensitivity, because the backing blocks one side of the gels for the solutions.

Colloidal Coomassie staining shows the same pattern like the hot staining procedure.
Görg A, Obermaier C, Boguth G, Harder A, Scheibe B, Wildgruber R, Weiss W. Electrophoresis 21 (2000) 1037–1053.

Silver staining often produces a pattern different from the pattern achieved with Coomassie Blue and other staining procedures like with Sypro Ruby (Görg *et al.* 2000). Figure 39 shows two separations of E.coli extracts: one stained with hot Coomassie Blue and the other one stained with silver staining. Particularly in the acidic pH and high molecular weight area Coomassie Blue picks up more proteins, and stains them more intensive than silver staining. Note: There is also a negatively stained spot in the silver stained gel at the low pI/low M_r corner.

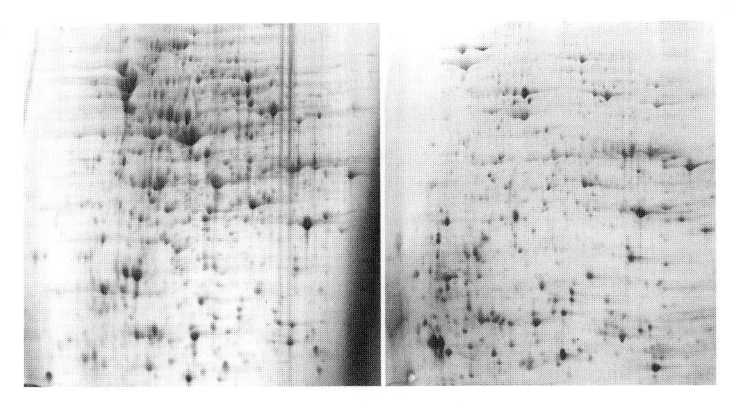

Fig. 39: 2-D electrophoresis of E.coli extract. 1st dimension: IPG 3–10 NL 24 cm, rehydration loading. 2nd dimension: SDS PAGE in 1 mm PPA gel 12.5 % *T*, 3% *C*. Left side: 1.2 mg protein, fast Coomassie staining; Right side: 180 μg protein, silver staining. A high number of proteins are stained with completely different intensity.

Double staining When the gel is first stained with Coomassie Blue and subsequently with silver staining, the pattern remains the same like with Coomassie Blue alone, it just detects more spots because of the higher sensitivity, see fig. 40.

When hot staining is used, the background has to be completely clear.

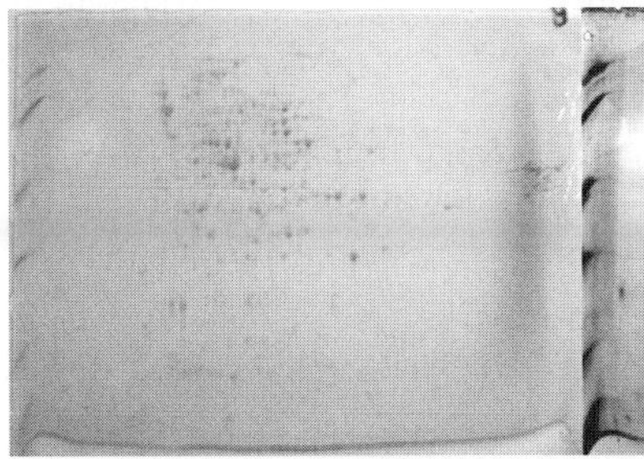

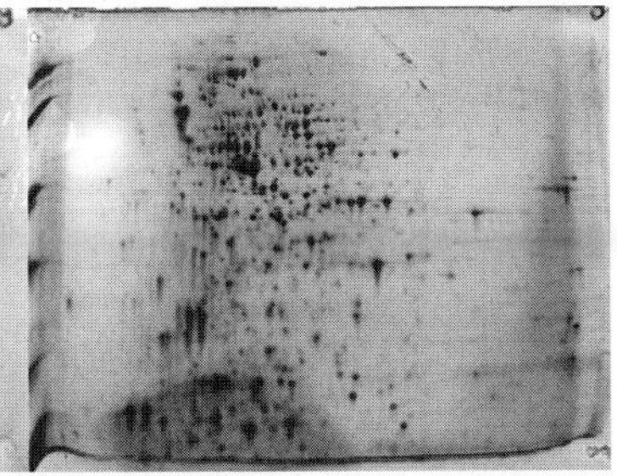

Fig. 40: 2-D electrophoresis of E.coli extract. 1st dimension: 24 cm IPG 3–10 L, rehydration loading of 180 μg protein. 2nd dimension: SDS PAGE in 1 mm PPA gel 12.5 % *T*, 3% *C*. Left side: hot Coomassie staining; Right side: gel stained afterwards with silver.

It is not unknown, that prestaining a gel with Coomassie Blue intensifies the signal of silver staining, resulting in improved detection sensitivity. Another advantage is, that double staining obviously prevents negative silver staining. When the Coomassie Blue staining is fast, it can almost compete with a conventional fixing procedure for silver staining.

Note: the best silver staining results are obtained when fixing is performed overnight.

Mass spectrometry compatibility Coomassie Brilliant Blue stained gels are usually compatible with mass spectrometry analysis, because the dye can be completely removed from the proteins. It is important to use a dye with good quality to avoid contaminants showing up in the mass spectrogram. For hot staining, shown in the two figures above, a very pure Coomassie Brilliant Blue R-350 dye was used, which is available in tablet form.

They use usually colloidal staining.

A spot, which is visible with Coomassie Blue staining, contains enough protein for identification and characterisation with mass spectrometry. Therefore many of the proteomics laboratories use almost exclusively this staining procedure in their routine work.

Shevchenko A, Wilm M, Vorm O. Mann M. Anal Chem 68 (1996) 850–858.
Yan JX, Wait R, Berkelman T, Harry RA, Westbrook JA, Wheeler CH, Dunn MJ. Electrophoresis 21 (2000) 3666–3672.
Gharahdaghi F, Weinberg CR, Meagher D, Imai BS, Mische SM. Electrophoresis 20 (1999) 601–605.

Silver staining can be modified for mass spectrometry compatibility by omitting glutardialdehyde from the sensitising solution, and formaldehyde from the silver solution (Shevchenko *et al.* 1996, Yan *et al.* 2000). The detection sensitivity decreases to about one fifth of the nonmodified procedure. It has also been reported, that removing the silver can improve the signals in mass spectrometry (Gharahdaghi *et al.* 1999). The silver-protein complexes are located only on the surface of the gel, the remaining proteins inside the gel layer can be further analysed.

In practice, however, it happens very frequently, that silver stained spots show no signals in mass spectrometry. There might be many reasons for this, for instance too long development time resulting in too much contact of proteins with formaldehyde, or the protein amount in the spot is simply too low.

At the present state of development it is a fact, that you are on the safe side with Coomassie Blue staining, whereas spot analysis of silver stained spots is still an adventure.

Strategy for visible spot detection and protein analysis A well practicable procedure is:

- stain the gel with colloidal or hot Coomassie Blue and scan it,
- perform image analysis to detect changing protein spots for further analysis,
- cut the spots out, digest the protein and analyse with mass spectrometry,
- stain the gel with silver and scan it again,
- find more interesting spots with image analysis,
- run narrow interval pH gradient gels of the interesting pH ranges with higher sample loads for staining with Coomassie Blue and so on.

2.1.5.3 Detection with fluorescence

Fluorescence staining A number of fluorescence dyes are available. They exhibit a wide dynamic range and they are therefore very well suited for quantification of proteins. Most of them show sensitivity like or below the Coomassie Blue dyes. Fluorescence staining is mass spectrometry compatible. Sypro ruby from Molecular Probes and a specially prepared RuBPS solution (Rabilloud *et al.* 2001) are very sensitive, almost as sensitive as mass spectrometry compatible silver staining (when the correct protocol has been applied). A fixing step, incubation overnight, and a few subsequent washing steps are very important to achieve maximum sensitivity. Dark polypropylene containers instead of glass or stainless steel trays must be used. A fluorescence scanner is required for visualization and detection. Most of the fluorescent dyes used here exhibit their highest excitation sensitivity at a wavelength apart from UV.

Rabilloud T, Strub J-M, Luche S, van Dorsselaer A, Lunardi J. Proteomics 1 (2001) 699–704.

Fluorescence labelling Fluorescence labelling with monobromobimane (Urwin and Jackson, 1993) or CyDyes prior to isoelectric focusing reaches similar sensitivities and dynamic ranges like fluorescence staining.

Urwin V, Jackson P. Anal. Biochem 209 (1993) 57–62.

2D-DIGE 2-D Fluorescence Difference Gel Electrophoresis (2D-DiGE) is a technique that labels complex protein mixtures with fluorescent dyes prior to conventional 2D electrophoretic separation (see figure 41). Up to three different samples can be labelled and mixed together and then separated on a single 2D gel. Cyanine dyes are used to label the proteins from different samples with dyes of different excitation and emission wavelengths (Ünlü *et al.* 1997, Zhou *et al.* 2002, Gharbi *et al.* 2002).

Ünlü M, Morgan ME, Minden JS. Electrophoresis 18 (1997) 2071–2077.
Zhou G, Li H, DeCamp D, Chen S, Shu H, Gong Y, Flag M, Gillespie J, Hu N, Taylor P, Buck ME, Liotta LA, Petricoin III EC, Zhao Y. Mol Cell Proteomics (2002) in press.
Gharbi S, Gaffney P, Yang A, Zvelebil MJ, Cramer R, Waterfield MD, Timms JF. Mol Cell Proteomics (2002) in press.

The dyes are matched for charge and molecular weight ensuring that the same protein found in each sample will migrate to the same position on a 2-D gel. One advantage of this method over conventional methods, is that since the samples are exposed to the same chemical environments and electrophoretic conditions, co-migration is guaranteed for identical proteins from the separate samples and analysis of sample differences is therefore simplified. The ratio of protein expression is always obtained from a single gel and an internal standard can be used in each gel significantly reducing gel to gel variation of protein ratio measurements. Matching proteins between gels allows ratio measurements to be compared from a number of different samples. A specific software package (DeCyder) has been developed to exploit the unique advantages of this technology allowing an automated approach to the analysis of differences found within and between gels.

The 2D-DIGE fluorescent dyes and associated DeCyder image analysis software have been developed and optimized for sensitive detection of differences in expression of proteins resolved in this format. DeCyder uses ratio measurements obtained within gels rather than between to derive its statistical data, and can routinely detect <10% differences in expression between samples with >95% confidence, within minutes.

The labelling method used is lysine labelling (minimal labelling), which labels the proteins via the epsilon amino group of lysine.

The lysine labelling dyes contain a single positive charge to replace the charge removed from lysine during the labelling procedure ensuring that the pattern of separation during isoelectric focusing (IEF) is unchanged. Lysine labelling is referred to as minimal labelling because the ratio of dye to protein is kept very low to ensure that the only protein visualised on the gel is that which contains a single dye molecule. This means that the visualised pattern on the 2D gel remains the same as those from gels that have been stained. The majority of protein contains no dye at all and therefore, the major non-labelled protein portion of a spot is slightly offset from the visualised spot, because the labelled protein has an increased size. For

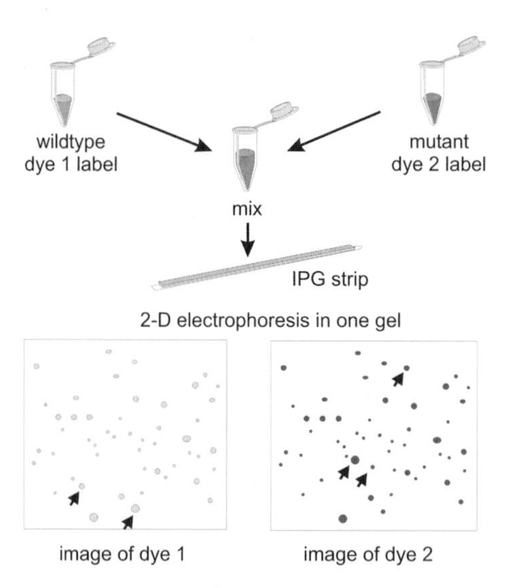

Fig. 41: Difference gel 2-D electrophoresis. The sample proteins are labelled with different fluorescent tags. The images are acquired with a fluorescence scanner at different wavelengths.

accurate spot picking the gel is stained with a fluorescent dye, for example SyproRuby. This image is matched to analysed cyanine dye labelled gels after analysis and automatic picking of the proteins of interest can be performed. Labelling has to be performed at alkaline pH and the presence of any primary amines or carrier ampholytes has to be avoided at this stage. Excessive amounts of thiols will reduce the labelling sensitivity and it is advised to avoid these until after labelling is complete. The best imaging results are achieved by using low fluorescent glass plates for the second dimension SDS-PAGE separation and the gel imaged while it gel is still between the glass plates. Silane is used on one plate to attach the gel to the glass allowing automated accurate spot picking.

2.1.6 Image analysis

The evaluation and comparison of the complex 2-D patterns with the eye is impossible. Therefore, the gel images have to be converted into digital data with a scanner or camera, and analysed with a computer. For a proper evaluation with the image analysis software it is important to acquire the image as a grey-scale TIFF file with adequate resolution and preferably as much as 16 bit intensity.

> Note
> ***If the resolution is too high, the image files will become too big to become processed in a reasonable time.***

Preparing the gels for spot picking When the gels should be placed into an automatic spot picker after image analysis, the gels must be fixed to a glass plate or plastic film support. Two self-adhesive reference markers – available also with fluorescence and radioactive signals – are glued to the bottom of this support. The positions of these markers are roughly predetermined to make it easier for the spot-picker camera to find them automatically. The reference markers are scanned together with the spots.

2.1.6.1 Image capturing

It has been mentioned before, that only large gels in the size range of 20 × 20 cm are used for proteome analysis. The resolution of the gel must be maintained for proper image analysis. At the present state of the art there is no CCD camera on the market, which can compete with the resolution of a scanner. This can change rapidly.

After the technology will have taken this step of development, you will find the update at the website indicated at the first page of this book.

Scanners for visible dyes Gels with visible spots have to be scanned in transmission mode. Otherwise quantification of spots would be incorrect. Blot membranes are scanned in reflectance mode. It is usually the scanning area, which sets the limit for the 2-D gel sizes. An A3 format scanner costs considerably more than a standard A 4 format instrument.

For the definition of O.D. see glossary.

The scanner must be calibrated – with the software – using a grey step tablet. The measured dimensionless intensity is converted into O.D. values (or AU, absorbance units).

New desktop scanners afford quick scanning with wide linear range of 0 to 3.6 O.D. in transmission mode and with high spatial resolution. In practice the offered resolution of down to 20 μm cannot be applied, because the image file would become too big. Scanning a 20 × 20 cm gel with 300 dpi takes 30 seconds to 1 minute.

Film- or glass-plate backed gels are placed on the scanning area with the gel surface down, with a thin layer of water in between. Therefore the standard desktop scanners have to be modified by sealing the scanning area against liquid leakage.

For accurate picking of spots according to the data of image analysis the x/y positions need to be absolutely correct in the μm tolerance range. Usually the x/y data have to be calibrated for each scanner with the help of a grid. These calibration data have to be imported into the spot picker computer for each scanner used.

Scanners for storage phosphor screens and fluorescent dyes The functions for storage phosphor imaging, multicolour fluorescence detection, and chemiluminescence can be combined in one instrument.

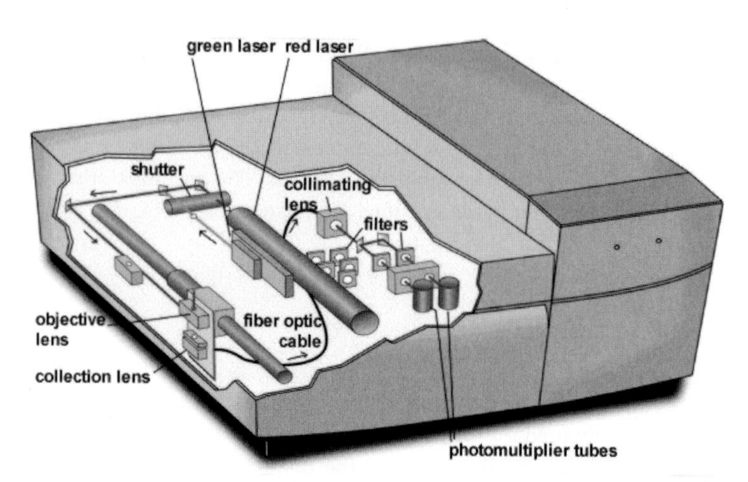

Fig. 42: Variable mode laser scanner Typhoon™: With kind permission from Amersham Biosciences, Sunnyvale, USA.

Lasers with different wavelengths are combined with different filters for the various scanning modes. The detectors are usually very sensitive photomultipliers. Figure 42 shows a drawing of such an instrument.

Radiolabelling provides the most sensitive signals. Storage phosphor screen scanners have a much wider linear dynamic range than X-ray films. The detection is much faster. After the exposure of a dried gel or a blotting membrane, the storage screen is scanned with a HeNe laser at 633 nm. For reuse, the screen is exposed to extra-bright light to erase the image.

Differentially labelled samples can be analysed in the following way: with direct exposure both ^{35}S and ^{32}P signals are recorded; with a second exposure through a thin copper foil only 32 P labelled proteins are detected.

Staining or labelling proteins with fluorescent dyes show less sensitivity than silver staining, but a much wider linear dynamic range. Modern instruments have a confocal scan head to cancel signals from scattered excitation light, and to reduce fluorescence background coming from glass plates and other supporting material. The laser light excites the fluorescent label or bound fluorescent dye; the emitted light of an offset wavelength is bundled with a collection lens and transported to the detector through a fibre optic cable (see figure 43). Signals emitted from bands or spots, which are excited by stray light, are focused out, they will not hit the "peep hole", and will thus not be conducted by the fibre optic cable.

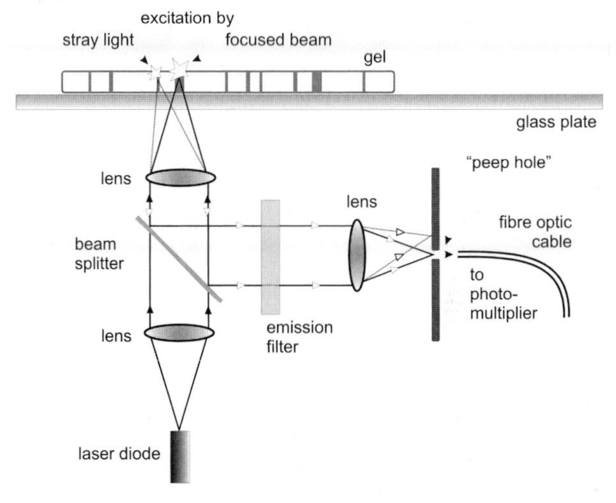

Fig. 43: Schematic diagram of the principle of fluorescence detection with confocal optics.

Amersham Biosciences Handbook: Fluorescence imaging: principles and methods (2000) 63-0035-28.

State of the art instruments can be equipped with up to three lasers with wavelengths of, for instance, 532 nm (green), 633 nm (red), and 457 and 488 nm (blue). Many aspects of fluorescence imaging are described in detail in a Handbook by Amersham Biosciences (2000).

Scanners with confocal optics deliver such exact data, that x/y calibration for accurate spot picking is not required.

2.1.6.2 **Image analysis software**

As proteomics is more an industrial approach than academic laboratory research, software, which runs on the Windows or UNIX® background, is mostly preferred to Apple Macintosh.

Image analysis is done for various purposes:

- Comparison of 2-D pattern of treated with non-treated samples.
- Construction of averaging gels, which are composed from replicate runs of the same sample.
- Detection of novel, missing, or modified proteins.
- Quantification of protein spots.
- Detection of up- or down-regulated proteins.
- Definition of spot positions for spotcutting.
- Detection and characterization of protein families and networks.
- Statistical analysis of experimental results.
- Enabling database queries.
- Linkage of 2-D data to mass spectrometry data.
- Integration of the image data with a laboratory workflow system.

The development of software for 2-D electrophoresis gel image analysis is a continuously ongoing process. The functions become more reliable, reproducible and automated from year to year. With the latest developed program it is already possible to compare gels of different sizes, shapes, and even damaged gels. However, it seems, that irreproducible results cannot be fixed, even not by the most sophisticated software. On the contrary: the highly developed programs recognise inconsistencies between different patterns. Exact and tidy laboratory work is still requested. Bad separation results cannot be turned into good results by the software.

There are three types of 2-D imaging software on the market:

- Cheap to free of charge tools for basic image analysis.
- Professional programs, which evaluate with high reproducibility and offer many valuable functions for image analysis, statistical evaluation, and reporting.

- Highly sophisticated, fully automated software and hardware solutions for high-throughput hands-free analysis of large gel packages.

Not all laboratories need and can afford to purchase the high-end solution. On the other side, the low-end programs are practically useless for proteomics analysis.

The following chapter describes the major features of an up-to-date image analysis procedure and some of the possibilities with the high-end solution.

Evaluation of a 2-D gel

The new scanned gels, or an experimental series of gels have to be placed into a common folder.

Spot detection The most important step is the spot detection, where different parameters have to be optimised:

- Size of pixel matrix used for detection around every pixel in the image
- Sensitivity to include small spots, but prevent too much spot merging
- Background factor to eliminate false weak signals
- Noise level

The settings are usually applied to a series of gels within an experiment.

This is nowadays done with the help of a spot detection "wizard", which is set up similar to image optimisation tools in some digital photo processing programs. A manual selected small area of the image is displayed in a 9-windows matrix, where the center image shows the result of the actual parameter adjustment, surrounded by the eight neighbouring settings (see figure 44). When the parameters are set, the spot detection occurs automatically and the result for the entire gel will be displayed on the screen.

Drawing spots is much more critical than deleting and splitting spots.

It is important, that the small and weak spots are included; large merging spots can be manually edited afterwards.

Spot filtering Before editing the spots, filtering should be employed to reduce the time for spot editing:

- Circularity: removes streaks and other non-spot signals.
- Minimum area: removes spikes caused by colloidal and Sypro Ruby staining.
- Minimum volume: excludes weak spots at the limit of detection.

With a choice of more filtering tools, certain groups of protein spots can be defined for further analysis. Filtering can also be used for other purposes, for instance to display only certain protein groups.

Spot editing With an editing toolbox spots can be drawn, erased and split. Dependent on the expertise of the operator and the complexity of the pattern, spot editing can take up to several hours per gel.

Setting parameters for spot detection and manual editing of merged spots is nowadays much easier and faster than in the past. But it requires still some time and is strongly influenced by the individual user.

> Note
> ***These are the most critical and time-consuming steps, and because of human interference the reproducibility can suffer considerably.***

Background correction Because the background intensity is never even over the gel, it has to be subtracted before spot quantification. Using the manual option will always lead to very subjective results. Several different algorisms are in use: according to practical experience the option "mode of non-spot" leads to the most reliable and reproducible results, because it works fully automatic. A box is drawn around each spot. This box is expanded by a number of pixels, which build an examination area. The intensity of those pixels in this area, which are not part of any spot, is used as the basis for background subtraction. As a control of the result the image should be displayed as background without spots.

Comparison of two and more 2-D patterns

Reference gel For gel comparisons it is important to create a reference gel, a virtual gel, which will contain all detected spots of an experiment. A new reference gel is created on the basis of an existing gel image of an experiment. All additional spots from other gels can be added to it.

Gel matching Now all gel patterns can be matched to the reference gel. This step is necessary, because the images of different gels are usually distorted due to shrinking or swelling of non-backed gels, and little concentration differences in the gels. The gels can be matched automatically, or with the help of seed matches that have been entered by the operator. The software will connect the reference spots and the "slave" spot with a vector. When the box size at the end of the vector is chosen large enough, it will automatically find other spots to be matched, even when the gel is distorted. Sometimes a few seed matches are not sufficient, more seed matches have to be added. The gel will be warped to the seed matches. In the gel overlay image the matched and unmatched spots are indicated as open or solid circles in different colours.

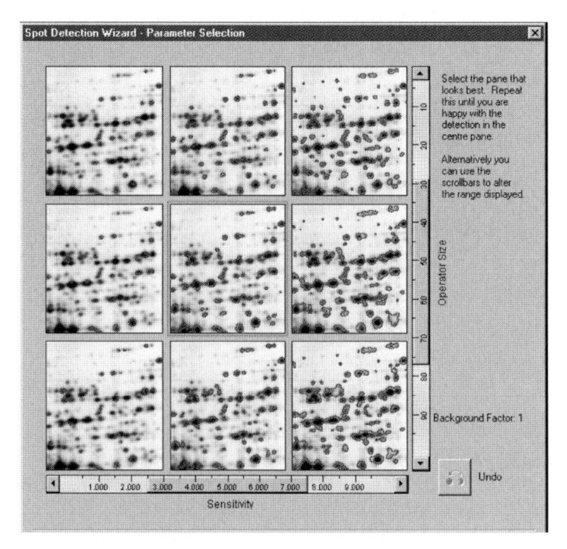

Fig. 44: Spot detection wizard for parameter optimisation. The result of the actual settings is displayed in the center, marked with a green frame.

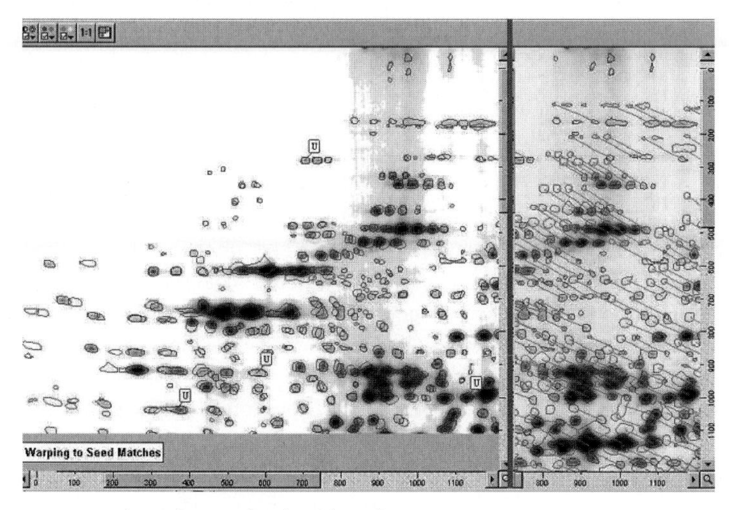

Fig. 45: Gel matching. Left side: Gel overlay image showing matched and unmatched spots. Right side: Unwarped overlay indicating the vectors between reference and slave spots.

The gel matching process can be visually controlled by displaying the gel overlay unwarped: vectors connecting the matched spots are shown (see figure 45). When these vectors look like "combed", the matching was correct; should they lie around with varying lengths and directions, this matching would not make any sense.

Updating the reference gel When the gels in an experiment are matched to the reference gel, it will turn out, that it will not contain all spots found in each of the gels. This additional spots need to be added now to the reference gel.

Normalisation Before gels are compared for differences in the spot pattern, the spot volumes of the different gels have to be adjusted by normalisation. This step corrects different protein loads and staining effectiveness. There is a choice to normalise the gels according to the total spot volume or to the volume of a single prominent reference protein, for instance *actin* in cell lysate patterns.

pI and M_r *calibration* There are two ways to calibrate the image for pI and M_r annotation (see also pages 53 and 79):

- *1-D calibration:* Calibration curves are derived from IPG pH gradient graphs for pI determination for one dimension and from the positions of co-run molecular weight markers for the second dimension by importing a ladder.
- *2-D calibration:* pI and M_r values are calibrated at the same time by interpolating between known values. These values usually come from mass spectrometry analysis. A number of spots with known protein properties are selected, the pI and M_r information are imported from the respective protein lists belonging to these spots.

Difference maps The changes between different gel images are displayed in a difference map, where unmatched, unchanged, increased and decreased spots are displayed in different colours. The settings for increase and decrease of spots must be selected carefully, because some detection procedures – like silver staining – do not allow too exact conclusions. A difference map delivers many of that information, which cannot be recognized with inspection of 2-D gels with the eye. Figure 46 shows an example.

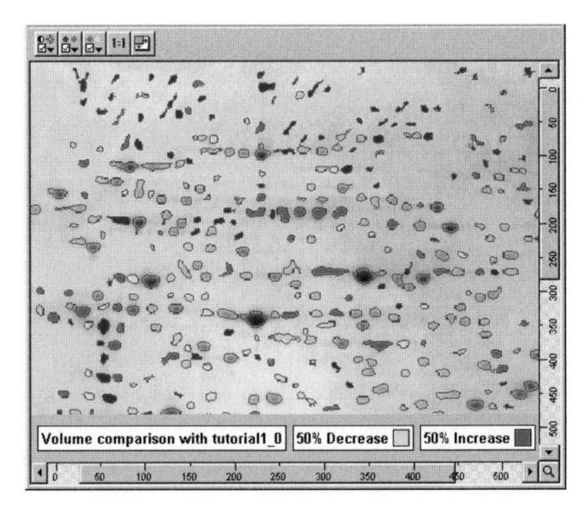

Fig. 46: Difference map of two gels. Matched spots are shown with a blue boundary, not matched spots are solid blue.

Averaged gels It is a good practice in proteomics to run replicate gels of the same sample under the same conditions, in order to average out the noise of the system. Artificial gels can be created from the statistical combination of several gel images, which can be compared with each other. These averaged gels are created after separate analysis of each image, including matching of the spots to the reference gel.

Quantification After all these steps the quantitative analysis of spot volumes becomes more reliable. However, one should not forget that it is very important first to apply a detection technique that is well suited for quantitative analysis. Selected spots from a series of gels can be displayed together in montage windows and in tables for quantitative comparisons of expression levels.

Preparing the spot picking list The x/y positions of the selected spots, including the reference markers, are exported into a table. The numbers of the reference markers are automatically named “IR1” and “IR2” (internal reference).

Reporting and exporting of data The results can be printed out in images with or without annotations and tables. The files can also be exported into Excel and other office software.

Fully automated image analysis

As already mentioned above, spot detection and editing is a cumbersome and time-consuming procedure. In fact, image analysis has been identified as a bottleneck in the proteomics workflow. And because of human interference in this process, the reproducibility and reliability of the data are limited.

Lately a completely automated software has been introduced, which can process large data sets hands-free and fast, and handle also imperfect gel images. New algorithms find spot boundaries at the very base of the spots, split merged spot packets and spots on smearing background detect and split faint shoulders and weak spots next to abundant, saturated protein groups. All these effects were causing huge problems and were demanding compromises in evaluation.

Whole image warping corrects differences generated by electrophoresis conditions, differences between scanners, and laboratory accidents.

Practice has shown, that with this automated software more changes between gels are detected than with the manually supported procedure described above.

2.1.6.3 Database software for 2-D electrophoresis

When a higher number of gels have to be evaluated, data base software has to be employed. The database contains all the information of the analysed gels including details about each protein that had been entered in the protein lists. Besides statistical tools like Student's t-test and correspondence analysis, there are possibilities to formulate complex queries in order to detect and characterize protein families and networks. With the help of an animation window it is possible to browse through a series of gels like a movie in order to detect changes in the patterns, which seem to show some systematic.

The database can also administrate data imported from mass spectrometry analysis and assign the spectra to the respective spots.

2.1.6.4 Use of 2-D electrophoresis data

Most of the image analysis software includes a web browser to check results of other laboratories in the World Wide Web. Because there is no standardization for sample preparation and for running 2-D electrophoresis, the spot patterns of different laboratories are not easy to compare in practice.

A protein cannot be identified on the basis of its position in the gel alone. Identification is only possible by further analysis of the protein, for instance with mass spectrometry. It is furthermore incorrect to conclude, that a protein is up or down regulated on the basis of growing or shrinking spot intensity. It could also have happened, that a different protein has changed its pI due to phosphorylation or another post-translational modification and is now sitting on top of the protein observed.

Therefore proteins of interest have to be analysed further. This is done mainly with mass spectrometry. The gel plugs containing the protein are cut out of the gel slab subsequently to image analysis.

2.2
Spot handling

Between image analysis and further analysis the gels can be kept in a refrigerator several weeks. It is also possible to dry the gels for storage and soak them in water before spot cutting.

In high-throughput proteomics it is very important to keep track of each sample and of each protein to be analysed. Before the proteins contained in the spots can be analysed with mass spectrometery a series of steps have to be performed:

- Spot cutting.
- Destaining and washing of the gel plug.
- In-gel digestion.
- Peptide extraction.
- Spotting of peptides on MALDI target slides.

During this procedure mix-ups and contaminations have to be prevented to avoid wrong conclusions.

In order to keep control over all these steps and parallel processes, as well as the results, two things are necessary:

- Automation.
- Workflow database.

Therefore a workstation has been developed, which handles all steps automatically. Figure 47 displays a schematic drawing of such an integrated system. In the fully automated system a robotic arm transports of gels, microtiterplates and target slides between the stations. The workstation is controlled by a laboratory workflow system software. This software can control the entire proteome analysis workflow including sample traceing.

The gels are identified with the help of barcodes on the support films and the trays. After spot picking the gel is transported back to the gel hotel. Many moves are needed for the digestion step. Spot cutting and spotting are combined to be carried out by the same instrument.

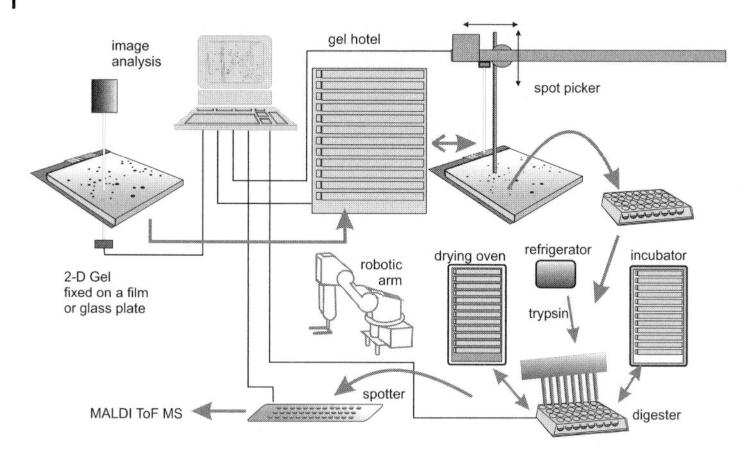

Fig. 47: Schematic drawing of an automated workstation for spot picking, in-gel digestion, and spotting peptides on a MALDI target slide.

The laboratory workflow system can be extended to include all instruments and software involved in the entire proteomics analysis, from sample preparation to mass spectrometry.

In most of the laboratories the different steps are performed by stand-alone instruments.

2.2.1 Spot cutting

Cutting spots from a gel manually is a very cumbersome job. The spots, marked in the printout of the image analysis for further analysis, have to be found in the gel again and transferred to the correct reaction tube or well of a microtiter plate. Sample tracing is not easy and errors can easily occur. Additionally, contaminations with keratin and other stuff must be avoided. This is a typical work for a laboratory robot.

Robotic spot pickers are much more reliable for excising selected spots from the gel slab and transferring them to defined wells of microtiter plates. This automated picking can be controlled with a CCD camera or by transferring the pixel coordinates from image analysis into the machine coordinates of the picker instrument.

The camera in this system has only the function to find the positions of the two reference markers.

For the second technique, the gel has to be immobilized on a rigid support – a glass plate or a plastic film – prior to scanning. Two self-adhesive reference markers, glued onto the support plate or film, are necessary to enable the machine to recalculate the x/y positions of the proteins from the imported picking list. The two markers are automatically recognised by the camera and used as calculation points for the spot positions. This procedure provides a very high picking efficiency and accuracy, because it utilises the high resolution and sensitivity of the scanning device. Furthermore, using the image analysis data, also fluorescent, nonstained and radioactive labelled spots can be excised.

Figure 48 shows the assembly of a robotic spot picker working according this concept. The gel is placed into a liquid-containing tray to prevent drying during the picking process. The glass plate or film support is fixed at the bottom of the tray.

A picking file has to be created to assign the spot position to the respective microtiter plate well.

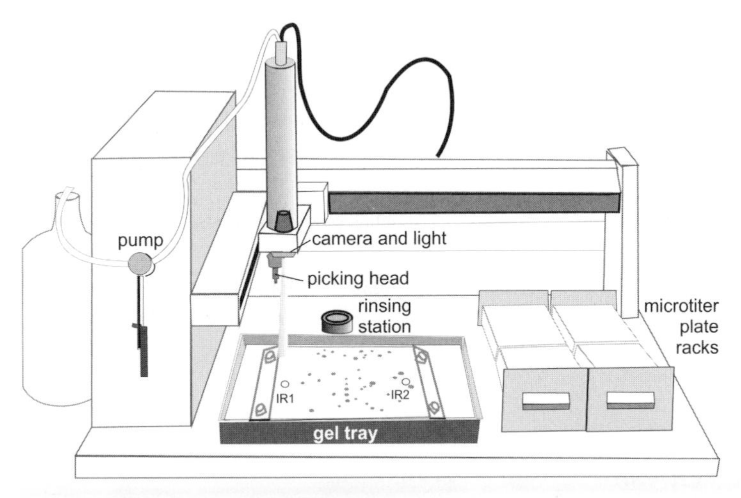

Fig. 48: Schematic drawing of an automatic spotpicker, which cuts the spots according to an imported picking list and transfers the gel plugs to microtiter plates.

The picking head is coated with a hydrophobic layer to prevent gel plugs sticking to it and protein cross-contamination. The gel plugs are cut out by punching down to the backing plate or film, followed by two side movements to shear the plug off the backing. Between each spot removal the cutting head cleans itself in a rinsing station to avoid cross contamination of proteins.

Such an instrument can also be operated in a semi-automated way in order to cut out a few spots with eye control. In this case the camera is also employed to find the spot to be removed from the gel.

2.2.2
Protein cleavage

As the mass spectrometers become ever more sensitive, delivering the sample of interest to the mass spectrometer in a manner which takes advantage of the innate high sensitivity of the mass spectrometer will become increasingly crucial.

Langen H. Takács B, Evers S, Berndt P, Lahm HW, Wipf B, Gray C, Fountoulakis M. Electrophoresis 21 (2000) 411–429.

The most commonly used strategies for protein identification by mass spectrometry require the protein of interest to be cleaved, proteolytically or chemically, into its constituent peptides (assembly of amino acids, see table 7). In the pathway highlighted in this book protein cleavage is performed by proteolysis after the protein has been separated by two-dimensional gel electrophoresis. This has been successfully applied to a wide range of proteome projects such as the analysis of the Haemophilus influenzae proteome (Langen *et al.* 2000). Other methods have described the use of proteolysis without separation of the proteins by 2-D electrophoresis (see section 2.5).

Tab. 7: Twenty common amino acids detailing structure, molecular weights, symbols and common modifications.

Residue (3 & 1 letter symbols)	***Residue structure (R = side chain)***	***Average mass***	***Monoisotopic Mass***	***Common modifications (and nominal molecular weight additions)***
Glycine (Gly & G)	OC—CH—NH; R = H	57.05	57.02146	
Alanine (Ala & A)	OC—CH—NH; R = CH_3	71.08	71.03711	
Serine (Ser & S)	OC—CH—NH; R = CH_2OH	87.08	87.03203	Phosphorylation (80) Glysosylation[a]
Proline (Pro & P)	OC—CH—N— (ring)	97.12	97.05276	
Valine (Val & V)	OC—CH—NH; R = $CH_2CH_2CH_3$	99.13	99.06841	
Threonine (Thr & T)	OC—CH—NH; R = $CH(OH)CH_3$	101.11	101.04768	Phosphorylation (80) Glysosylation (*O*-linked sugars)

Residue (3 & 1 letter symbols)	*Residue structure (R = side chain)*	*Average mass*	*Monoisotopic Mass*	*Common modifications (and nominal molecular weight additions)*
Cysteine (Cys & C)	OC–CH(R)–NH; R = CH_2SH	103.15	103.00919	Carbamidomethyl[b] (57) *S*-propionamide[c] (71) *S*-pyridylethyl[d] (105)
Leucine (Leu & L)	OC–CH(R)–NH; R = $CH_2CH(CH_3)_2$	113.16	113.08406	
Isoleucine (Ile & L)	OC–CH(R)–NH; R = $CH(CH_3)CH_2CH_3$	113.16	113.08406	
Asparagine (Asn & N)	OC–CH(R)–NH; R = $CH_2C(=O)NH_2$	114.10	114.04293	N-linked glycosylation[e] (Sequon Asn-X-Ser/Thr)
Aspartic acid (Asp & D)	OC–CH(R)–NH; R = $CH_2C(=O)OH$	115.09	115.02694	Esterification, methyl ester[f] (14)
Glutamine (Gln & Q)	OC–CH(R)–NH; R = $CH_2CH_2C(=O)NH$	128.13	128.05858	
Lysine (Lys & K)	OC–CH(R)–NH; R = $(CH_2)_4NH_2$	128.17	128.09496	Acetylation[g] (42) Homo-arginine (42)
Glutamic acid (Glu & E)	OC–CH(R)–NH; R = $CH_2CH_2C(=O)OH$	129.16	129.04259	Esterification, methyl ester[f] (14)
Methionine (Met & M)	OC–CH(R)–NH; R = $CH_2CH_2SCH_3$	131.20	131.04049	Oxidation[h] (16)

Residue (3 & 1 letter symbols)	Residue structure (R = side chain)	Average mass	Monoisotopic Mass	Common modifications (and nominal molecular weight additions)
Histidine (His & H)	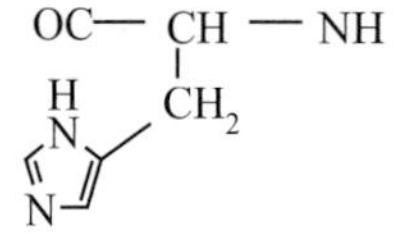 OC—CH—NH, CH₂-imidazole	137.14	137.05891	
Phenylalanine (Phe & F)	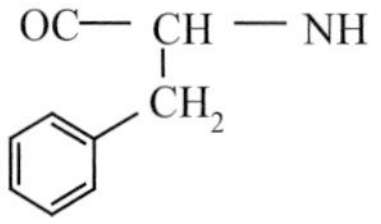 OC—CH—NH, CH₂-phenyl	147.18	147.06841	
Arginine (Arg & R)	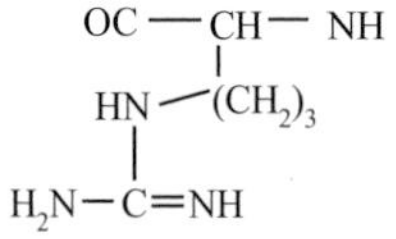 OC—CH—NH, $(CH_2)_3$—HN, H_2N—C=NH	156.19	156.10111	
Tyrosine (Tyr & Y)	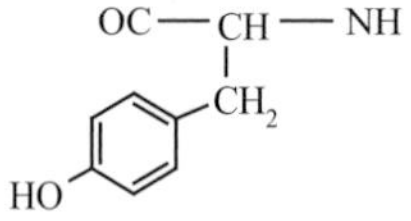 OC—CH—NH, CH₂-phenyl-HO	163.18	163.06333	Phosphorylation (80)
Tryptophan (Trp & W)	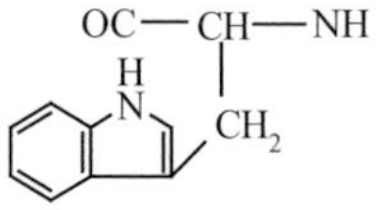 OC—CH—NH, CH₂-indole	186.21	186.07931	Oxidation (16)

Key:

- **a** *O*-linked glycosylation – a range of sugars can link to Ser/Thr, including pentoses ($C_5H_{10}O_5$), hexoses ($C_6H_{12}O_6$) and *N*-acetylhexosamines
- **b** Modification of cysteine residue with iodacetamide
- **c** Modification of cysteine residue with acrylamide monomer
- **d** Modification of cysteine residue with 4-vinylpyridine
- **e** *N*-linked glycosylation – a pentasaccharide core linked to the Asn residue, with a variety of structures extending from this core
- **f** Also esterification of the C-terminal carboxyl group of each peptide
- **g** Also acetylation of the α-amino group at the n-terminus of each peptide
- **h** Oxidation of methionine to the sulfoxide and the sulphone

2.2.2.1 Protein cleavage – proteolysis

Proteolysis has become the routine method of protein cleavage used in proteomics with a range of enzymes available (see table 8). Proteolysis offers several practical advantages; including high specificity, minimal side reactions and good cleavage efficiency. Important practical considerations for protein digestion include the digestion buffer and its pH, the enzyme:substrate ratio, temperature and time.

Tab. 8: Enzymes commonly used in proteomics.

Method of protein cleavage	***Site of cleavage***	***Exception***	***pH range***
Trypsin	C-terminus of R-X, K-X	If X = P	7–9
Endoproteinase Glu-C (V8-DE)	C-terminus of E-X, D-X	If X = P	4–8
Chymotrypsin	C-terminus of F,Y,W,L,I,V,M	If X = P	7.5–8.5
Endoproteinase Lys-C	C-terminus of lysine, K-X	If X = P	8.5–8.8
Arg-C	C-terminus of arginine, R-X	If X = P	7.5–8.5
Elastase	Not very specific. C-terminal side of G, A, S, V, L and I.		8–8.5
Pepsin	C-terminus of F,L and E		2–4
Pronase	Pronase is a mixture of endo- and exo-proteinases. It cleaves almost any peptide bond.		7–8, dependent on proteases present

Of all the enzymes available, the most commonly used enzyme is trypsin. Trypsin has well defined specificity, yields tryptic peptides of an appropriate size for efficient MS analysis and locates the basic residues at the terminus of the peptide. Though trypsin is commonly used there may be occasions where digestion with an alternative enzyme will be advantageous, specifically post-translational studies such as phosphorylation analysis. The specific site of phosphorylation may not reside on a tryptic fragment that is appropriate size for efficient MS analysis.

Restricting the basic residues to the c-terminus enables efficient peptide fragmentation during an MS/MS product ion experiment, see section 2.3.3.

The simplest digestion would be an in-solution digestion. However, optimal electroelution of a protein from the polyacrylamide gel matrix into solution for digestion is difficult and highly variable from protein to protein (Aebersold and Patterson, 1995). The electrotransfer (blotting) of the protein to a membrane (of which there were many types) with subsequent digestion on the membrane surface (Aebersold *et al.* 1987; Pappin *et al.* 1995).

Patterson S, Aebersold R. Electrophoresis 16 (1995) 1791–1814.
Aebersold R, Leavitt J, Saavedra RA, Hood LE, Kent BS. Proc Natl Acad Sci USA 84 (1987) 6970–6974; Pappin DJC, Rahman D, Hansen HF, Jeffery W, Sutton CW. Methods in protein structure analysis (1995) 161–173.

A further development of the blotting technology was described by Bienvenut *et al.* 1999. In this method the proteins are blotted through a membrane of immobilised trypsin, onto a support membrane where the constituent peptides are trapped and scanned directly by MALDI peptide mass fingerprinting. This technology is ultimately envisaged as the basis of a clinical scanner (Binz *et al.* 1999; Bienvenut *et al.* 1999; Schleuder, Hillenkamp and Strupat, 1999).

Bienvenut WV, Sanchez JC, Karmime A, Rouge V, Rose K, Binz PA, Hochstrasser DF. Anal Chem 71(1999) 4800–4807.

Wilm M, Shevchenko A, Houthaeve T, Breit S, Schweigerer L, Fotsis T, Mann M. Nature 379 (1996) 466–469.
Shevchenko A, Wilm M, Vorm O, Mann M. Anal Chem 68 (1996) 850–858.

Despite these methods the in-gel digestion methodology has become routine for proteins separated by 2-D electrophoresis. Wilm and co-workers, 1996 and Shevchenko *et al.* 1996a; reported the in-gel digestion of proteins from Coomassie Blue and silver stained gels which was compatible with protein identification by mass spectrometry.

75mM ammonium bicarbonate in 40% Ethanol is an effective CBB destain.

The basic theory is summarised in the following paragraphs, with some novel developments reported. Following visualisation of the gel, the protein spots of interest can be excised from the gel manually or automatically with commercially available gel spot pickers and be subsequently destained. The destaining of the protein spot is particularly important. For instance, Coomassie Brilliant Blue binds to proteins via ionic and hydrophobic interactions. The ionic interactions between the sulphonic acid group of the CBB and the basic residues of proteins (arginine and lysine residues) affect the trypsin digestion efficiency, and the presence of CBB in the final sample can hamper MS performance. This destaining step also allows the removal of unwanted detergents such as SDS from the digestion procedure and the MS analysis (in which detergents ionise very efficiently).

Yan JX, Wait R, Berkelman T, Harry RA, Westbrook JA, Wheeler WH, Dunn MJ. Electrophoresis 21 (2000) 3666–3672.

The presence of glutardialdehyde in many silver stain procedures precluded the stained protein from subsequent analysis, as the glutardialdehyde reacted with free amino groups in the protein. However, Shevchenko *et al.* 1996; demonstrated that this component could be omitted from the staining procedure and still realise high sensitive visualisation and successful MS analysis; Yan *et al* 2000; also published a silver stain method compatible with mass spectrometry.

Gharahdaghi F, Weinberg CR, Meagher DA, Imai BS, Mische SM. Electrophoresis 20 (1999) 601–605.
Additionally, the negative zinc/imadazol stain is compatible with mass spectrometric analysis.

The residual silver ions from the silver stain procedure can also be removed using the method reported by Gharahdaghi *et al.* 1999.

At this stage many methods have reported the reduction and alkylation of cysteine residues contained within the protein embedded in the gel piece (Rosenfield *et al.* 1992; Moritz *et al.* 1996; Shevchenko *et al.* 1996b; Gevaert and Vandekerckhove 2000). The derivatisation comprises two steps: reduction of the disulphide bonds and alkylation of the subsequent thiol side chain of the cysteine. This step is included to improve the detection of cysteine containing peptides and hence improve the potential protein coverage. However, the cysteine thiol group can become modified as it passes through the PAGE gel with free acrylamide monomer. Sechi and Chait, 1998; noted that the methods referred to above do not generally label the cysteine residues post electrophoresis with the same mass addition that occurs within the gel, hence potentially producing a heterogeneous derivatistion of cysteine residues making protein identification complicated, as the cysteine residues have effectively been labelled with different reagents.

Rosenfield J, Capdevielle J, Guillemot JC, Ferrara P. Anal Biochem 203 (1992) 173–179.
Shevchenko A, Jensen ON, Podtelejnikov A, Sagliocco F, Wilm M, Vorm O, Mortensen P, Shevchenko A, Boucherie H, Mann M. Proc Natl Acad Sci USA 93 (1996) 1440–1445;
Moritz RL, Eddes JS, Reid GE, Simpson RJ. Electrophoresis 17 (1996) 907–917.
Gevaert K, Vandekerckhove J. Electrophoresis 21 (2000) 1145–1154.
Sechi S, Chait BT. Anal Chem 70 (1998) 5150–5158

However, the 2D-electrophoresis method described in this book includes a quantitative reduction and alkylation derivatisation step of the cysteine residues prior to the second dimension SDS-PAGE. If the procedure is repeated post-electrophoresis and prior to digestion, then the same reagent should be used.

The cysteine residues become modified with acrylamide monomer to form the β-propionamide derivative, a mass difference of 71 Da. The modification of the cysteine with iodoacetamide (a mass difference of 57 Da) and 4-vinylpyridine, a mass difference of 105 is generally performed (see table 7).

The in-gel digestion of the embedded protein can be performed either manually or automatically. Staudenmann *et al.* 1998; reported that the digestion of proteins above 10 pmols was routine, giving few problems. However, once the concentration dropped below the 5 pmol level the number and yield of the peptides drops significantly, suggesting sample loss onto vessel surfaces and digestion efficiency as potential reasons for the disparity.

Staudenmann W, Dainese-Hatt P, Hoving S, Lehman A, Kertesz M, James P. Electrophoresis 19 (1998) 901–908.

Some major considerations involved with in-gel digestion that ultimately affect MS performance may include

- digestion efficiency
  - Substrate-enzyme kinetics.
  - Delivering the enyme to the protein whilst minimising autolysis
- extraction efficiency of peptides from the polyacrylamide gel matrix after digestion
- sample handling

- losses during the digestion procedure, for example onto the walls of vessels
- presence of detergents or staining agent
- minimising contamination with other proteins, such as keratin.

Shevchenko A, Jensen ON, Podtelejnikov A, Sagliocco F, Wilm M, Vorm O, Mortensen P, Shevchenko A, Boucherie H, Mann M. Proc Natl Acad Sci USA 93 (1996) 1440–1445.

Before addition of the enzyme to the gel plug, the gel plug (containing the embedded protein) is dehydrated using successive washes of acetonitrile and subsequent dried under vacuum. The dehydration of the gel plug to a completely dry dust-like material will enable efficient uptake of the enzyme when the gel plug is rehydrated. A sufficient volume of the enzyme in the digestion buffer (volatile buffer at the optimum pH for enzyme activity) is applied to rehydrate the gel plug. The enzyme has to passively diffuse through the gel matrix and digest the protein before autolysis has a significant effect. The rehydration step of the digestion procedure can be performed on ice (Shevchenko *et al.*1996) presumably to minimise autolysis of the trypsin whilst it diffuses through the gel during the rehydration step. Additionally, the modified form of the enzyme can be purchased to minimise autolysis.

Upon completion of the digest reaction, between 3 and 24 hours, the resultant digested peptides need to be extracted from the gel matrix. To enhance peptide recovery, the inclusion of detergents such as SDS, Tween 20, Triton X-100 or NP-40 in the extraction buffer has been reported. However, the presence of detergent was detrimental to MS analysis, both for matrix-assisted laser desorption ionisation and electrospray ionisation (see section 2.3.1). Instead, high peptide recovery can be obtained from the gel matrix by passive diffusion using acidic and organic solvents, namely solutions of trifluoroacetic acid and acetonitrile.

Once extracted the peptides have to be delivered to the mass spectrometer in a form, which enables optimal MS performance. As described in section 2.3.1 though MALDI and ESI MS are highly applicable to the analysis of biomolecules they are considerably different techniques, requiring different sample preparation approaches. MALDI involves the analysis of the sample from a remote, solid surface after co-crystallisation with the matrix, whilst ESI is a liquid inlet system requiring the sample to be presented in the liquid state. Thus ESI is effectively coupled with separation techniques such as capillary electrophoresis and particularly high performance liquid chromatography (HPLC).

Unfortunately, the extracted peptides tend to be in a large volume after extraction (>50 μL) and this is not ideal for MALDI or ESI analysis, especially if the sample is low in abundance. Hence the sample needs to be concentrated. However, extraction of the digested peptides also extracts unwanted material from the gel which may hamper MS ionisation and reduce MS performance. Simply concentrating the gel extracts is insufficient, as both the contaminants and analyte will be concentrated. Also, concentrating the sample by drying in vacuum or by lyophilisation can be problematic because of irrecoverable sample loss on the walls of the digestion vessel; this is particularly relevant for high sensitivity analysis.

Extensive washing of the sample at the destaining stage to effectively remove any residual detergent from the electrophoresis is very important. Further a low salt, volatile buffer can be used as the digestion buffer, minimising any detrimental effect observed in the MS analysis

An efficient approach to removing the salts, buffers and unwanted contaminants from the digestion process is to use reversed phase chromatography (RP-HPLC) off line prior to MALDI MS. In such a method using a RP 300 μm × 15 cm column, the peptides of interest can generally be eluted into two or three fractions in a concentrated volume (~4–5 μL) for subsequent MALDI MS analysis. In addition, the eluate, if eluted with the correct solvent, is ideally suited for ESI MS analysis.

A quicker solution for both MALDI and ESI analysis can be achieved by binding the digested peptides to small amounts of reversed phase resin packed into gel-loader pipette tips in a micro-scale purification. This not only has the desired effect of desalting, but also concentrates the sample in a suitable volume for the MALDI analysis (Kussmann *et al.* 1997). Similarly, this approach can be applied to on-target desalting, a method described by Gevaert *et al.* 1998.

Kussmann M, Nordhoff E, Rahbek-Nielsen H, Haebel S, Rossel-Larsen M, Jakobsen L, Gobom J, Mirgorodskaya E, Kroll Kristensen A, Palm L, Roepstorff P. J Mass Spectrom 32 (1997) 593–601

Gevaert K, De Mol H, Sklyarova T, Houthaeye T, Vandekerckhove J. Electrophoresis 19 (1998) 909–917.

Stensballe and Jensen, 2001, described a methodology where the digestion and MALDI analysis was performed on the MALDI target, minimising sample manipulation and losses.

Stensballe A, Jensen ON. Proteomics 1 (2001) 955–966.

Alternatively a small aliquot (<0.5 μL) can be spotted directly from the digestion mixture onto a pre-formed layer of matrix (α-cyano-4-hydroxy-cinnamic acid) and nitrocellulose. After the sample has dried, the salts from the digestion reaction can be effectively removed by washing with 0.1%TFA (Vorm *et al.* 1994; Jensen *et al.* 1996).

Vorm O, Roepstorff P, Mann M. Anal Chem 66 (1994) 3281–3287.

Jensen ON, Podtelejnikov A, Mann M. Rapid Commun Mass Spectrom 10 (1996) 1371–1378.

2.2.2.2 **Protein cleavage – chemical methods**

Chemical methods of protein cleavage are complimentary to proteolysis, though not used as commonly. They are often used for specific applications where no enzyme is available or proteolysis is not appropriate, such as cyanogen bromide cleavage of insoluble or membrane

Washburn MP, Wolters D, Yates JR. Nature Biotech 19 (2001) 242–247.

proteins. In this instance, a chemical fragmentation in an organic solvent capable of solubilising the proteins is an attractive alternative (Wasburn *et al.* 2001). Cyanogen bromide cleaves specifically at methionine residues yielding relatively large peptides. Similarly large peptides can be expected if cleavage at tryptophan residues is performed with 2(2-nitrophenylsulphonyl-3-indolenine) (BNPS-skatole).

Li A, Sowder RC, Henderson LE, Moore SP, Garfinkel DJ and Fisher RJ. Anal Chem 73 (2001) 5395–5402.

Acid hydrolysis of proteins with a mild formic acid solution has been reported to be an effective method. The formic acid solution is a good solvent for a whole range of proteins, and yields a specific cleavage at aspartyl residues.

See table 9 for further examples of chemical methods of protein cleavage

Tab. 9: Chemical reagents for protein cleavage.

Chemical reagent	*Site of cleavage*
Cyanogen bromide	C-terminus of Methionine
Acid hydrolysis	Cleavage at aspartyl residues
Hydroxylamine	Cleavage of asparagine-glycine bonds
BNPS-skatole	Cleavage at tryptophan

2.3 Mass spectrometry

Pappin DJC, Rahman D, Hansen HF, Bartlet-Jones M, Jeffery W, Bleasby AJ. Mass Spectrometry in the biological sciences. (1996) 135–150.

As has already been explained in previous chapters, large format two-dimensional gel electrophoresis enables the resolution of several thousand proteins in a reproducible fashion in a relatively short period of time. The instrumental developments in two-dimensional gel electrophoresis have offered the momentum for proteomics. A similar ground shift in the methods of protein identification was essential if proteomics was going to expand into the dominating field we know today. Edman sequencing was the principal method of protein identification in the eighties; combining the derivatisation of the n-terminal amino acid of a protein or peptide with the subsequent cleavage of the derivatised residue. This two step procedure was continued for each amino acid through the peptide sequence. The retention times of the cleaved derivatised amino acids were compared, by HPLC, with the retention times of a series of standards to determine the peptide/protein sequence. Though Edman was used with considerable success for routine protein identification, the method is relatively slow (one or two peptides per day) and relatively insensitive (upper fmol – low pmol amounts) (Pappin *et al.* 1995). It is still useful for certain aspects of applications in proteomics such as phosphorylation analysis, but is has generally been replaced with mass spectrometry for routine protein identification.

The introduction of the soft ionisation mass spectrometry techniques, matrix assisted laser desorption ionisation (MALDI) and electrospray ionisation (ESI) mass spectrometry in the late eighties provided the ground shift. Subsequent hardware and software developments have enabled the automated acquisition of data with high sensitivity and high accuracy in a robust and sensitive manner. The simultaneous development of software algorithms to utilise the mass spectrometry data in an efficient manner has established MS as the routine method for protein identification.

Principle

Mass spectrometry is an analytical technique that measures the molecular weight of molecules based upon the motion of a charged particle in an electric or magnetic field. The sample molecules are converted into ions in the gas phase and separated according to their mass : charge ratio (m/z). Positive and negatively charged ions can be formed. This technique is performed by a mass spectrometer.

Mass spectrometry enables identification and is complimentary to techniques such as NMR. Generally MS can be used in two different ways for identification: it can be used to indicate the composition of a particular analyte and secondly the structure of a particular analyte. The first instance is an example of the instrument being run in MS mode, where the actual mass of the analyte is measured and this mass is indicative of the composition of the analyte. For example, peptide mass fingerprinting utilises the mass of each of the tryptic peptides in the spectrum for a database search. In the second instance the mass spectrometer is used in MS/MS mode to obtain structural information relating to the analyte. In this mode, the mass of the analyte is determined and the analyte fragmented within the mass spectrometer to yield structural information. This method is common for the determination of small molecule structure and peptide sequence (see section 2.3.3).

In the field of protein identification and proteomics, a mass spectrometer is judged upon sensitivity, mass accuracy and tandem mass spectrometry. The former is always of significant importance and needs no further explanation. The latter, tandem mass spectrometry, is explained in section 2.3.3. The accuracy of mass measurement, an important parameter for protein ID, is closely related to resolution.

Amino acids consist of the elements carbon, hydrogen, nitrogen and oxygen, and to a lesser sulphur. Each of these elements exists naturally as a mixture of isotopes. For instance, carbon exists as a mixture of the ^{12}C isotope (98.9%) and the ^{13}C isotope (1.1%). As such the abundance of the two isotopes in a given compound will be reflected in a mass spectrum by the isotope envelope of the compound.

There are two types of mass measurement for a given compound

- Average mass. The measurement which reflects the contribution of the isotopes in an isotopic envelope. The average mass is taken at the centroid of the isotopic envelope.
- Monoisotopic mass. The mass of the first peak in a peptide isotopic envelope. With respect to carbon, this is termed the ^{12}C peak.

Figure 49 illustrates the predicted isotope distribution of the peptide HLKTEAEMK at resolution 5000 (FWHM).

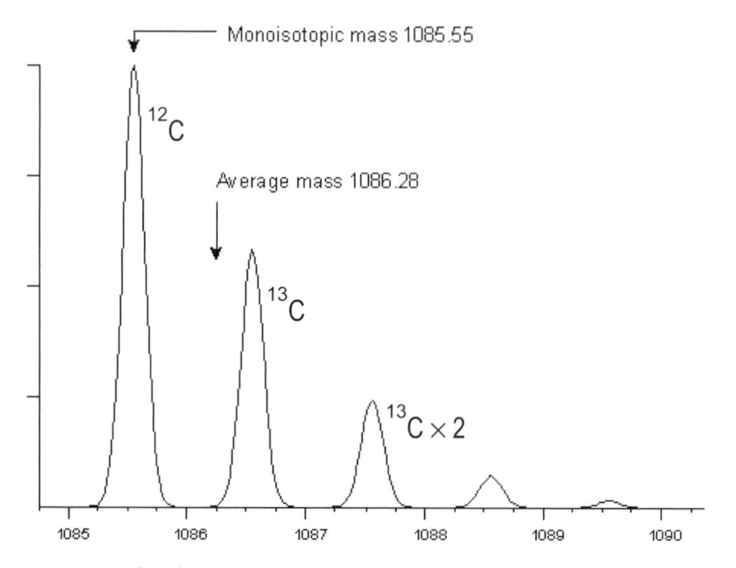

Fig. 49: Predicted isotope distribution of the peptide HLKTEAEMK at resolution 5000 (FWHM).

Thus the mass of the peptide which contains a ^{13}C atom is 1 Da higher.

The spectrum clearly demonstrates that with sufficient resolution the ^{12}C and ^{13}C isotopes are clearly resolved. As a result the monoisotopic mass, ^{12}C peak, can be clearly assigned. The second peak in the envelope represents that one of the carbon atoms in the peptide is a ^{13}C atom. The third peak represents that two of the carbon atoms in the peptide are a ^{13}C atoms and so on. The mass difference between the average and monoisotopic masses is 0.73 Da. This is a significant difference which can be capitalised on for peptide mass fingerprinting.

However, the ^{12}C monoisotopic peak is not always the most abundant peak in the isotope envelope. As the mass of a given compound increases, the contribution of the ^{13}C isotope increases (see figure 50). At a particular mass the monoisotopic ^{12}C peak is no longer observed, at this point the average mass is recorded for the given compound. This is the measurement used for proteins.

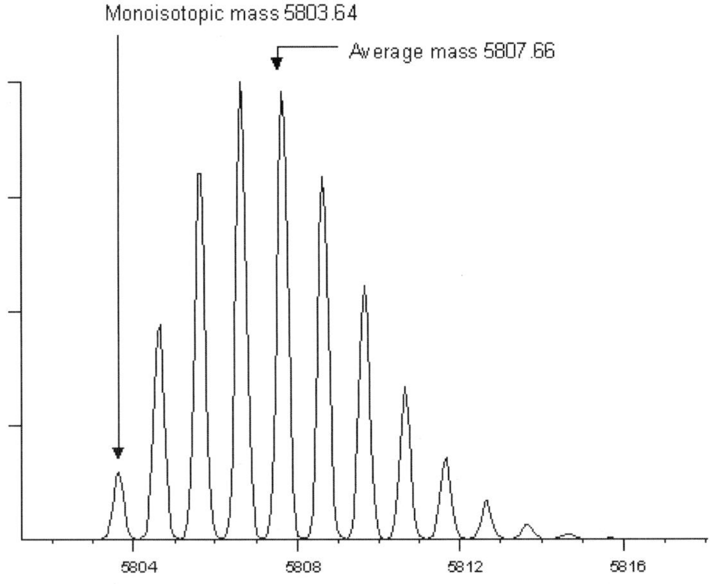

Fig. 50: The predicted resolved isotopic envelope for insulin. Note the ^{12}C peak is clearly no longer the most abundant peak in the envelope. The various ^{13}C contributions have become the most abundant peaks.

With respect to peptides the monoisotopic mass can only be measured if the ^{12}C and ^{13}C isotopes of the peptide envelope are sufficiently resolved and if the ^{12}C peak of the envelope is actually observed. If the resolution is sufficient to provide unambiguous assignment of the monoisotopic peak then mass accuracy is increased. Resultantly, the peptide monoisotopic mass is almost always used for protein identification by mass spectrometry.

Does not have to be baseline resolution as above.

A mass spectrometer can be effectively split into three constituent parts; the area of ion production, the ion source; the area of ion separation, the analyser and the area of ion detection, the detector (figure 51).

Ion production	Ion separation	Ion detection
MALDI Electrospray	Time-of flight (ToF) Triple quadrupole Ion trap Hybrid ToF FT-ICR	

Fig. 51: A schematic indicating the three regions of a mass spectrometer.

2.3.1
Ionisation

McFarlane R, Torgerson DF. Science 191 (1976) 920–925. Barber M, Bordoli RS, Sedgwick RD, Tyler AN. Nature 293 (1981) 270–271.

The ion source is the region of the mass spectrometer where the gas phase ions are produced from sample molecules, the area of ion production. The method of ion production is termed the ionisation technique. Several ionisation techniques have been developed, the earliest incarnations include electron impact and chemical ionisation which are useful for ionising small molecular weight molecules, but less applicable for larger (bio)molecules. The first example of ionising larger biomolecules was reported by MacFarlane and Torgerson in 1976 using the technique plasma desorption. The introduction of fast atom bombardment in 1981 (Barber *et al.* 1981) revolutionised biomass spectrometry enabling the ionisation and detection of a range of intact biomolecules with relatively good sensitivity. The FAB source was coupled with magnetic sector analyser, enabling bio-molecule analysis with relatively good sensitivity to be coupled with high resolution and MS/MS information for the first time.

Karas M Hillenkamp F. Anal Chem 60 (1988) 2299–2301. Fenn JB, Mann M, Meng CK, Wong SK, Whitehouse C. Science 246 (1989) 64–71.

However, FAB was supplanted by two ionisation techniques developed in the late eighties. Karas and Hillenkamp introduced matrix assisted laser desorption ionisation (MALDI) in 1988 as a technique that could readily ionise (large) biomolecules in a very sensitive manner MALDI is a pulsed ionisation technique which utilises the energy from a laser to desorb and ionise the analyte molecules in the presence of a light absorbing matrix. As this is a pulsed ionisation technique, generating packets of ions with each laser pulse, a pulsed analyser is typically required for separation and resolution of the ions. Consequently, MALDI is routinely coupled with a time-of flight analyser (TOF). In another breakthrough, Fenn and co-workers demonstrated that electrospray ionisation could also ionise large biomolecules with high sensitivity. Electrospray is typically coupled to ion trap, quadrupole, and ion cyclotron (ICR) analysers. These two ionisation techniques became the techniques of choice for protein and peptide ionisation.

2.3.1.1 Matrix Assisted Laser Desorption Ionisation (MALDI)

Principle MALDI ions are created by mixing the analyte with a small, organic molecule which absorbs light at the wavelength of the laser, the matrix. The analyte becomes incorporated into the crystal lattice of the matrix and is then irradiated with a laser. The laser causes the desorption and ionisation of the matrix and analyte, either by protonation or cationation (positively charged ions) or by deprotonation (negatively charged ions) (figure 52). The ions are then accelerated into the MS analyser (see section 2.3.2).

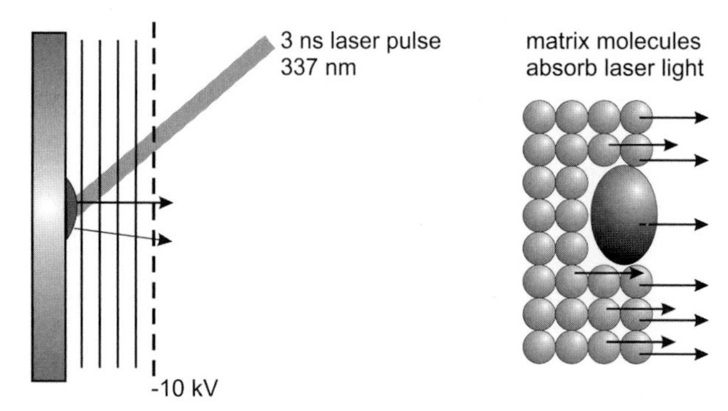

Fig. 52: Schematic of the MALDI process.

The typical wavelength of the UV lasers utilised is 337 nm. The precise mechanisms of desorption and ionisation is still unclear. Generally, irradiation of the matrix-analyte solid causes sublimation of the matrix-analyte solid and ejection of a plume of matrix and analyte ions into the gas phase. MALDI ions are generated under high vacuum (5×10^{-6}). Packets of ions are generated with each laser pulse and each packet is pulsed into the analyser. As MALDI is a pulsed ionisation technique, which is ideally coupled with a TOF analyser (see section 2.3.2.1).

Though new developments have seen MALDI ions produced at intermediate pressure (Loboda et al., 2000) and at atmospheric pressure (Laiko et al., 2000) similar to ESI.
Lobada A, Krutchinsky A, Bromirski M, Ens W, Standing KG. Rapid Commun Mass Spectrom 14 (2000) 1047–1057.
Laiko VV, Baldwin MA, Burlingame AL. Anal Chem 72 (2000) 652–657.

A wide range of matrices for bio-mass spectrometry applications have been adopted for use with UV lasers (see table 10). The three widely used matrices for peptides and proteins are α-cyano-4-hydroxy cinnamic acid, 2,5-dihydroxybenzoic acid (2,5-DHB) and sinapinic acid. Beavis and Chait reported the use of α-cyano, 4 hydroxy cinnamic acid for peptides and it is still widely used today affording high sensitivity and negligible matrix adduction. Many different procedures for the sample preparation for α-cyano exist. Figure 53 shows a

MALDI-ToF spectrum of a BSA tryptic digest acquired with α-cyano-4-hydroxy cinnamic acid.

2,5-dihydoxybenzoic acid DHB was described by Strupat *et al*, 1991. Further applications include detection of high molecular weight proteins and the analysis of oligosaccharides released from glycoproteins (Mock *et al.* 1991; Stahl *et al.* 1991; Harvey 1993).

Strupat K, Karas M, Hillenkamp F. Int J Mass Spectrom Ion Proc 111 (1991) 89–102.
Mock KK, Davy M, Cottrell JS. Biochem Biophys Res Commun 177 (1991) 644–651.
Stahl B, Steup M, Karas M, Hillenkamp F. Anal Chem 63 (1991) 1463–1466.
Harvey DJ. Rapid Commun Mass Spectrom 7 (1993) 614–619.

MALDI produces predominantly singly charged ions.

Tab. 10: Common MALDI matrices used in biological applications.

Matrix	Matrix Structure	Application
α-cyano-4-hydroxy-cinnamic acid	CH=C(COOH)(CN), OH	UV laser Peptide analysis & Protein digests. Analytes < 10 kDa
Sinapinic acid (4-hydroxy-3,5-dimethoxycinnamic acid)	CH=CH–COOH, CH_3O, OMe, OH	Analysis of large poly-peptides & proteins > 10 kDa
2,5-Dihydroxybenzoic acid (2,5 DHB)	COOH, OH, HO	UV laser Protein digests & Proteins Oligosaccharides released from Glycoproteins
2,4,6-trihydroxyacetophenone (THAP)	O, CH_3, HO, OH, OH	UV laser Oligonucleotides < 3 kDa
3-hydroxy picolinic acid	OH, N, COOH	UV laser Oligonucleotides > 3kDa

Beavis RC, Chaudhary T, Chait BT. Org Mass Spectrom 27 (1992) 156–158.
Beavis RC, Chait BT. Rapid Commun Mass Spectrom 3 (1989) 432–435.
Strupat K, Karas M, Hillenkamp F. Int J Mass Spectrom Ion Proc 111 (1991) 89–102.
Pieles U, Zurchner W, Schar M, Moser HE. Nucleic Acids Res 21 (1993) 3191–3196.

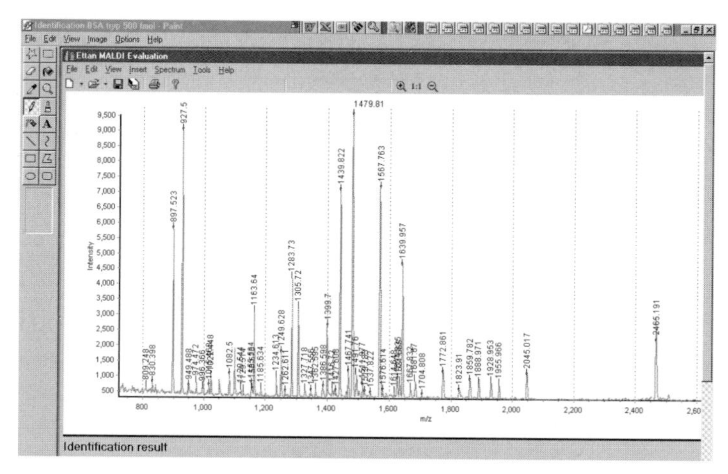

Fig. 53: MALDI in positive ion mode for the analysis of a tryptic digest of myoglobin. The spectrum was acquired using α-cyano-4-hydroxycinnamic acid as the matrix. Each of the peptide signals observed in the mass range m/z 800–2500 exist as the $[M+H]^+$ molecular ion, where M is the mass of the peptide. A MALDI-ToF spectrum of a BSA tryptic digest.

2.3.1.2 Electrospray ionisation (ESI)

Fenn and co-workers described the use of electrospray for ionising large biomolecules in 1989.

Taylor G. Proc R. Soc Lond A280 (1964) 383–397.

Principle ESI ions are produced at atmospheric pressure by applying sample dissolved in solvent to a narrow capillary tube, which is under the influence of an electric field. A potential difference is created between the capillary and the inlet of the mass spectrometer generating a force extending the liquid to form a cone from the capillary tip, from which a fine mist of droplets will emerge (Taylor cone, Taylor, 1964). The cone is formed as a result of repulsive coulombic forces between the like charges. Subsequently, evaporation of the droplets reduces droplet size whilst the charge on the droplet remains constant. Eventually, the surface coulombic forces exceed the surface tension and the droplets fission into smaller droplets (the Rayleigh limit). This process continues until nanometer sized droplets are produced (figure 54).

The charges are statistically distributed over the analyte's potential charge sites, enabling the formation of multiply charged ions. Each multiply charged ion can be termed a charge state, and a distribution of charge states is characteristic of large macromolecules during ESI analysis.

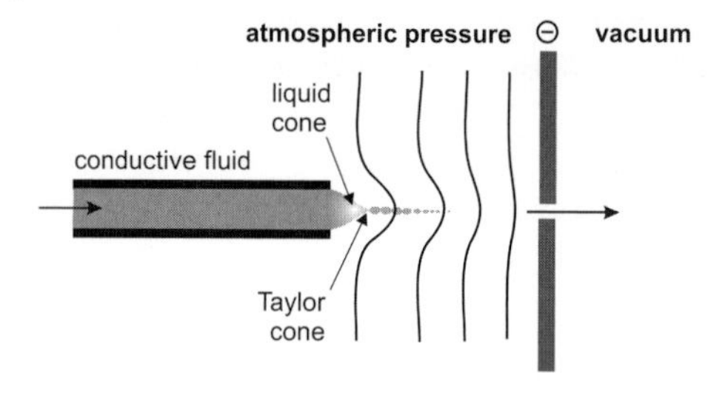

Fig. 54: Schematic of ESI.

ESI produces predominantly multiply charged ions

Note the mass range and the number of charges (z value in m/z) on the protein in the spectrum.
Its mass is known from the spectrum, but without knowledge of the charge state, the real mass of the analyte remains unknown.

This uniquely distinguishing feature of ESI (as opposed to singly charged ions produced by MALDI), yields an effectively reduced mass for the analyte of interest. Simply, the observable mass range is lowered offering significant advantages for the analysis of large molecular weight analytes, particularly proteins. Proteins become multiple charged to such an extent that the protein can be observed on the

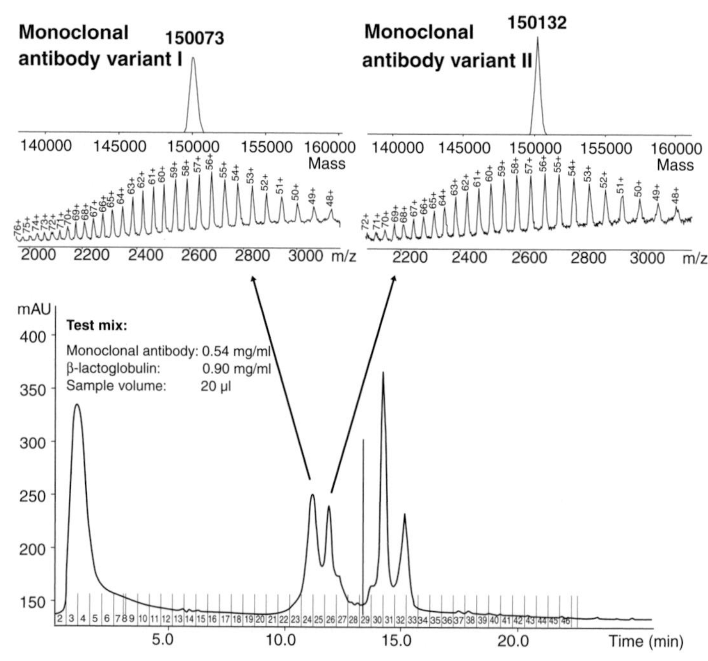

Fig. 55: Characteristic multiply charging of full length protein acquired with the Ettan ES-ToF. Upper panel – true mass of protein after deconvolution; lower panel – multiply charged spectrum. Note the number of charge states observed for this 150kDa protein.

simplest of mass spectrometers. Figure 55 demonstrates the multiply charging of 150 Kda protein. Generally, with respect to proteins the resolution of each multiply charged ion is insufficient to determine the number of charges the ion possesses, from which can be determined its true mass. Therefore, the masses of at least two of the multiply charged ions in the envelope are used in a deconvolution algorithm to give the actual mass of the whole protein (lower panel). Due to the formation of the multiple charge states and a large number of estimates of the mass (one per charge state), the mass accuracy afforded for the protein is generally very good.

For small biomolecules such as peptides, the charge state of the multiply charged ion can be readily determined if the resolution is sufficient. As tryptic peptides tend to produce doubly or triply charged ions the charge state can be readily be observed (see figure 56). The mass difference between the adjacent ^{12}C and ^{13}C istopes is indicative of the charge state. For a singly charged ion the mass difference between the ^{12}C and ^{13}C isotopes of a peptide isotopic envelope will be 1 Da. In the case of a doubly charged peptide ion (m/z = m/2) the mass difference between the ^{12}C and ^{13}C isotopes will be 0.5 Da, for a triply charged peptide (m/z = m/3) the mass difference between the ^{12}C and ^{13}C will be 0.33 Da and so on. With the knowledge of the charge state the true mass of the peptide can be determined. In the case of the peptide ion in figure 55, the charge states are 0.33 Da apart indicating that the peptide is triply charged. Together with the knowledge of the mass of the first ion in the envelope the mass of the peptide is easily determined.

In this example, the mass of triply charged ^{12}C ion was measured at 558.312 Da (mass of peptide + 3 (3 ionising protons)/3); a mass accuracy of 2.5.

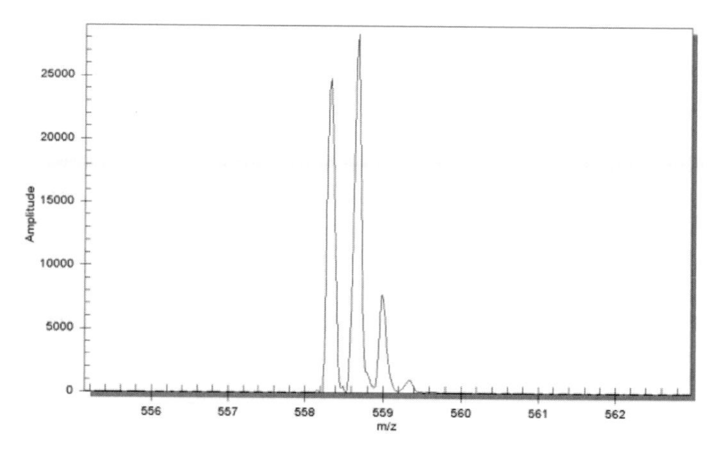

Fig. 56: Characteristic multiply charging of tryptic peptides: triply charged neurotensin protein acquired with the Ettan ESI-ToF.

Emmett MR, Caprioli RM. J Am Soc Mass Spectrom 5 (1994) 605–613.
Wilm M, Mann M. Anal Chem 68 (1996) 1–8.

The electrospray ionisation technique is readily coupled with separation techniques, principally HPLC and capillary electrophoresis. Electrospray, coupled with LC or not, generates flow rates ranging from tens of microlitres-hundreds of microlitres per minute. This high flow rate can be problematic for protein identification, particularly as the analysis time is very quick with all the sample being consumed in a short amount of time. Hence there is insufficient time to perform all the necessary experimentation (for example multiple MS/MS spectra), a particular problem for lowly abundant samples. In the early nineties reductions in ESI low rates were reported; sub-microlitre flow rates by continuous infusion affording significant increases in sensitivity (Emmett and Caprioli, 1994) Wilm and Mann reported the use of a nanospray ionisation source, electrospray without continuous infusion. (Wilm and Mann, 1996) This miniaturisation of electrospray generated flow rates in the order of tens of nanolitres per minute (typically 25–40 nL/min), by spraying the solvent from a narrow bore coated needle (sub 10 μm tip diameter). Furthermore, in addition to the longer analysis times observed enabling multiple experiments to be performed, increased sensitivity is also attained.

DE-MALDI-TOF and electrospray have transformed protein ID methods.

2.3.2
Ion separation

After the gas phase ions have been produced, they are accelerated from the ion source and guided into the analyser, the region of ion separation. Here the ions are separated according to their mass to charge ratio (m/z). Many types of analyser exist and those commonly used for proteomic applications are briefly described below.

2.3.2.1 Time-of-flight analyser

Time-of-flight (ToF) analysers are one of the simplest MS analysers in use today. Ions produced in the ion source are accelerated by high voltage into the ToF analyser, acquiring an initial velocity that is dependent on their mass. The analyser affords a high sensitivity and theoretically an unlimited mass range; the practical limit is determined by the detection efficiency of the ions arriving at the detector. The analyser is a pulsed analyser and routinely coupled with MALDI.

Principle Mass measurement is recorded by the time of flight of an ion in the ToF flight tube. The time-of-flight of an ion is proportional to the square root of its mass/charge ratio, given a constant accelerating voltage.

Thus, the smaller the molecule the faster it will travel the distance of the flight tube to the detector.

$$\text{time of flight} = k\sqrt{m/z}$$

The time of flight of an ion is typically measured as the time it takes to traverse the flight tube to the detector tube from when the ion leaves the source.

The simplest example is a linear ToF analyser, where each ion is accelerated into the field-free region (the flight tube) and maintains the constant velocity it acquired by the acceleration until it hits the detector. With respect to resolution and hence mass accuracy, the performance of a linear ToF analyser is related to the ionising conditions within the ion source and to the length of the flight tube. An inherent problem of MALDI is the initial kinetic energy distribution that is observed within the source for ions of the same mass.

The longer the flight tube the better the resolution.
Ions of the same mass are formed in different places within the ion source and hence are slightly differently affected when the voltage is applied, and will be accelerated from the source with slightly different kinetic energies into the flight tube.

Though the linear ToF analysers are useful for some proteomic applications, the resolution of peptides in the range of interest (FWHM ~100–500), and subsequent mass accuracy, obtained with this analyser is poor. A linear ToF instrument is incapable of resolving the ^{12}C and ^{13}C isotopes of the peptide isotopic envelope across the range of the protein digest, 700–3500 Da. As the initial velocity distributions are fairly constant with mass the width of the kinetic energy distribution will increase as the mass increases, giving rise to further peak broadening and worsened resolution and mass accuracy (figure 57).

$KE = ½ mv^2$

With the increasing size of the protein databases, it was recognised that greater stringency would be required to search these databases effectively using peptide mass fingerprinting. Improved resolution would enable the peptide isotopic envelope to be resolved allowing accurate mass measurement of the mono-isotopic ^{12}C isotope, rather than the average mass of the peptide. Instrument resolution can be improved by extending the flight tube of the analyser (but other parameters begin to take effect as the flight tube length is increased dramatically). Two more practical measures can be used. Two more practical measures can be used.

Protein database search confidence will be improved if mass accuracy is increased.

1. Incorporation of a reflectron into the flight tube (Mamyrin, 2001).
A reflectron acts as an ion mirror reversing the trajectory of the ions back down the flight tube, effectively increasing the length of the flight tube. Furthermore the reflectron accommodates for the initial

Mamyrin BA. Int J Mass Spectrom 206 (2001) 251–266.

kinetic energy spread the ions of the same mass experience in the source. Ions of the same mass formed in the source can have different kinetic energies when they leave the source depending upon their position in the source when the accelerating voltage was applied. This results in some ions of the same mass arriving at the detector before the slightly less energetic ions of the same mass, causing a spread of arrival times, thus reducing resolution and mass accuracy. The reflectron can accommodate these small differences in kinetic energy; the ions with slightly higher kinetic energy travel further into the reflectron than the ions with slightly lower energy. Resultantly, the ions of the same mass are better time focused at the detector, greatly improving resolution and subsequently mass accuracy. Resolution obtained with a reflectron instrument can exceed 5000 (FWHM) (figure 57).

Cornish TJ, Cotter RJ. Rapid Commun Mass Spectrom 8 (1994) 781–785.
Anderson UN, Colburn AW, Makarov AA, Raptakis EN, Reynolds DJ, Derrick PJ, Davis SC, Hoffman AD, Thomson S. Rev Sci Instrum 69 (1998) 1650–1660.

A number of reflectron designs are commonly used in commercial mass spectrometers; including conventional linear field reflectron, the curved field reflectron (Cornish and Cotter, 1994) and the harmonic or quadratic field described by Anderson *et al.* 1998. In a quadratic field an increasing voltage is applied, non-linearly, to create a perfect quadratic field. As with the other reflectron designs the kinetic energy correction leads to a better time focus at the detector and gives mass spectra of high resolution and excellent mass accuracy. Such a design has significant advantages for performing post-source decay, a technique used to determine peptide sequence (see section 2.3.3.5).

2. Timed delay ion extraction

Wiley WC, McLaren IH. Rev Sci Instrum. 26 (1955) 1150–1157.
Brown R, Lennon JJ. Anal Chem. 67 (1995) 1998–2003.
Vestal ML, Juhasz P, Martin SA. Rapid Commun Mass Spectrom 9 (1995) 1044–1050.
Whittal RM, Li L. Anal Chem 67 (1995) 1950–1954.

Secondly, resolution can be further improved by delaying the extraction of the ions (or timed delay ion extraction) from the source for a short period of time. Wiley and McLaren, 1953 reported such a method, time lag focusing, and noted the increased resolution. This was revisited by Brown and Lennon, Vestal *et al.*, Whittal and Li in 1995 and applied to MALDI. They observed that by delaying the extraction of ions from the source the ions were allowed to spread out before the accelerating voltage was applied This allows for the initially faster ions to drift further away from the sample plate than the initially slower ions, which after the electric field is pulsed on means that the ions further away from the sample plate will be at lower potential energy than the slower ions. Thus, after pulsed acceleration, the initially slower ions, which are lagging behind will come out at a higher velocity than the initially faster ions, and will catch up at a position in space. Resultantly, at this position a much tighter packet of ions are produced giving significantly higher resolution and hence mass accuracy (figure 57). Timed delay ion extraction coupled with a reflectron can achieve greater than 10,000 resolution (FWHM) (figure 58).

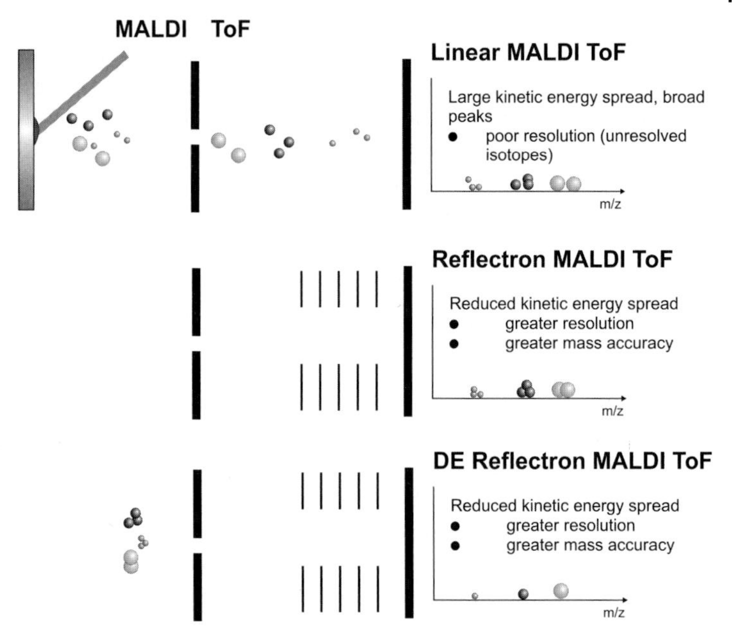

Fig. 57: Schematic demonstrating the effect of a reflectron and timed delay ion extraction on peak width and hence resolution and mass accuracy.

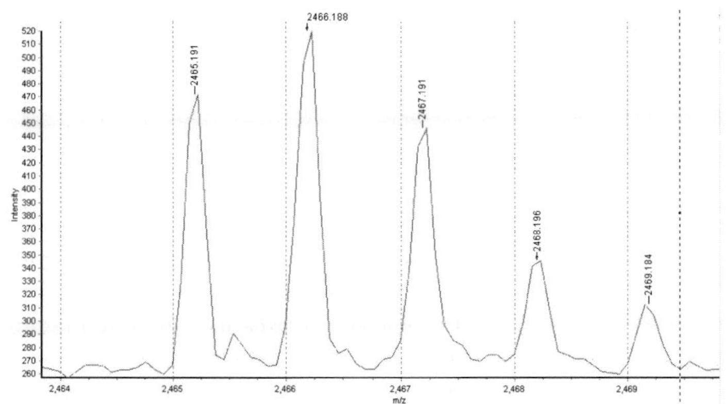

Fig. 58: Spectrum demonstrating the resolution capable with a MALDI-ToF equipped with a reflectron and timed delay ion extraction. Resolution >12000 at 3657.922 (hACTH clip 7–38).

Although a ToF analyser is commonly coupled with a pulsed MALDI source it has recently been combined with electrospray, the ESI-ToF instrument (Boyle and Whitehouse, 1992; Mirgorodskaya *et al*, 1994; Verentchikov *et al.* 1994).

Boyle JG, Whitehouse CM. Anal Chem 64 (1992) 2084–2089.
Mirgorodskaya OA, Shevchenko AA, Chernushevich V, Dodonov AF, Miroshnikov AI. Anal Chem. 66 (1994) 99–107.
Verentchikov N, Ens W, Standing KG. Anal Chem 66 (1994) 126–133.

2.3.2.2 **Triple quadrupole analyser**

The quadrupole analyser consists of four parallel hyperbolic rods, through which the gas phase ions have to achieve a stable trajectory. The analyser is operated by the application of a voltage (dc) and an oscillating voltage (radio frequency, rf) to one pair of rods and DC voltage of opposite polarity and rf voltage of different phase to the opposite pair of rods. The alternating electric field helps to stabilise and destabilise the passing ions. The ions traverse through the space between the rods, and only at specific voltages applied to the rods will certain m/z values be allowed to pass through the rods and reach the detector. The voltages are scanned to allow a wide mass range to be observed.

von Haller PD, Donohoe S, Goodlett DR, Aebersold R, Watts JD. Proteomics 1 (2001) 1010–1021.

A single quadrupole analyser has negligible benefits for proteomic analysis, but if three quadrupoles are arranged in sequence, the subsequent triple quadrupole analyser coupled with ESI has been demonstrated to be very useful for proteomics (von Haller *et al.* 2001) This configuration is almost exclusively used to provide structural information by tandem mass spectrometry (see section 2.3.3). Here the first and third quadrupoles operate as the mass filters described above, with the second quadrupole operating as a collision cell (where the structural information is produced). The analyser can be operated in three different ways to perform MS/MS experiments (see 2.3.3.1–3).

This configuration has recently been overtaken for proteomic applications by the hybrid quadrupole ToF instrument. In this configuration the third scanning quadrupole is replaced with a non-scanning reflectron ToF analyser, thereby increasing sensitivity and resolution in both MS and MS/MS modes (see section 2.3.2.4).

2.3.2.3 **Quadrupole ion trap**

Stafford GC, Kelley PE, Syka JEP, Reynolds WE, Todd JFJ. Int J Mass Spectrom Ion Proc 60 (1984) 85–98.

The quadrupole ion trap is based on the same theory as the quadrupole analyser, with the quadrupole field generated within a three dimensional trap. The ion trap itself is filled with helium and comprises a ring electrode and two end cap electrodes, creating the three dimensional electric field by applying a large RF voltage to the ring electrode. The orbiting motion of the ions in the trap is governed by the large rf voltage and the cooling effects of collisions with the helium gas (the helium gas reduces the kinetic energy of the ions and helps focus the ions in the centre of the trap). A mass spectrum is acquired by sequentially ejecting fragment ions from low m/z to high m/z. This is performed by scanning the rf voltage to make ion trajectories sequentially become unstable, the mass selective instability mode developed by Stafford *et al.* 1984; Ions are ejected and detected at the detector.

A product ion MS/MS spectrum is obtained via resonance excitation. The precursor ion is isolated by the application of a waveform signal to the endcap electrodes. Fragmentation of the precursor ion is caused by the application of a small voltage across the endcap electrodes and collisions with helium gas. The resultant fragmentation is similar to that observed from a triple quadrupole mass spectrometer; a low energy collisional regime with fragmentation predominantly at the peptide amide bonds (see section 2.3.3). A drawback of the instrument for the acquisition of product ion MS/MS spectra is that the fragment ions in the lower third of the mass range (approximately 1/3 rd of the mass of the precursor ion) will not be detected. Depending on the mass of the precursor ion, several fragment ions from both the b- and y-ion series may not be observed (see section 2.4.5).

An additional advantage of quadrupole ion traps (and FT-ICR MS, see section 2.3.2.6), is their unique ability to perform multiple stages of mass spectrometry (MS^n) (see section 2.3.3.4). For a detailed review see Jonscher and Yates, 1997.

Jonscher KR, Yates JR III. Anal Biochem 211 (1997) 1–15.

The performance characteristics of a quadrupole ion trap make them a highly attractive instrument for proteomic applications, particularly for tandem mass spectrometry applications, either interfaced with nanospray (Hoess *et al.* 1999) LC-MS/MS (Ho *et al.* 2002) 2DLC-MS/MS (Washburn *et al.* 2001) or MALDI (Qin *et al.* 1996).

Hoess M et al. 1999;
Washburn MP, Wolters D, Yates JR. Nature Biotech 19 (2001) 242–247.
Ho et al., Nature 415 (2002) 180–183.
Qin J, Steenvorden RJJM, Chait BT. Anal Chem. 68 (1996) 1784–1791.

2.3.2.4 **Hybrid analysers (quadrupole time-of flight)**

Initially, described in 1996 (Morris *et al.* 1996) for oligosaccharide analysis and more recently by Lobada *et al.*, 2000; these instruments have rapidly become the instrument standard for MS/MS applications within the theatre of proteomics. By combing a mass filtering quadrupole analyser and a collision cell (rf only quadrupole) with a non-scanning reflectron ToF analyser, the user is able to acquire MS and, most notably MS/MS data with high mass accuracy, resolution and sensitivity. The instrument is generally with interfaced with HPLC (Bell *et al.* 2001, Gavin *et al.* 2002).

Morris HR, Paxton T, Dell A, Langhorne J, Berg M, Bordoli RS, Hoyes J, Bateman RH. Rapid Commun Mass Spectrom 10 (1996) 889–896.
Shevchenko A, Chemushevich IV, Ens W, Standing KG, Thomson B, Wilm M, Mann M. Rapid Commun Mass Spectrom 11 (1997) 1015–1024.
Bell AW, Ward MA, Blackstock WP, Freeman HNM, Choudhary JS, Lewis AP, Chotai D, Fazel A, Gushue JN, Paiement J. J Biol. Chem 276 (2001) 5152–5165.
Gavin et al., Nature 415 (2002) 141–147.

Lobada A et al. Rapid Commun Mass Spectrom 14 (2000) 1047–1057.
Shevchenko A et al. Ana. Chem 72 (2000) 2132–2141.
Baldwin MA et al. Anal Chem 73 (2001) 1707–1720.

More recently, a MALDI ion source has been fitted to such a hybrid instrument enabling similar MS/MS performance to be attained. (Lobada *et al.* 2000; Shevchenko *et al.* 2000 Baldwin *et al.* 2001).

This configuration enables a peptide mass fingerprint and MS/MS data with high specifications data to be acquired using the same instrument, removing the need for a second tier of mass spectrometry, as is commonly currently used. The MS/MS spectra will be derived from singly charged MALDI precursor ions as opposed to the doubly or triply charged precursor ions observed when the analyser is combined with ESI.

2.3.2.5 TOF/TOF

Medzihradszky KF, Campbell JM, Baldwin MA, Falick AM, Juhasz P, Vestal ML, Burlingame AL. Anal Chem 72 (2000) 552–558.

A TOF/TOF analyser has recently been coupled with a MALDI ion source (Medzihradszky *et al.* 2000) generating peptide mass fingerprint data and peptide sequence derived by high energy collision induced dissociation (as opposed to low energy CID obtained with the analysers listed above). Basically, two ToF analysers are separated by a collision cell, with the first ToF analyser, used for precursor ion selection. High energy collisions occur within the collision cell, and the second ToF analyser resolves the ions. The configuration allows for high sensitivity and high resolution in both MS and MS/MS modes, and is capable of reducing protein identification to a one tier process.

2.3.2.6 Fourier transform ion cyclotron (FT-ICR)

Martin SE, Shabonowitz J, Hunt DF, Marto JA. Anal Chem 72 (2000) 4266–4274.
Goodlett DR, Bruce JE, Anderson GA, Rist B, Pasa-Tolic L, Fiehn O, Smith RD, Aebersold R. Anal Chem 72 (2000) 1112–1118.
Shi SDH, Hemling ME, Carr SA, Horn DM, Lindh I, McLafferty FW. Anal Chem. 73 (2001) 19–22.

Similarly to an ion trap, an FT-ICR is capable of trapping and storing ions. The ICR resides in a strong magnetic field and consists of three pairs of three parallel plates arranged in a cube. In the cell, ions of a given m/z ratio have a given cyclotron frequency of a given orbit radius. On applying an rf voltage at the same frequency as the cyclotron frequency, the respective ions absorb energy and are accelerated to a larger orbit radius. When the rf voltage is removed the energised, accelerated ions still rotate at a constant radius. Ions which have a different cyclotron frequency remain unexcited and hence ions of differing mass can be seperated.

An FT-ICR is the top performing mass spectrometer in terms of resolution and mass accuracy. In addition, its compatibility with both MALDI and ESI, high sensitivity and MS/MS and MS^n capability ensure proteomic applications are increasing rapidly (Martin *et al.* 2000) The high mass accuracy combined with novel chemistry allows very stringent searches of the protein database (Goodlett *et al.* 2000). Furthermore electron capture dissociation (ECD), a type of tandem mass spectrometry which is unique to FT-ICR instruments, enables the unambiguous assignment of sites of phosphorylation (Shi *et al.* 2001).

2.3.3
Tandem mass spectrometry (MS/MS)

In the introductory paragraphs to mass spectrometry in section 2.3, we briefly described the two modes of operation of a mass spectrometer. Firstly, the MS mode where molecular weights are measured and composition inferred; and secondly where the analyte ion of interest is mass measured, specifically selected and fragmented in the mass spectrometer generating structural information. Fragmentation is generated by collision induced dissociation (CID) within the mass spectrometer. This second technique is termed tandem mass spectrometry or MS/MS.

The technique is performed with instruments capable of selecting ions of a particular *m/z* value and subjecting the selected ions to fragmentation within the mass spectrometer. Generally, these experiments are performed successfully on two types of instrument; those where analysers are in series (tandem in space) such as the triple quadrupole and hybrid quadrupole-ToF configurations described earlier in this section; and secondly those instruments which employ ion trapping mechanisms such as the quadrupole ion trap and FT-ICR analyser (tandem in time).

Three types of MS/MS experiments are performed routinely within proteomics. The configuration used in triple quadrupole and hybrid quadrupole-ToF instruments will be used to explain the three types of experiment (see figure 59). Effectively the analyser region of these instruments can be regarded in three parts; MS1, the collision cell and MS2.

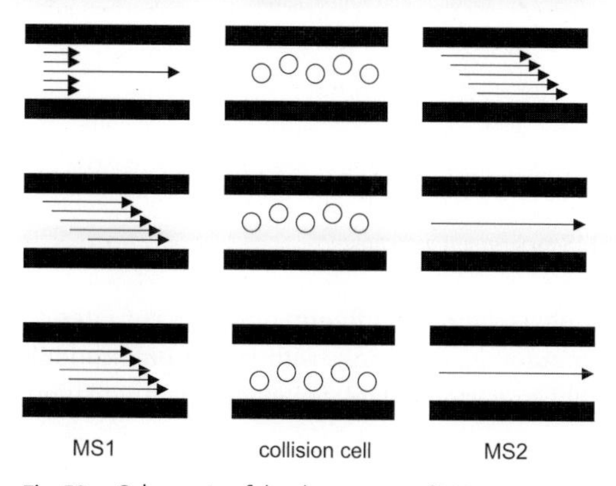

Fig. 59: Schematic of the three types of MS/MS experiment. Top panel – product ion scan Middle panel – parent ion scan, and bottom panel – neutral loss scan.

2.3.3.1 Product ion scan

In a product ion scan, the first part of the analyser, MS1, is used to specifically select the ion of interest, the precursor ion (i.e. a peptide). The precursor ion is allowed into the collision cell where it will undergo CID (figure 59 upper panel). Here, the peptide precursor ions collide with molecules of the collision gas (typically argon or nitrogen) causing the precursor ion to fragment yielding a distribution of fragment ions, or product ions. The resultant product ions are resolved by the third part of the analyser, MS 2, before detection at the detector producing a product ion spectrum. With respect to a triple quadrupole analyser, MS 2 is a mass filtering quadrupole and hence scans the selected mass range to enable the detection of the product ions. In the case of a hybrid quadrupole ToF analyser, MS 2 is a reflectron ToF analyser, in which the entire mass range is not scanned, but collected simultaneously with significant improvements in sensitivity and resolution. With respect to the ToF/ToF configuration, the first ToF acts as MS1 and the second ToF acts as MS2. Figure 60 demonstrates the experiment workflow, where an ion of interest is selected and isolated by MS1 and fragmented yielding the product ion MS/MS spectrum.

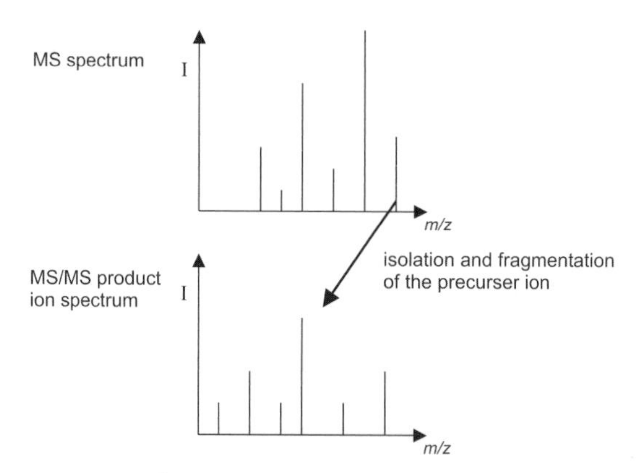

Fig. 60: Product ion MS/MS experiment workflow.

Spahr CS, Susin SA, Bures EJ, Robinson JH, Davis MT, McGinley MD, Kroemer G, Patterson SD. Electrophoresis 21 (2000) 1635–1650.

The process of selecting, isolating and fragmenting ions of interest can be performed in an automated fashion, with recent publications demonstrating the ability to perform data dependant fragmentation (Spahr *et al.* 2000).

CID can be further defined by the energy of the collisions observed in the collision cell. Fragmentation within triple quadrupole, quadrupole ion trap and hybrid quadrupole ToF analysers occurs at collisional energies in the order of 10–100 eV range, whilst fragmentation within a magnetic sector or ToF/ToF analyser occurs at collisional energies at least an order of magnitude higher in the keV range. The former is described as low energy CID whilst the latter is described as high energy CID.

As a result of the low energy collisional fragmentation the peptide precursor ion fragments predictably at each peptide amide bond along the peptide backbone yielding a distribution of product ions in complimentary ion series forming a ladder which is indicative of the peptide sequence. This fragmentation nomenclature was described by Roepstorff and Fohlman, 1984 (figure 61).

Roepstorff P, Fohlman J. Biomed Mass Spectrom 11 (1984) 601.

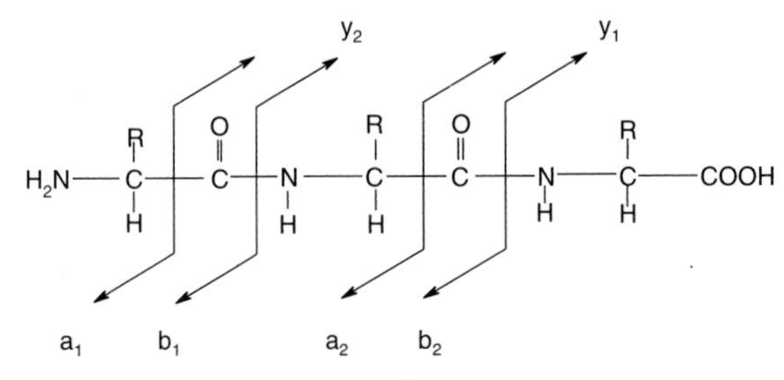

Fig. 61: Fragmentation nomenclature described by Roepstorff and Fohlman. Only the ions typically observed in low energy CID are included (a-, b- and y-type ions).

The peptide is fragmented forming two complimentary ion series (figure 61)

- the n-terminal ion series, or b-ion series. The ions of the n-terminal ion series will contain the n-terminal amino acid and extensions from this residue (figure 62).
- the c-terminal ion series, or y-ion series. The ions of the c-terminal ion series will contain the c-terminus of the peptide and extensions from this residue (figure 62).

The precursor ion and subsequent product ions can undergo additional fragmentation through the loss of small molecules such as

- ammonia (typically from basic residues such as arginine and lysine, table 7)
- water (typically from the alcohol containing residues serine and threonine; and the acid containing residues, glutamic and aspartic acid)
- carbon monoxide, which is lost from a b-ion to yield an a-ion, a second type of n-terminal ion.

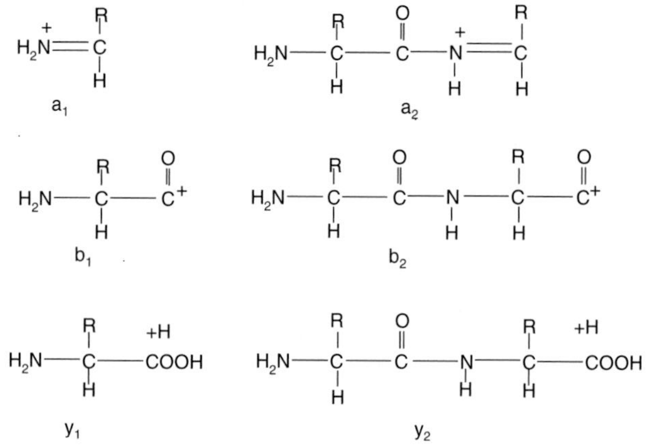

Fig. 62: Structure of the peptide fragment ions commonly formed during a low energy CID experiment.

Tab. 11: Common fragment ions formed during low energy CID and the respective formulae to calculate the mass of each ion.

Fragment ion	*Formula*
a-ion	Total residue mass of the amino acids present – 28 (for the loss of CO)
b-ion	Total residue mass of the amino acids present
y-ion	Total residue mass +18+1

Johnson RS, Martin SA, Biemann K. Int J Mass Spectrom Ion Proc 86 (1988) 137–154.

At higher collisional energies, the peptide is additionally fragmented at the amino acid side chains (Johnson *et al.* 1988). Though this type of experiment yields extra information, such as the specific differentiation of the leucine and isoleucine residues it produces more complex spectra and is generally no more useful than low energy CID for the sequence analysis of peptides.

The product ion MS/MS spectrum can be used to identify a protein by manually determining the peptide sequence and using the determined sequence to search the sequence databases or by automatically correlating the native product ion MS/MS spectra with the sequence database (see section 2.4.4).

2.3.3.2 **Precursor ion scan**

In this experiment the first part of the MS analyser, MS1, is set to allow all the components of the mixture to enter the collision cell and undergo CID. The third part of the analyser, MS2, is fixed at a specific mass value, so that only analytes which fragment to give a fragment ion of this specific mass will be detected at the detector (see figure 59 middle panel). In this manner the precursor ion scan can be used to selectively identify certain species in a complex mixture, including sites of phosphorylation.

For example, the presence of serine, threonine or tyrosine phosphorylated peptides in a complex peptide digest can be investigated using this type of MS/MS experiment. A phosphorylated peptide fragments during low energy collisional fragmentation to give a fragment ion of 79 Da (the PO_3^- group). Therefore to specifically identify phosphorylated peptides in a mixture the third part of the analyser is set to 79. Thus, only the species which fragment to give a fragment ion of 79 reach the detector and hence indicating of phosphorylation (Wilm *et al,* 1996; Carr *et al.,* 1996; Annan *et al.,* 2001).

Wilm M, Neubauer G, Mann M. Anal Chem. 68 (1996) 527–533; Carr SA, Huddleston MJ, Annan RS. Anal Biochem 239 (1996) 180–192.

Annan RS, Huddleston MJ, Verna R, Deshaies RJ, Carr SA. Anal Chem 73 (2001) 393–404.

This experiment has been historically restricted to a triple quadrupole instrument, but recent applications have been demonstrated using a hybrid quadrupole ToF instruments (Steen *et al.* 2001). The method has been exploited for the specific detection of tyrosine phosphorylated peptides in complex mixtures

Steen H, Küster B, Fernandez M, Pandy A, Mann M. Proceedings of the - Proceedings of the 49th ASMS conference on mass spectrometry and allied topics, Chicago (2001).

2.3.3.3 **Neutral loss scan**

In this experiment the 1st and 3rd parts of the analyser are scanned synchronously but with a specific m/z offset. Once more the second part of the analyser is used as the collision cell and the entire mixture is allowed to enter the collision cell, only those species which fragment to yield a fragment with the same mass as the offset will be observed at the detector (see figure 60). For instance, serine and threonine phosphorylated peptides readily lose phosphoric acid during low energy CID. Phosphoric acid has a mass of 98 Da, thus the offset for a doubly charged peptide would be set at 49. Hence any species which loses 49 Da from a doubly charged ion would be observed at the detector and be indicative of phosphorylation (Covey *et al.* 1991; Schlosser *et al.* 2001).

Covey TR, Huang EC, Henion JD. Anal Chem 63 (1991) 1193–1200.

Schlosser A, Pipkorn R, Bossemeyer D, Lehman WD. Anal Chem 73 (2001) 170–176.

2.3.3.4 **MS^n**

This type of experiment is restricted to the quadrupole ion traps and FT-ICR MS. A fragment ion produced in the product ion MS/MS spectrum can be selected, isolated and fragmented a second time generating even further information in a third spectrum. This spectrum would be called an MS/MS/MS product ion spectrum, or MS^3 (see figure 63) (Louris *et al.* 1989). This type of experiment is potentially useful for phosphorylation or glycosylation analysis if sufficient material exists.

Louris JN, Amy JW, Ridley TY and Cooks RG. Int J Mass Spectrom Ion Proc 88 (1989) 97–111.

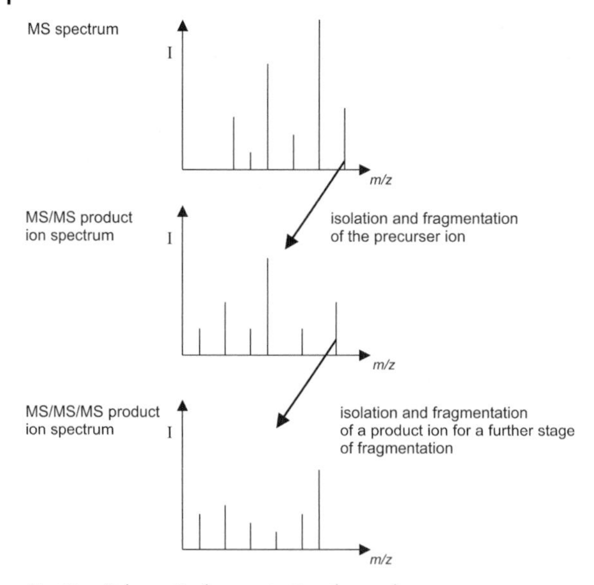

Fig. 63: Schematic demonstrating the workflow in an MS^3 experiment.

2.3.3.5 MALDI-TOF Post Source Decay (PSD)

Spengler B. J Mass Spectrom. 32 (1997) 1019–1036. Chaurand P, Luetzenkirchen F, Spengler B. J Am Soc Mass Spectrom 10 (1999) 91–103.

This technique was described by Spengler and co-workers in the early nineties and used as a method of generating peptide sequence information with a MALDI reflectron ToF instrument (Spengler, 1997; Chaurand *et al.* 1999).

$KE = 1/2mv^2$. Thus a fragment ion will have a lesser mass than the precursor ion, hence lower KE.

Spengler and co-workers observed that MALDI was not such a soft ionisation technique as first reported, with MALDI ions decaying in the flight tube prior to the detector. After the analyte is accelerated from the source it continues to undergoes dissociation or post-source decay forming product ions in the field free region before the detector. Since the fragmentation occurs in the field free region after the ions have been accelerated, each product ion has the same velocity as the precursor ion, and in a linear ToF instrument they will arrive at the detector at the same time, and cannot be differentiated. However, because the product ions have a lower mass than the precursor ion they have a lower kinetic energy. Fortunately the ion mirror properties of a reflectron enable it to act as an energy filter accommodating to some extent these kinetic energy differences (earlier in this section, the ability of the reflectron to minimise initial energy distributions was discussed, this describes how it can be used to analyse PSD ions)

The peptide of interest, the precursor ion, can be specifically selected from a complex mixture for post-source decay analysis using an ion gate; typically an electrical deflector.

However, a conventional linear reflectron (one stage or two stage) can only accommodate a small range of KE differences i.e the reflectron will only be able to focus those ions that are similar in mass (figure 64). In this instance, only PSD ions that are close in mass to the precursor ion will be mass measured correctly. Thus, to perform PSD with a conventional reflectron the voltage on the reflectron has to drop sequentially to allow the fragment ions of lower mass (and lower KE) to become focussed at the detector. This is termed stepping the reflectron. In a conventional reflectron, several different voltages or segments, (as many as 12 segments may be required) are required to focus the entire mass range of product ions (figure 65a). The individual segments are then "stitched" together using the instrument software, generating the complete product ion spectrum. (figure 65b).

Conventional linear reflectron

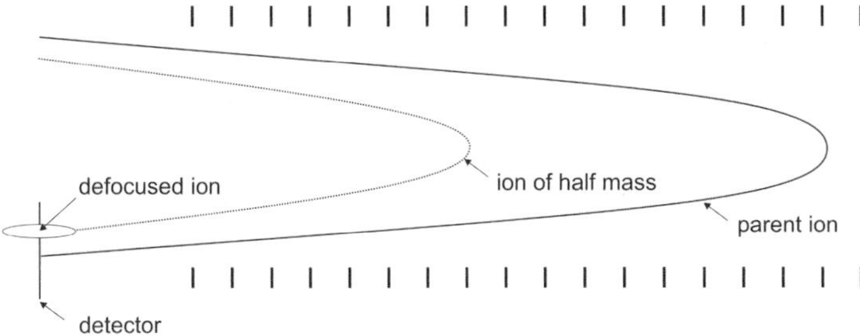

New harmonic reflectron

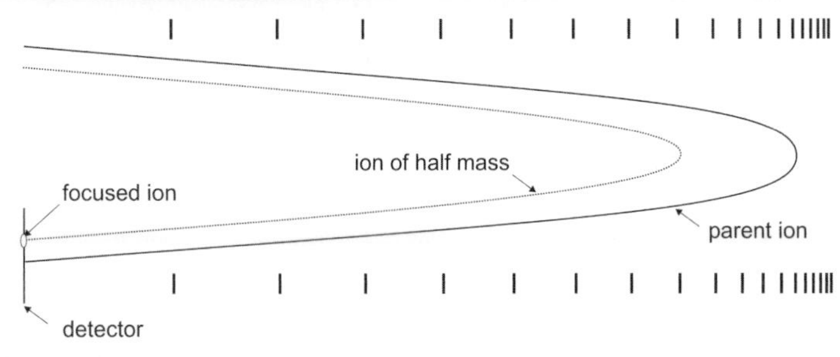

Fig. 64: Schematic demonstrating the passage of the precursor ion and product ion through two types of reflectron. Upper panel – conventional linear reflectron, lower panel – quadratic field reflectron.

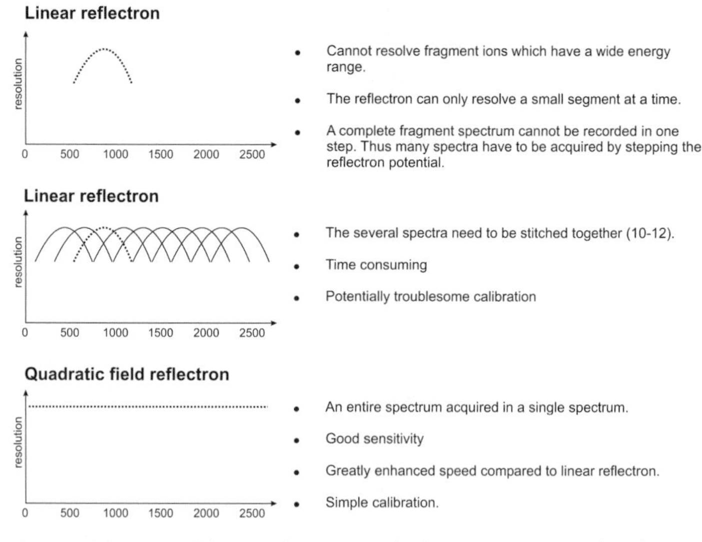

Fig. 65: Schematic of the time focusing of a conventional and quadratic field reflectron. Upper panel – only a small segment of the mass range can be focussed at one time with a conventional linear reflectron. Middle panel – multiple segments acquired to obtain the whole mass range of fragment ions using a conventional linear reflectron. Lower panel – quadratic field reflectron allows the focussing of the mass range in a single ToF spectrum.

The use of a quadratic field reflectron for PSD analysis significantly reduces these drawbacks, because it is capable of accommodating ions of widely different kinetic energies in a single spectrum (figures 64–65). This is because in a quadratic field reflectron an increased voltage is applied, non-linearly, to create a perfect quadratic field across the reflectron. Resultantly, all the product ions created during the post source decay of a particular precursor ion are focused at the detector, irrespective of their energy, over the entire range of m/z. Hence a complete PSD spectrum is obtained under the same experimental conditions with each pulse of the laser, without the need for data stitching, in one or two minutes.

Silles E, Mazon MJ, Gevaert K, Goethals M, Vandekerckhove, Lebr R, Sandoval IV. J Biol Chem. 275 (2000) 34054–34059.
Gevaert K, Demol H, Martens L, Hoorelbeke B, Puype M, Goethals M, Van Damme J, De Boeck S, Vandekerckhove J. Electrophoresis 22 (2001) 1645–1651.

The fragmentation data obtained from a PSD experiment is similar to that obtained with an MS/MS product ion spectrum. Cleavages along the peptide backbone yielding b-type ions from the n-terminus and y-type ions from the c-terminus are prominant, with small molecule losses from these ions, such as a-ions, additionally observed. The data generated from such an experiment can readily be used for peptide sequence analysis and protein identification. (Silles *et al.* 2000; Gevaert *et al.* 2001).

Under typical timed ion delay MALDI conditions, peptide fragmentation observed during PSD is reduced. (Kauffman *et al.* 1996). Keough and co workers described a derivatisation method for peptide sequence analysis by MALDI PSD that facilitated fragmentation during PSD, increasing sensitivity significantly. The tag was designed to promote efficient charge-site initiated fragmentation of the peptide bonds. Simultaneously only a single ion series, the y-ion series, is observed in the PSD spectrum, simplifying interpretation.

Kauffman R, Chaurand P, Kirsch D, Spengler B. Rapid Commun Mass Spectrom 10 (1996) 1199–1208.
Keough T, Youngquist RS, Lacey MP. Proc Natl Acad Sci USA 96 (1999) 7131–7136.
Keough T, Lacey MP, Fieno AM, Grant RA, Sun Y, Bauer MD, Begley KB. Electrophoresis 21 (2000) 2252–2265.

In the original method the peptide digest was derivatised with 2-sulphonyl acetyl chloride, labelling the α-amino group at the n-terminus of each peptide (and the ε-amino group of the each lysine side chain) with the very strong acid, sulphonic acid (pKa<2).

The theory proposed that under MALDI ionisation conditions, the strong basic residue at the c-terminus of tryptic peptides (arginine) would be protonated and the strong acidic group at the c-terminus would be deprotonated (figure 66). The additional proton (almost exclusively MALDI ions in the peptide mass range are singly charged) would then be free to randomly ionise and subsequently fragment the peptide amide bonds of the peptide backbone. Hence when the fragment ions are formed only the y-ions will be observed because, the formal positive charge typical of a b-ion will be neutralised by the negative charge on the sulphonic acid group, hence b-ions will not be observed

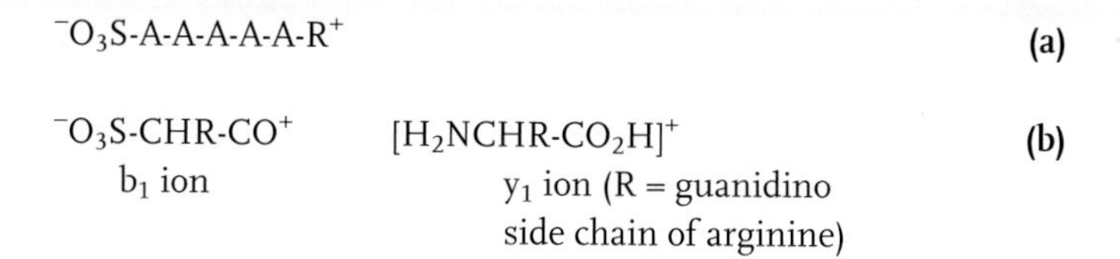

Fig. 66: The effect of a strongly acidic group at the n-terminus of an arginine c-terminating peptide.

Labelling of arginine-terminating tryptic peptides greatly facilitates fragmentation, particularly fragmentation by MALDI PSD. Figure 67 demonstrates the extent of fragmentation during a PSD experiment. In the case of this synthetic peptide, 16 consecutive residues can be readily determined. The derivatistion technique is compatible with protein digests derived from 2D gels (see section 2.4.3–2.4.4).

Brancia FL, Oliver SG, Gaskell SJ. Rapid Commun Mass Spectrom 14 (2000) 2070–2073.
Keough T, Lacey MP, Youngquist RS. Rapid Commun Mass Spectrom 14 (2000) 2348–2356.

As the derivatisation strategy also causes the derivatisation of the ε-amino group of the lysine side chain, therefore labelling the lysine with a strong acidic group, PSD of lysine c-terminating peptides becomes inefficient. To extend the applicability of this method to lysine terminating peptides, the lysine must first be modified to the homo-arginine derivative. This has recently been reported by a number of groups to increase lysine tryptic peptide detection in peptide mass fingerprinting (Brancia *et al.*, 2000; Keough *et al.*, 2000). If this modification of the lysine peptides is performed prior to the sulphonation reaction then the sulphonation methodology is equally applicable to lysine c-terminating peptides.

Liminga M, Borén M, Åström J, Carlsson U, Keough T, Maloisel JL, Palmgren R, Youngquist S. Proceedings ASMS conference on mass spectrometry and allied topics, Chicago, USA (2001).

The sulphonation method has further been improved with the development of a water stable sulphonic NHS ester reagent, figure 68 (Liminga *et al.* 2001). Subsequent PSD spectra acquired from peptides derivatised in this manner is termed chemically assisted fragmentation (CAF).

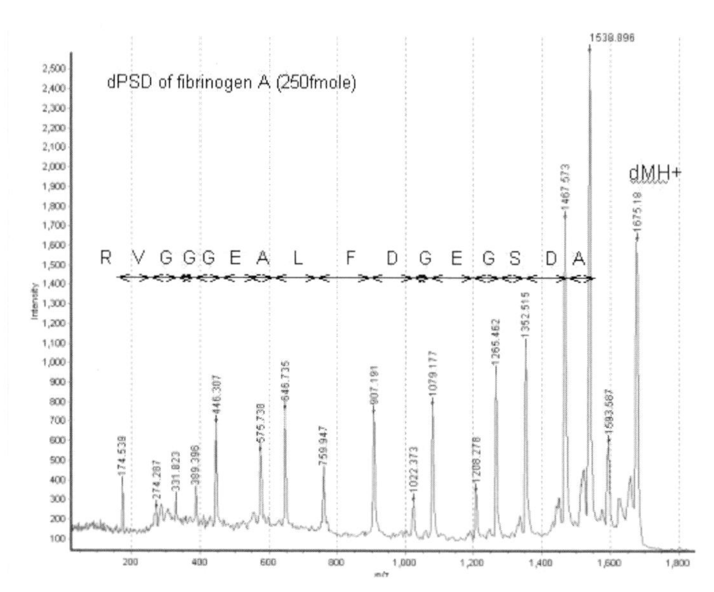

Fig. 67: PSD spectrum of a CAF derivatised synthetic peptide acquired with the Ettan MALDI-ToF Pro (250 fmol applied to the target). The 16 consecutive residues are readily determined. Though Leucine (L) is specified in the sequence, this cannot be determined from the experiment (Isoleucine and leucine are isobaric). dMH^+ is the precursor ion. Note the intensity of the precursor ion. In a typical PSD experiment, fragment ion abundance is a very low percentage of the precursor ion abundance.

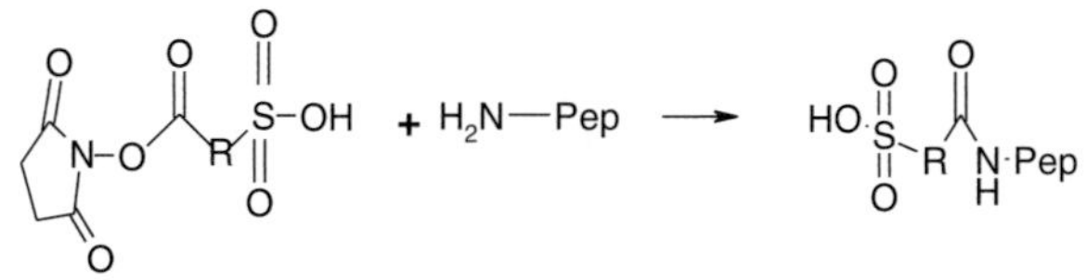

Fig. 68: Reaction scheme for the water stable sulphonic NHS ester reagent.

One of the drawbacks of the original method was the hygroscopic nature of the acetyl chloride reagent, requiring the derivatisation procedure to be performed under stringent conditions in a non-aqueous environment. The novel watersoluble reagent allows the quantitative derivatisation of the peptide digest at sub pmol quantities of starting material and is compatible with the lysine modification step affording a one pot derivatisation (see step 10). The derivatisation step is amenable to automation. This derivatisation approach enables peptide mass fingerprinting data and subsequent sequence information to be acquired on the same instrument with good sensitivity (see sections 2.5.1).

This derivatisation strategy is equally applicable to any MS/MS instrument, which employs MALDI as the ionisation source, enabling the simple interpretation of the peptide sequence from the fragmentation of a singly charged ion.

2.4 Protein identification by database searching

As stated earlier, MS has become the preferred method of choice for protein identification. However, the development of several software programs and continuous updating of sequence databases have been crucial to the success of mass spectrometry in this field.

Generally, MS data can be used in four approaches for protein identification

- Peptide mass fingerprint (PMF) – MS mode. The mass measurement of each peptide derived from the enzyme digestion or chemical cleavage of the protein.
- Peptide mass fingerprint and composition information. The molecular weights of each of the peptides derived from the enzyme digestion or chemical cleavage of the protein can be used alongside some composition information relating to one or more of the peptides.

- Peptide mass fingerprint and sequence information The molecular weights of each of the peptides derived from enzyme digestion or chemical cleavage of the protein can be used along side some direct sequence information relating to one or more of the peptides.
- Product ion MS/MS sequence data from one or more peptide – MS/MS mode.

2.4.1 Peptide mass fingerprint

Pappin DJC, Højrup P, Bleasby A. Curr Biol 3 (1993) 327–332.
Henzel WJ, Billeci TM, Stults JT, Wong SC, Grimley C, Watanabe C. Proc Natl Acad Sci USA 90 (1993) 5011–5015.
Mann M, Hojrup P, Roepstorff P. Biol Mass Spectrom 22 (1993) 338–345.
Yates JR III, Speicher S, Griffin PR, Hunkapiller T. Anal Biochem 214 (1993) 397–408.
James P, Quadroni M, Carafoli E, Gonnet G. Biochem Biophys Res Commun 195 (1993) 58–64.

The technique originally described in 1993 comprises protein digestion, MALDI ToF analysis and sequence database search algorithms. (Pappin *et al.* 1993; Henzel WJ *et al.* 1993; Mann *et al.* 1993; Yates *et al.* 1993 and James *et al.* 1993).

Simply, every protein in the protein database (or in the genome under investigation) is theoretically digested with the cleavage reagent used in the digestion reaction, generating many hundreds of thousands of theoretical peptides. The experimental peptide masses derived from the MS spectrum, the peptide mass fingerprint (PMF) is then compared to the theoretical peptide masses and a score is calculated and assigned. The score reflects the match between the theoretically and experimentally determined masses; the protein identified as the most probable is the one that gives the best match between the experimental and the theoretical peptides. The number of peptides observed in the PMF and the accuracy to which they are measured determines the confidence of the protein identification.

Jensen ON, Podtelejnikov A, Mann M. Rapid Commun Mass Spectrom 10 (1996) 1371–1378.
Clauser KR, Baker P, Burlingame AL. Anal Chem 71 (1999) 2871–2882.

The incorporation of reflectron technology and delayed extraction (timed delay ion extraction) into MALDI-ToF instrumentation has enhanced the performance of peptide mass fingerprinting considerably. The ^{12}C isotope of the peptide isotopic envelope, the mass used in a PMF search, can be unambiguously assigned across the peptide mass range of interest, significantly enhancing mass measurement of this peak (sub 50 ppm mass accuracy is routine). The improved mass accuracy further constrains the database search, reducing the potential for ambiguous protein identifications (Jensen *et al.* 1996; Clauser *et al.* 1999).

Several programs are available to perform this type of search, varying in the execution of the task (including MASCOT at www.matrixscience.com profound at www.prowl.com; MS-FIT at www. prospector.ucsf.edu/) Accuracy, reliability and speed will determine the program of choice. Regardless which programme is used, four user variables are important for a PMF search.

- Peptide mass list.
- Specification of the cleavage agent.
- Error tolerance. The accuracy of mass measurement is determined by the calibration, the higher the mass accuracy the greater the specificity.
- Knowledge of peptide modifications i.e. methionine oxidation.

For this approach to give an unambiguous result a significant number of experimentally determined peptide masses have to match the experimental masses, the peptides have to be a result of the correct cleavage specificity of the cleavage agent used and the protein in question has to exist in the protein database. Resultingly, the PMF approach is highly suited to genetically highly characterised genomes and is commonly used. Shevchenko *et al.* 1996; demonstrated that up to 90% of proteins selected from a 2D gel of an E coli lysate were identified by peptide mass fingerprinting.

Shevchenko A, Jensen ON, Podtelejnikov A, Sagliocco F, Wilm M, Vorm O, Mortensen P, Shevchenko A, Boucherie H, Mann M. Proc Natl Acad Sci USA 93 (1996) 1440–1445.

Although an unambiguous match may be returned from a database search, not all the peptides within the peptide mass fingerprint may be identified. It may prove important to identify the remaining unidentified peptide peaks, which may arise from

- The presence of other digested proteins within the sample (in the case of 2D gel, comigration to the same spot).
- Peptides arising from the simultaneous digestion of contaminant proteins such as keratin.
- Incorrect cleavage, protease acting in a non-predicted fashion
- Post-translationally modified peptides, altering the mass of the native peptide rendering it unrecognisable in the database search.
- The identified protein may be (highly) homologous to the protein under study (or be a splice variant), but not exactly the same and thus unmatched peptides will exist.

Though many keratin peaks are commonly known and these peptides, if present, can be subsequently omitted from the search.

MALDI-ToF peptide mass fingerprinting is fast and simple method of protein identification, but the success of the method can be compromised in a number of ways:

- Insufficient peptides are observed in the peptide mass fingerprint to submit to a database search. i.e. insufficient information to identify the protein.

Common for small molecular weight proteins, basic proteins if digested with trypsin low abundant proteins.

- The sample maybe a mixture of proteins. Although it is possible to analyse simple mixtures (Axelsson *at el.* 2001) there may not be a significant number of peptides from each protein component in the mixture to yield a successful identification. As several peptides are required to give a statistical match, this approach is not readily suited to complex mixtures.

Axelsson J, Boren M, Naven TJP, Fenyö D. Proceedings of the 49th ASMS conference on mass spectrometry and allied topics, Chicago (2001).

- Excessive post-translational modifications of the protein can result in masses, which are not predicted in the theoretical digestion of the proteins in the database as well as precluding peptides from the fingerprint.
- Very little homology can be found with another protein in the database or the protein in question may not actually exist in the protein database.
- If the above is true, peptide mass fingerprinting cannot be used with confidence for searching against the EST databases (which is the next stage of protein identification if a search of the protein database is unsuccessful). The probability that a significant number of peptides from the peptide mass fingerprint will find matches with a single EST yielding a unique protein identification is small.

If any of these situations arise, then more specific information is required to identify the protein.

2.4.2 Peptide mass fingerprinting combined with composition information

Fenyö D, Qin J, Chait BT. Electrophoresis 19 (1998) 998–1005.
Pappin D, Rahman D, Hansen HF, Bartlet-Jones M, Jeffrey W, Bleasby AJ. Methods in Biological Sciences (1996) 135–150.

Composition information can supplement the PMF data providing extra specificity for searching protein databases (Fenyö *et al.* 1998). For instance, this publication demonstrated that with the knowledge of the presence of a cysteine residue in a tryptic peptide of mass 2000 Da (measured to 0.5 Da mass accuracy) the number of matching proteins in S cerevisiae was reduced by a factor of five. A number of approaches have been employed to gain compositional information. In one example reported by Pappin, a PMF is acquired and the peptide masses noted. A subsequent methyl esterification reaction is performed on a small aliquot of the sample esterifing the carboxyl side chain of the acidic residues, glutamic and aspartic acid, and the carboxyl group of the c-terminal residue present in each peptide of the digest. A PMF spectrum is re-acquired. The subsequent mass changes between corresponding peptides are indicative of the number of acidic groups each peptide contains. This information was combined with the original PMF using a MOWSE composition search potentially increasing search discrimination by orders of magnitude (Pappin *et al.* 1995). This type of search can be performed using MASCOT at www.matrixscience.com.

The number of exchangeable hydrogens within a peptide offers composition information. Again, this procedure requires the acquisitionof a PMF, subsequent labelling of the mixture a deuterium solution (D_2O) and finally re-acquisition of a second spectrum. The mass increase of each peptide by the number of exchangeable hydrogens is indicative of amino acid composition (Sepetov *et al.*, 1993; James *et al.* 1994).

Sepetov NF, Issakova OL, Lebi M, Swiderek K, Stahl DC, Lee TD. Rapid Commun Mass Spectrom. 7 (1993) 58–62; James P, Quadroni M, Carafoli E, Gonnet G. Protein Sci 3 (1994) 1347–1350.

More recently Goodlett *et al.* (2000) demonstrated composition information and high mass accuracy can be very specific information for unambiguously identifying proteins. By labelling cysteine residues with a specific isotopic label, IDEnT (the tag contained chlorine, which has a specific isotopic profile owing to the relative abundance of ^{35}Cl and ^{37}Cl) the distinctive isotopic pattern of the labelled peptide could be simply recognised. With this specific composition information and mass accuracy measured to within 1 ppm using an FT-ICR MS, the mass of a single peptide was sufficient to unambiguously identify a protein from the whole yeast database (1 peptide from a possible in ~345,000 peptides).

Goodlett DR, Bruce JE, Anderson GA, Rist B, Pasa-Tolic L, Fiehn O, Smith RD, Aebersold R. Anal Chem 72 (2000) 1112–1118.

Cysteine was chosen as the labelled residue because it is one of the rarest amino acid residues, allowing constrained database searching.

The isotopically labelling of cysteine residues has also been exploited for extra composition information. Sechi and Chait (1998) reported a method for the modification of the cysteinyl thiol group with isotopically labelled acrylamide.

Sechi S, Chait BT. Anal Chem 70 (1998) 5150–5158.

When a PMF is combined with composition information inferred from a MALDI PSD spectrum, highly discriminating information can be obtained. Immonium ions present in a MALDI PSD spectrum indicate the presence of certain amino acid residues; again this can be included in a composition search as described above. (Clauser *et al.* 1999). This information is also available in a product ion MS/MS spectrum acquired with a MALDI QToF or ToF/ToF instrument.

Clauser KR, Baker P, Burlingame AL. Anal Chem 71 (1999) 2871–2882.

2.4.3
Peptide mass fingerprint combined with partial sequence information

Supplementing PMF data with some partial sequence is an attractive approach. If the two pieces of data can be acquired on the same instrument and the resultant information combined in the same search, sample throughput and identification rates can be enhanced significantly. PMF and sequence information can be used in the same search using the sequence query function with the MASCOT search engine. An unsuccessful search of the protein database with the peptide mass fingerprint data may be overturned with some partial sequence if the protein exists in the protein database.

The data required for this type of database search can be readily obtained with a MALDI-ToF instrument capable of post-source decay. A worked example is demonstrated below whereby a PMF is

Including only selecting the ^{12}C isotope from each peptide isotopic distribution for the database search.

acquired, derivatisation of the peptide digest and subsequent CAF-PSD is performed (section 2.3.3.5) using MALDI-ToF instrument fitted with a quadratic field reflectron. The PMF, spectrum processing and database search were performed automatically.

An unsuccessful result was obtained from the PMF search (figure 69). Subsequently PSD was performed on two of the derivatised peptides. As the spectra demonstrate (figures 70a and 71a) partial amino acid sequence can be inferred for both peptides from the single y-ion series of product ions (figures 61 and 62, section 2.3.3.1). Each sequence is combined with the peptide mass fingerprint in a sequence query search using MASCOT. Combining the interpreted sequence derived from the PSD spectrum of peptide m/z 1569.79 with the original PMF in a database sequence query search, yields an unambiguous match (figure 70b). Similarly, an unambiguous match is returned when the determined sequence from the second peptide (peptide m/z 1157.57) is combined with the PMF in a sequence query search (figure 70b). When both sets of sequence data are combined with the peptide mass fingerprint the confidence in the match returned from the database search increases dramatically (figure 70c). As the data was acquired with a quadratic field reflectron the calibration of the PSD spectrum was reasonably good enabling successful protein identification.

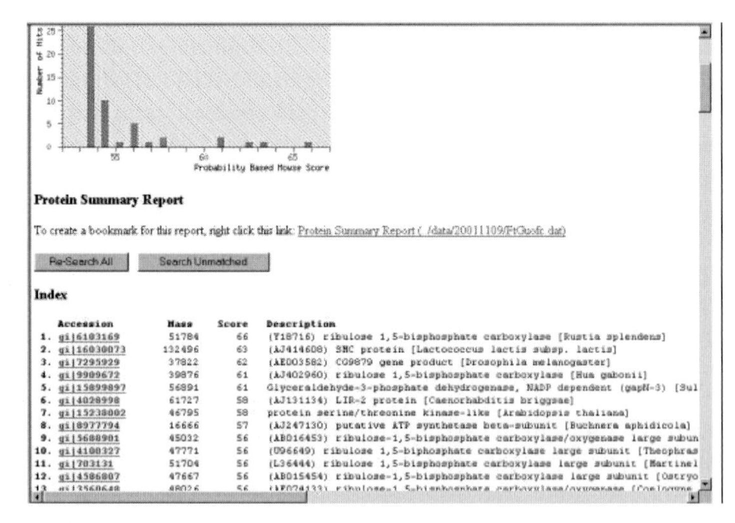

Protein Summary Report

To create a bookmark for this report, right click this link: Protein Summary Report (/data/20011109/FtGusfc.dat)

Re-Search All | Search Unmatched

Index

	Accession	Mass	Score	Description
1.	gi|6103169	51784	66	(Y18716) ribulose 1,5-bisphosphate carboxylase [Rustia splendens]
2.	gi|16030073	132496	63	(AJ414608) SMC protein [Lactococcus lactis subsp. lactis]
3.	gi|7295929	37822	62	(AE003582) CG9879 gene product [Drosophila melanogaster]
4.	gi|9909672	39876	61	(AJ402960) ribulose 1,5-bisphosphate carboxylase [Hua gabonii]
5.	gi|15899897	56891	61	Glyceraldehyde-3-phosphate dehydrogenase, NADP dependent (gapN-3) [Sul
6.	gi|4028998	61727	58	(AJ131134) LIR-2 protein [Caenorhabditis briggsae]
7.	gi|15238002	46795	58	protein serine/threonine kinase-like [Arabidopsis thaliana]
8.	gi|8977794	16666	57	(AJ247130) putative ATP synthetase beta-subunit [Buchnera aphidicola]
9.	gi|5688901	45032	56	(AB016453) ribulose-1,5-bisphosphate carboxylase/oxygenase large subun
10.	gi|4100327	47771	56	(U96649) ribulose 1,5-biphosphate carboxylase large subunit [Theophras
11.	gi|703131	51704	56	(L36444) ribulose 1,5-bisphosphate carboxylase large subunit [Martinel
12.	gi|4586807	47667	56	(AB015454) ribulose-1,5-bisphosphate carboxylase large subunit [Ostryo

Fig. 69: An unsuccessful PMF search of the protein database.

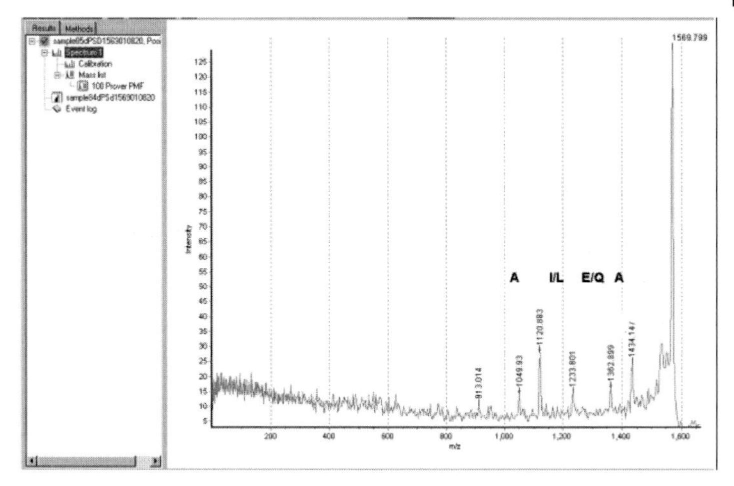

Fig. 70a: PSD spectrum of the derivatised peptide, m/z 1569.79. The sequence – A[IL]EA was determined from the spectrum and combined with the PMF as before. The result of the combined search is demonstrated in figure 70b. An unambiguous result is returned from the search.

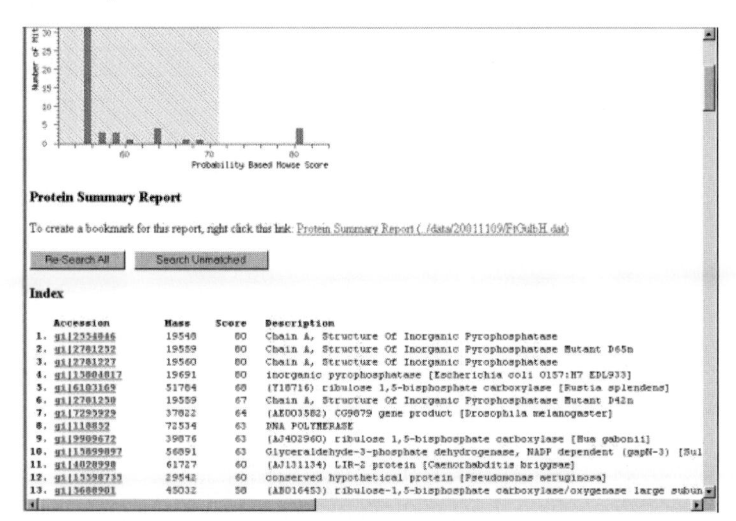

Protein Summary Report

To create a bookmark for this report, right click this link: Protein Summary Report (../data/20011109/FtGulbH.dat)

Re-Search All   Search Unmatched

Index

	Accession	Mass	Score	Description
1.	gi|2554846	19548	80	Chain A, Structure Of Inorganic Pyrophosphatase
2.	gi|2781232	19559	80	Chain A, Structure Of Inorganic Pyrophosphatase Mutant D65n
3.	gi|2781227	19560	80	Chain A, Structure Of Inorganic Pyrophosphatase
4.	gi|15804817	19691	80	inorganic pyrophosphatase [Escherichia coli O157:H7 EDL933]
5.	gi|6103169	51784	68	(Y18716) ribulose 1,5-bisphosphate carboxylase [Rustia splendens]
6.	gi|2781230	19559	67	Chain A, Structure Of Inorganic Pyrophosphatase Mutant D42n
7.	gi|7293929	37822	64	(AE003582) CG9879 gene product [Drosophila melanogaster]
8.	gi|118832	72534	63	DNA POLYMERASE
9.	gi|9909672	39876	63	(AJ402960) ribulose 1,5-bisphosphate carboxylase [Mua gabonii]
10.	gi|15899897	56891	63	Glyceraldehyde-3-phosphate dehydrogenase, NADP dependent (gapN-3) [Sul
11.	gi|4028998	61727	60	(AJ131134) LIR-2 protein [Caenorhabditis briggsae]
12.	gi|15598735	29542	60	conserved hypothetical protein [Pseudomonas aeruginosa]
13.	gi|3688901	45032	58	(AB016453) ribulose-1,5-bisphosphate carboxylase/oxygenase large subun

Fig. 70b: MASCOT search result from the combined PMF and partial sequence determined from the PSD spectrum in figure 70a.

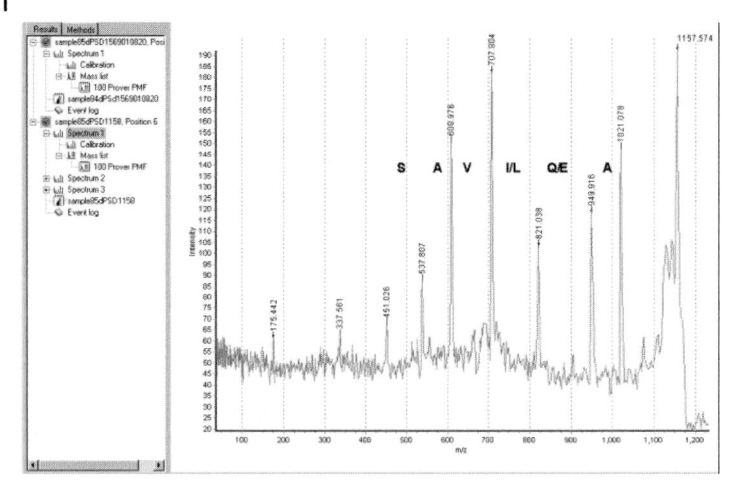

Fig. 71a: PSD spectrum of the derivatised peptide, m/z 1157.57. The sequence – SAV[IL]EA was determined from the spectrum and combined with the PMF as before. The result of the combined search is demonstrated in figure 71b. An unambiguous result is returned from the search.

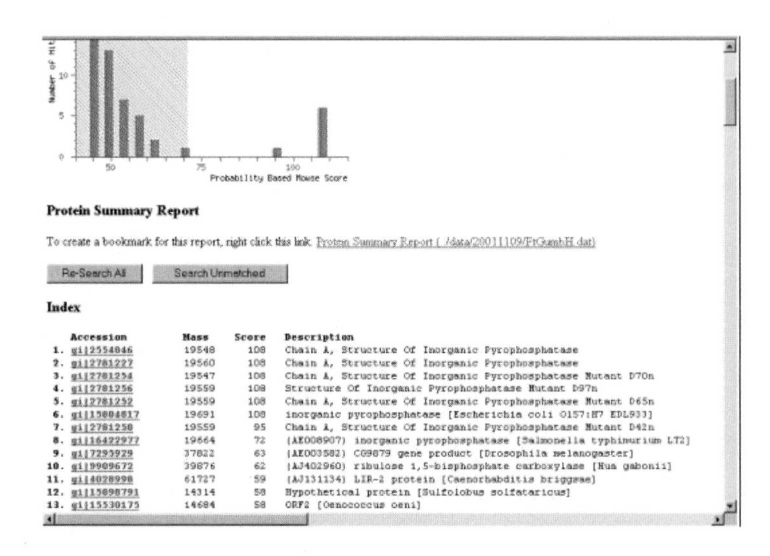

Protein Summary Report

To create a bookmark for this report, right click this link: Protein Summary Report (../data/20011109/FtGumbH.dat)

Re-Search All | Search Unmatched

Index

	Accession	Mass	Score	Description
1.	gi|2554846	19548	108	Chain A, Structure Of Inorganic Pyrophosphatase
2.	gi|2781227	19560	108	Chain A, Structure Of Inorganic Pyrophosphatase
3.	gi|2781254	19547	108	Chain A, Structure Of Inorganic Pyrophosphatase Mutant D70n
4.	gi|2781256	19559	108	Structure Of Inorganic Pyrophosphatase Mutant D97n
5.	gi|2781252	19559	108	Chain A, Structure Of Inorganic Pyrophosphatase Mutant D65n
6.	gi|15804817	19691	108	inorganic pyrophosphatase [Escherichia coli O157:H7 EDL933]
7.	gi|2781250	19559	95	Chain A, Structure Of Inorganic Pyrophosphatase Mutant D42n
8.	gi|16422977	19664	72	(AE008907) inorganic pyrophosphatase [Salmonella typhimurium LT2]
9.	gi|7293929	37822	63	(AE003582) CG9879 gene product [Drosophila melanogaster]
10.	gi|9909672	39876	62	(AJ402960) ribulose 1,5-bisphosphate carboxylase [Hua gabonii]
11.	gi|4028998	61727	59	(AJ131134) LIR-2 protein [Caenorhabditis briggsae]
12.	gi|15898791	14314	58	Hypothetical protein [Sulfolobus solfataricus]
13.	gi|15530175	14684	58	ORF2 [Oenococcus oeni]

Fig. 71b: MASCOT search result from the combined PMF and partial sequence determined from the PSD spectrum in figure 70 a.

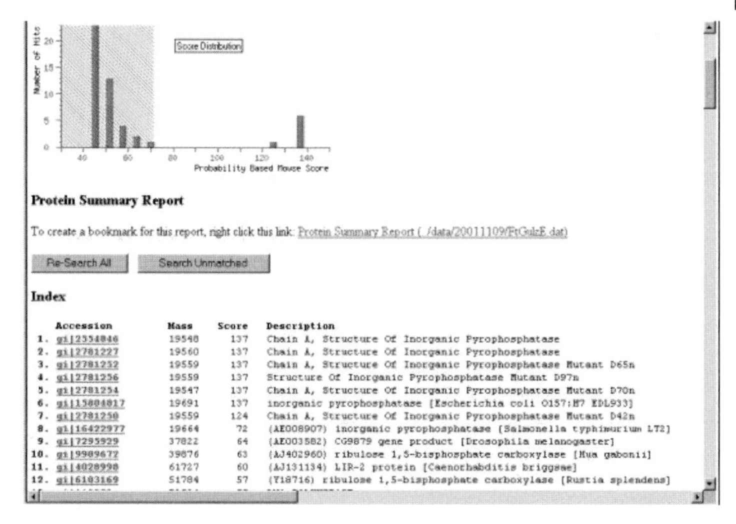

Protein Summary Report

To create a bookmark for this report, right click this link: Protein Summary Report (../data/20011109/FtGulzE.dat)

Re-Search All | Search Unmatched

Index

	Accession	Mass	Score	Description
1.	gi|2354846	19548	137	Chain A, Structure Of Inorganic Pyrophosphatase
2.	gi|2781227	19560	137	Chain A, Structure Of Inorganic Pyrophosphatase
3.	gi|2781252	19559	137	Chain A, Structure Of Inorganic Pyrophosphatase Mutant D65n
4.	gi|2781256	19559	137	Structure Of Inorganic Pyrophosphatase Mutant D97n
5.	gi|2781254	19547	137	Chain A, Structure Of Inorganic Pyrophosphatase Mutant D70n
6.	gi|15804817	19691	137	inorganic pyrophosphatase [Escherichia coli O157:H7 EDL933]
7.	gi|2781250	19559	124	Chain A, Structure Of Inorganic Pyrophosphatase Mutant D42n
8.	gi|16422977	19664	72	(AE008907) inorganic pyrophosphatase [Salmonella typhimurium LT2]
9.	gi|7295929	37822	64	(AE003582) CG9879 gene product [Drosophila melanogaster]
10.	gi|9989672	39876	63	(AJ402960) ribulose 1,5-bisphosphate carboxylase [Hua gabonii]
11.	gi|4028998	61727	60	(AJ131134) LIR-2 protein [Caenorhabditis briggsae]
12.	gi|6103169	51784	57	(Y18716) ribulose 1,5-bisphosphate carboxylase [Rustia splendens]

Fig. 71c: MASCOT search result of the combined PMF and partial sequence from two peptides.

Close inspection of the PSD spectra demonstrated that the sequence can readily determined using the CAF chemistry and the quadratic field reflectron. In the case of the second peptide, there is some ambiguity whether the residue is a glutamine or a glutamic acid (experimentally measured difference 128.7 Da) in part of the determined sequence. However, the MASCOT search query allows both amino acid residues to be inputted into the search query (see step 14). As the figures demonstrate a short stretch of determined sequence combined with the PMF is a powerful approach for reducing ambiguous protein identifications.

2.4.4 Product ion MS/MS sequence data

The final approach is to use sequence data derived from product ion MS/MS or PSD spectra from one or more peptides. As described in section 2.3.3, fragmentation of peptides within the mass spectrometry can yield (long) stretches of sequence information and a number of search engines can correlate this information to protein sequences within the databases. As a result database searching in this fashion overcomes many of the difficulties encountered with peptide mass fingerprinting. Again a number of search engines can be used to perform this analysis including SEQUEST (fields.scripps.edu/sequest/index.html) MASCOT, SONAR (www.proteometrics.com) MS-TAG at prospector.ucsf.edu/.

Sequence information derived from product ion MS/MS or PSD experiments can be used in three ways.

Perkins DN, Pappin DJC, Creasy DM, Cottrell JS. Electrophoresis 20 (1999) 3551–3567.

- Manual interpretation of the sequence (generally from the b- and y-ion series of fragment ions) where the determined sequence can be inputted as a search query (Perkins *et al.* 1999).

Mann M, Wilm M. Anal Chem 66 (1994) 4390–4399.

- Manual interpretation of a short stretch of sequence used together with the accurate mass of the precursor ion and the masses of the first and last fragment ions in the stretch of the sequence (knowledge of these two masses, gives the masses of the uninterpreted sequence in the peptide), a method termed the peptide sequence tag (Mann and Wilm; 1994).

Eng JK, McCormack AL, Yates JR III. J Am Soc Mass Spectrom. 5 (1994) 976–989; Yates JR III, Eng JK, McCormack AL, Schieltz D. Anal Chem 67 (1995) 1426–1436.

- The automatic, uninterpreted MS/MS search (Eng *et al.* 1994; Yates *et al.* 1995; Perkins *et al.* 1999).

Susin SA, Lorenzo HK, Zamzami N, Marzo I, Brothers G, Snow B, Jacotot E, Constantini P, Larochette N, Goodlett DR, Aebersold R, Pietu G, Prevost MC, Siderovski D, Penninger J, Kroemer G. Nature 397 (1999) 441–446; Shevchenko A, Loboda A, Shevchenko A, Ens W, Standing KG. Anal Chem 72 (2000) 2132–2141.

Sequence information is significantly more discriminating than a molecular mass alone and as such, one peptide of sufficient length is sufficient to identify a protein from a well characterised genome (Susin *et al.* 1999; Shevchenko *et al.*, 2000).

Such information can be acquired using CAF chemistry and PSD. Figure 72 demonstrates the PSD spectrum of a peptide derived from a protein digest. In this spectrum twelve consecutive residues of sequence can be determined from the spectrum resulting with an unambiguous hit from the database search.

Once a CID spectrum has been matched, a theoretical tryptic digest of the identified protein can be generated and the presence of other tryptic peptides can be checked and confirmed.

As only one or two peptides are generally required to unambiguously identify a protein from a known genome, complex mixtures of proteins are readily analysed using product ion MS/MS spectra, without the need for PMF, see section 2.5.2.

Mann M. Trends Biol Sci. 21 (1996) 494–495.

A further tool for protein identification is the expressed-sequence tag databases. ESTs are short stretches of nucleotide sequence, which peptide sequence data can be used to search against. Peptide sequence data can be used to search the EST databases if a search against the protein database is unsuccessful (Mann, 1996). (A PMF search against the EST database is inappropriate, as it would not be sufficiently discriminating for unambiguous identification.)

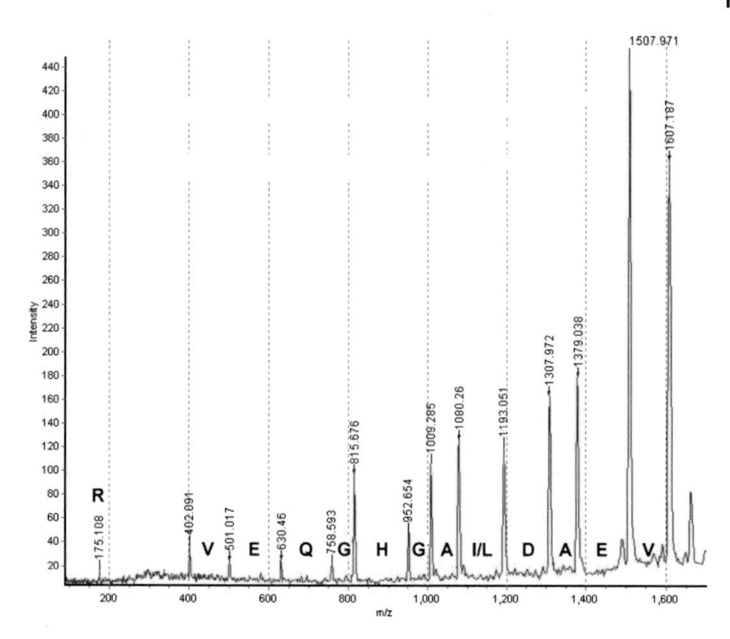

Fig. 72: MALDI-PSD spectrum of a derivatised peptide, m/z 1743.6 derived from the tryptic digest of myoglobin. The sequence VEQGHGAI/LDAEV with a c-terminal R can be readily interpreted from the sequence. If the protein is in the database, a unique match will be returned for this stretch of sequence.

2.4.5 *De novo* sequencing

Despite the automated searches having overtaken the manual methods as the choice for searching product ion MS/MS data, manual peptide sequencing still has to be performed, particularly for *de novo* sequencing and sequencing of MHC class peptides. Manual peptide sequencing is time-consuming and requires considerable familiarity with interpretation of MS/MS product ion spectra.

De novo peptide sequencing is required if the peptide mass fingerprint and EST database searches are unsuccessful or ambiguous. This is often the case if the genome of the organism in question is not fully sequenced or poorly characterised. Further, *de novo* sequencing can be applied to the characterisation of splice variants. Though the peptide sequences probably exist in the database, they may not exist in a consecutive order. Thus the sequencing of peptides, which span two exons, could be very useful for determining splice variants (B Küster, personal communication).

The fragmentation pattern of a peptide in a product ion MS/MS spectrum is indicative of its sequence, if a significant number of contiguous residues can be determined from the spectrum (greater than 12 residues) oligonucleotide probes can be developed. However, under low energy CID conditions the peptide precursor ion fragments to yield y- and b-ion series of varying lengths. Neither series may be long enough to generate the sufficient stretch of contiguous residues required for subsequent oligonucleotide probe construction. In such circumstances it is important to observe both ion series in the spectrum because they are complimentary. For instance, the residue indicated by the first b-ion, the n-terminal residue, b_1, is the same residue indicated by the final y-ion and vice versa. In this manner the complimentary ion series can be sued to fill in the gaps that may exist in the sequence of one of the ion series.

However, the presence of two ion series can complicate the interpretation of the spectrum, even though they are complimentary. Thus, for efficient *de novo* sequencing it is very important, if not essential, to be able to differentiate these two series. Once this has been achieved the presence of the two ion series is not a burden, the interpretation of the spectrum is simplified significantly.

Hunt DF, Yates JR III, Shabanowitz J, Winston S, Hauer CR. Proc Natl Acad. Sci USA 83 (1986) 6233–6237.

A range of different approaches have been taken to perform *de novo* sequencing by mass spectrometry. The first report of peptide *de novo* sequencing by ESI mass spectrometry was described by Hunt and colleagues as early as 1986, where tryptic peptides of aplipoprotein B were successfully sequenced using FAB-triple quadrupole mass spectrometry (Hunt *et al.* 1986). In this method, the group used differential modification of the peptides to differentiate each ion series, simplifying interpretation and confirming the peptide sequence. Specifically, product ion MS/MS spectra were acquired of the native peptides. Subsequently the peptide fractions underwent methyl esterification, and the product ion MS/MS spectra of the corresponding derivatised peptides were acquired. This reaction esterified the c-terminus of each peptide (and all subsequent acidic resides in each peptide, aspartic and glutamic acids). Resultantly, the starting point of the y-ion series (c-terminal containing ions), y_1, was shifted by 14 Da (and hence the remainder of the y-ion series) whilst the starting point of the b-ions series remained unchanged (unless the n-terminal residue, b_1-ion was a glutamic or aspartic acid). Hence, the y-ion series became recognisable and the two ion series could be differentiated.

However, with this method the sample has too be split and two product ion MS/MS experiments are required for each peptide. The presence of aspartic or glutamic residues internal to the sequence would also change by the esterifying mass, and subsequent y- and b-ions internal to the sequence would also alter by 14 Da.

A second approach described the incorporation of an isotopic label during the peptide digestion. Specifically, the protein of interest is digested as normal but in a 1:1 mixture of ^{16}O:^{18}O digestion buffer. Subsequently all peptides will appear as ^{16}O:^{18}O isotope doublets; with the label being incorporated into the carboxyl group of the c-terminal residue. When this isotopic doublet is fragmented during a product ion MS/MS experiment, all c-terminal product ions, the y-ion series, will exhibit this isotopic doublet and hence can be immediately differentiated from the b-ion series and other non c-terminal containing fragment ions.

Schnolzer M, Jedrzejewski P, Lehman WD. Electrophoresis 17 (1996) 945–953.
Shevchenko A, Chemushevich IV, Ens W, Standing KG, Thomson B, Wilm M, Mann M. Rapid Commun Mass Spectrom 11 (1997) 1015–1024.
Qin J, Herring CJ, Zhang X. Rapid Commun. Mass Spectrom 12 (1998) 209–216.

Wilm and co-workers developed this approach further, combining the isotopic labelling with a technique calling differential scanning enabling improved assignment of the y-ion series (Uttenweiller-Joseph *et al.* 2001). The method requires two product ion MS/MS spectra to be acquired; one where the whole ^{16}O:^{18}O peptide envelope is selected for fragmentation and the second where only the ^{18}O labelled ions are selected for fragmentation. Using a software algorithm the y-ion series can be filtered automatically.

Uttenweiler-Joseph S, Neubauer G, Christoforidis S, Zerial M, Wilm W. Proteomics 1 (2001) 668–682.

However, both approaches do not affect the fragmentation pattern and hence the b-and y-ion series still fragment in the same manner. Furthermore, both approaches are only applicable to enzymatic methods and to proteins which can be digested; it is not applicable to those samples that are presented such as MHC class peptides.

A wide range of peptide derivatisation protocols have been reported for altering the peptide fragmentation pattern and fragmentation efficiency to aid interpretation of product ion MS/MS spectra. Derivatisation of the peptide digest at the n-terminus with phosphonium quaternary tags, to improve sensitivity and yield a single ion series, the b-ion series, have been reported (Roth *et al.* 1998). Munchbach *et al.* 2000, also described a derivatisation protocol to aid the *de novo* sequencing of peptides.

Roth KDW, Huang ZH, Sadagopan N, Throck Watson J. Mass Spectrom Reviews 17 (1998) 255–274.

Munchbach M, Quadroni M, James P. Anal Chem. 72 (2000) 4047–4057.

Keough and co-workers developed a strategy to obtain a single y-ion series, particularly for PSD analysis, with excellent results (see section 2.3.3.5). Long stretches of consecutive y-ions from the y-ion series can be readily observed with this technique and can be useful for *de novo* sequencing. The data demonstrated earlier highlights how simple the fragment ion spectra are to manually interpret, a useful approach for *de novo* sequence interpretation.

Another derivatisation approach to *de novo* sequence analysis is the derivatisation of peptide digests with basic NHS esters. Pyridyl quaternary NHS ester reagents were reported for the derivatisation of peptide digests for PSD analysis facilitating charge localisation at the n-terminus, subsequently yielding only the b-ion series. (Spengler *et al.* 1997; Cardenas *et al.*, 1997). These NHS ester reagents were

Spengler B, Luetzenkirchen F, Metzger S, Chaurand P, Kaufmann R, Jeffrey W, Bartlet-Jones M, Pappin DJC. Int J Mass Spectrom Ion Proc 169 (1997) 127–140.
Cardenas MS, Van der Heeft E, De Jong APJM. Rapid commun Mass Spectrom. 11 (1997) 1271–1278.

further developed to produce a series of gas phase basic reagents to quantitatively label the peptide digest (N-succinimidyl morpholino acetate and N-succinimidyl pyridyl acetate; SMA and SPA respectively). Each of these tags defined the starting points for both the b- and y-ion series, enabled the observation of often complete y-ion series for tryptic peptides (the final y-ion is observed in all cases) and long stretches of the b-ion series. Hence, both the starting and end points of the peptide are known and the ion series are subtley differentiated. Armed with these pieces of information *de novo* sequencing is significantly simplified. Furthermore, because the two ion series are complimentary the determined sequence can be readily checked in each ion series. This proved highly beneficial for uninterpreted, automated MS/MS database searching and essential for de novo sequencing (Tugal *et al.* 1998; Kondo *et al.* 1997; Hoess *et al.* 1999).

In this approch the α-amine N-terminus of each peptide and the ε-amine group of the lysine side chain are modified with a basic functionality (figure 73). The derivatives, or tags, were specifically designed for high sensitivity derivatisation and enhanced sequence coverage in product ion MS/MS spectra. Over 30 potential reagents were synthesised, with two determined to be very useful for *de novo* sequencing. The two basic NHS ester reagents developed were synthesised with the following considerations

The basic tags have significant benefits for peptide sequencing using a quadrupole ion trap.

- water soluble, simple reaction in aqueous conditions
- polar, easily removed by RP chromatography
- yield quantitative derivatisation for both the α-amine N-terminus and ε-amine of the lysine side chain
- the mass of the tag to be unique, easily differentiated from the mass of amino acid residues
- the tag-peptide bond designed to fragment in a similar fashion to the other peptide bonds.

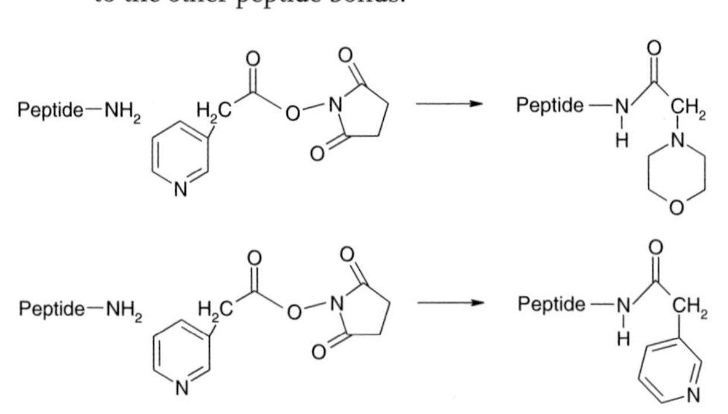

Fig. 73: Reaction scheme for peptide derivatisation with the basic NHS ester reagent. Upper reaction scheme using the SMA reagent. Lower reaction scheme using SPA reagent.

Furthermore, the protein digest does not need to be split as performed in the differential modification method, all the sample can be quantitatively derivatised from the beginning. A peptide mass fingerprint of the digest can be acquired as normal because the confidence in the derivatisation is high. If the peptide mass fingerprint is unsuccessful then the derivatised peptides can be fragmented immediately (unlike the ^{16}O:^{18}O approach). In addition, the method can be applied to any sample regardless of it nature (i.e. it doesn't matter what enzyme or chemical approach is used or if the enzyme is presented instead of being digested).

In order to simplify the *de novo* interpretation of the peptide sequence, the differentiation of the two ion-series and knowledge of both ends of the peptide are very important. Using the derivatisation chemistry developed in the protein sequencing group at Imperial Cancer Research Fund (M Bartlet-Jones, W Jeffries, T Naven, B Canas, D Pappin) both these criteria can be satisfied.

With respect to tryptic peptides

- The c-terminal residue is either an arginine or a lysine (trypsin cleavage rules, section 1.2, table 2), and in a product ion MS/MS spectrum the c-terminal residue is reflected in the y_1 fragment ion. Resultingly, the starting point of the y-ion series in the spectrum will always be known. With knowledge of the structure of a y-ion (residue amino acid mass + water + 1 Da for the ionising proton, tabel 11 & figure 62), if the tryptic peptide is arginine terminating and is present in the spectrum the y_1 ion will be observed at 175.1 Da. If the tryptic peptide is lysine terminating and is present in the spectrum, then y_1 will be observed at 147.1 Da. After location of the y_1-ion in the spectrum, the y-ion series can begin to be determined from this point. Luckily in the product ion MS/MS spectrum of an arginine terminating peptide the y-ion series is generally predominant
- Unfortunately, knowledge of the n-terminal residue is not defined from the trypsin cleavage laws and the n-terminal b_1 ion, reflecting the n-terminal residue, cannot be immediately identified in the spectrum in the same way.

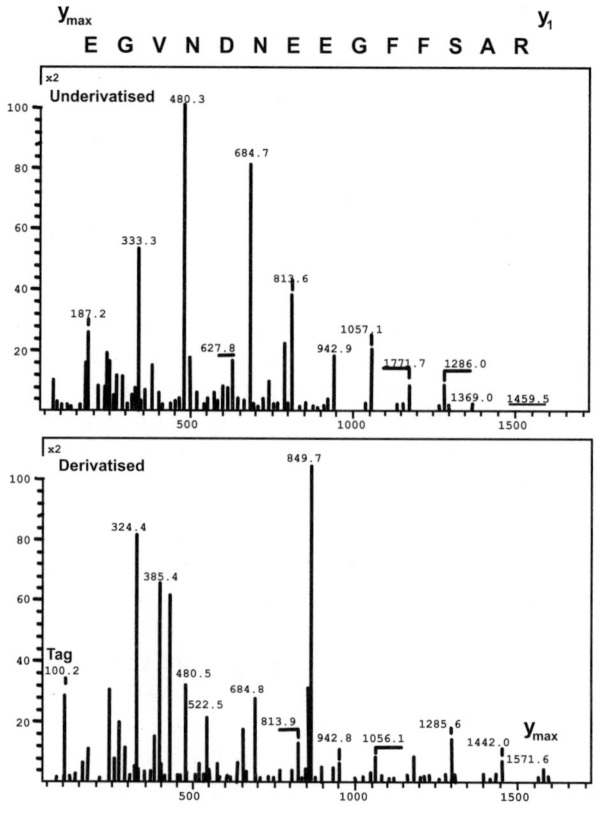

Fig. 74: Product ion MS/MS spectrum acquired with a triple quadrupole equipped with nanospray. Upper panel – underivatised [Glu]-fibrinopeptide b (doubly charged ion *m/z* 786.3), lower panel derivatised – [Glu]-fibrinopeptide b (doubly charged ion *m/z* 849.7).

Figure 74 demonstrates the effect of derivatisation on the fragmentation pattern of a doubly charged peptide. As the peptide is underivatised the n-terminal ion cannot unambiguously be determined from the spectrum. From the complimentary ion series the final y-ion is not observed in the spectrum (final y-ion expected at 1571.6). Hence the n-terminal residue cannot be determined from either ion series. Despite this the y_1 ion is observed at 175.1(arginine) and several consecutive residues of sequence can be determined from this point (175.1, 246.2, 333.3, 480.3, 627.8, 684.7, 813.6, 942.9, 1057.1, 1171.7, 1286.0; corresponding to the sequence RASFFGEENDN.) This length of sequence would be highly discriminating for searches against the protein and EST databases. However, there is still 285 Da still unaccounted for which cannot be inferred from the y-ion series and the b-ion series remains unclear.

Compare this spectrum to the spectrum of the corresponding derivatised peptide, molecular weight 1699.6 (doubly charged ion, 849.7). Following the steps listed above:

- 1699.6 – mass of the tag (128), will give the position of the final y-ion in the spectrum.
- 1571.6. This ion is clearly observed and labelled in the spectrum. From this point the sequence can begin to be interpreted. The next ion down from the final y-ion that co-incides with the mass of an amino acid is the ion at 1442.0, a difference of 129.6 (Glu, E). Thus the final y-ion, which is the first b-ion or the n-terminal residue, is a glutamic acid.
- The process continues, the next ion is observed at 1385.1, a difference of 56.9 (Gly, G).
- The next ion is the ion at 1285.6, a difference of 99.6 (Val, V).
- The remainder of the y-ion series can be followed as in the spectrum of the underivatised peptide (all the way through the sequence to the arginine at 175.1). From the y-ion series alone the remainder of the sequence has been determined.
- As the two ion-series are complimentary, the sequence determined from the y-ion series can be checked using the n-terminal b-ion series. We determined from the y-ion series that the n-terminal ion was a Glu (E, 129 Da). Thus the b_1 ion in the spectrum should be observed at 257 Da. The ion is readily observed along with the a_1 ion 28 Da below (as is the a-ion of the tag at 100.2 Da).
- The n-terminal ion series can be determined from this point. As the y-ion series has already been determined, the two ion series can be clearly differentiated. The next residue determined from the y-ion series was a Gly (G, 57 Da). Thus the addition of 57 Da to the ion at 257 Da, yields 314 Da. This ion at 314.4 Da, b_2, is clearly observed. The next residue determined from the y-ion series was a Val (V, 99). The corresponding n-terminal ion, b_3, at 413.2 is clearly observed along with the corresponding a_3 ion at 385.4. This process is continued for the remainder of the observable b-ion series.

Hence the derivatisation methodolgy has enabled the remainder of the sequence to be determined by locating the end points of the peptide and clear differentiation of the two complimentary ion series. The sequence can be corroborated in both directions removing all ambiguity; essential for de novo sequencing.

If either the c- or n-terminal residue is known, differentiation of the ion series is easier; if both are known then differentiation is greatly simplified.

This approach has significant advantages for de novo sequencing using a quadrupole ion trap. During an ion trap product ion MS/MS spectrum approximately the bottom third of the mass range is not observed in the final spectrum, therefore several fragment ions will not be present in the spectrum.

An example of this application is demonstrated below (Iglesias *et al.* 2000). A product ion MS/MS spectrum of a peptide precursor ion (2147 Da, m/z 1072.6) was acquired using a quadrupole ion trap (figure 75 and table 12). The bottom third of the spectrum is absent. However, following the rules listed above the mass of the tag can be subtracted from the total mass of the peptide (2146–128 = 2018). However, this mass is still above the mass range of the mass spectrometer, but there is an ion a further 101 Da below this mass; this ion is indicative of the final y-ion (as it is bound to the tag and is therefore indicative of the n-terminal ion). From this ion 17 consecutive y-ions can be interpreted from the latter end of the y-ion series. As the ion series are complimentary, these ions can be followed in the b-ion series.

Tab. 12: Fragment ion masses observed in the product ion spectrum of the doubly charged peptide *m/z* 1072.6.

Ion	***Observed m/z***	***residue***	***Ion***	***Observed m/z***	***residue***
y_1			b_1		
y_2	361.3	S	b_2		
y_3	474.5	I/L	b_3		
y_4	603.4	E	b_4		
y_5	716.6	I/L	b_5	613.1	S
y_6	803.5	S	b_6	700.4	S
y_7	890.7	S	b_7	813.1	I/L
y_8	1005.6	D	b_8	927.4	N
y_9	1134.9	E	b_9	1014.2	S
y_{10}	1220.7	S	b_{10}	1143.0	E
y_{11}	1334.6	N	b_{11}	1257.6	D
y_{12}	1447.4	I/L	b_{12}	1344.4	S
y_{13}	1534.8	S	b_{13}	1431.7	S
y_{14}	1621.5	S	b_{14}	1544.9	I/L
y_{15}	1734.7	I/L	b_{15}	1674.4	E
y_{16}	1821.4	S	b_{16}	1787.3	I/L
y_{17}	1918.6	P	b_{17}	1874.8	S
y_{18}		T	b_{18}		K

Iglesias T, Cabrera-Poch N, Mitchell MP, Naven TJP, Rozengurt E, Schiavo G. J Biol Chem. 275 (2000) 40048–40056.

Furthermore the ion at 1874.8 is 274 Da below the mass of the peptide, the mass of a derivatised lysine. Thus, from this ion it can be deduced from the spectrum that the tryptic peptide is lysine terminating. Thus the derivatisation process has enabled us to identify both ends of the peptide sequence. The next ion below this mass which corresponds to an amino acid is 1787.3, a mass difference of 87.5 (Ser, S) (the end of the b-ion series). From this point 12 consecutive b-ions can be interpreted. Thus if this sequence is now applied to the complimentary c-terminal ion series, KS, then y_2 should be present at 361.1 (274, mass of derivatised lysine, + 87), an ion of this mass is clearly observed in the spectrum. The entire y-ion series is sequenced from this point (further aided by working backwards from 1874). As a result the entire 18 residue peptide was sequenced despite the n-terminal and c-terminal ions of the peptide being excluded from the spectrum.

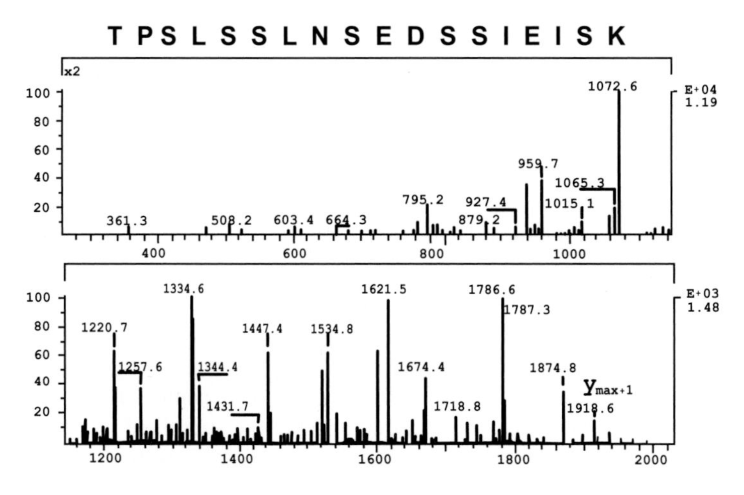

Fig. 75: Product ion MS/MS spectrum of the doubly charged derivatised tryptic peptide *m/z* 1072.6.

Software programmes have been written which aim to *de novo* sequence peptides automatically, without interpretation (Skilling *et al.* 1999). These methods would potentially benefit from knowledge of the starting and end points of both ion series.

Skilling J, Cottrell J, Green B, Hoyes J, Kapp E, Landgridge J, Bordoli B. Proc. 47th ASMS conference Mass spectrometry and allied topics, Dallas (1999).

Using this derivatisation approach the protein digest does not need to be split as performed in the differential modification method, all the sample can be quantitatively derivatised from the beginning. A peptide mass fingerprint of the digest can be aquired as normal because the confidence in the derivatisation is high. Further, if the peptide mass fingerprint is unseccessful then the derivatised peptides can be fragmented immediately (unlike the ^{16}O:^{18}O approach). In addition, the method can be applied to any sample regardless of its nature (i. e., it doesn't matter what enzyme or chemical approach is used or if the enzyme is presented instead of being digested).

2.5
Methods of proteome analysis

An array of approaches have been developed for addressing proteome analysis and protein identification. All of the approaches require mass spectrometry and database searching for protein identification, but differ in the way the proteins are separated and isolated. The methods listed in this section offer a brief view of the variety of approaches that can be employed.

2.5.1
2D-MS

Identification of proteins by mass spectrometry separated by two-dimensional gel electrophoresis is the most commonly used method in proteomics (figure 76). The methods that comprise this approach have been introduced in earlier chapters. Following the second dimension, enabling software allows the analysis of multiple gel images in an automated fashion, determining experimental differences between gels. Selection of protein spots is dependent on experimental design, with the spots of interest picked either manually or automatically, and subsequently digested with trypsin. The extracted tryptic peptides are analysed by MALDI-ToF MS. The peptide molecular masses observed in the discriminating PMF are submitted to a non-redundant protein database search. This approach lends itself to high throughput analysis, with the steps from 2D-MALDI (including database searching) approaching complete automation. If the protein is successfully and unambiguously identified, the next sample is similarly analysed. However, if the result is ambiguous or unsuccessful then more specific information is required, essentially sequence information derived from a product ion MS/MS (or MALDI-PSD) spectrum. Using CID, peptide ions of interest are individually fragmented within the mass spectrometer to reveal its partial or complete amino acid sequence. This information is very specific and can be used to not only search against the protein database, but also the EST database and ultimately for de novo sequencing if no known sequence can be matched. This second stage of MS analysis is typically performed by nanospray MS/MS or LC-MS/MS.

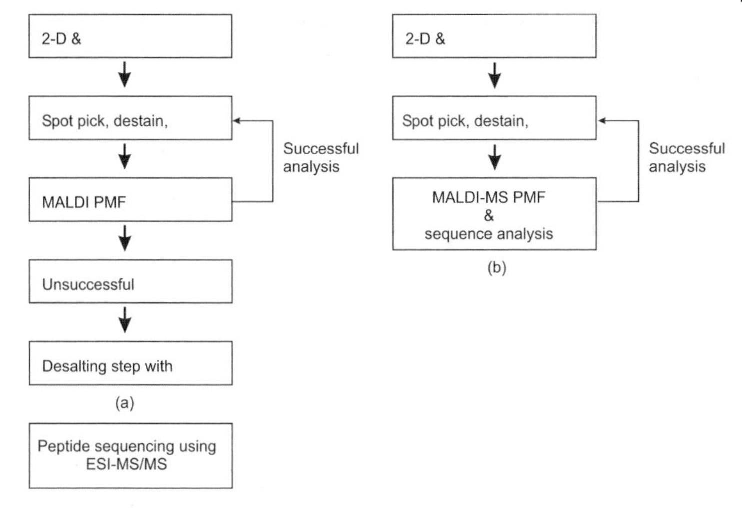

Fig. 76: Workflow for the identification of proteins separated by 2-D electrophoresis. (a) – two tier approach, (b) – one tier approach.

Although a highly successful approach, it is evidently a two tier identification system requiring two different mass spectrometers. With respect to throughput, automation and increased database search confidence, it would be highly advantageous if a single mass spectrometer could combine both peptide mass fingerprinting and peptide sequence analysis. The accessibility and throughput of a MALDI instrument, endear it to such an approach (Keough *et al.* 2000; Shevchenko *et al.* 2001).

Keough T, Lacey MP, Fieno AM, Grant RA, Sun Y, Bauer MD, Begley KB. Electrophoresis 21 (2000) 2252–2265.

Shevchenko A, Sunyaev S, Loboda A, Shevchenko A, Bork P, Ens W, Standing KG. Anal Chem 73 (2001) 1917–1926.

The one tier approach followed by Keough *et al.* 2000 employs PSD to acquire peptide sequence information following a chemical derivatisation of the peptide mixture. Further development of the reagent has enabled a water-soluble reaction to be performed with improved sensitivity, reproducibility, sample handling and automation. The simplified chemical reaction and subsequent facilitated PSD, title CAF MALDI, enables sequence information to be acquired in a sensitive and rapid fashion if the PMF yields an unambiguous or unsuccessful match from a database search (figure 77).

The combination of MALDI PMF and PSD sequence information using a single instrument is an attractive proteomics approach generally yielding unambiguous protein identification (Gavaert *et al.* 2000; Vandahl *et al.* 2001).

Gevaert K, Eggermont L, Demol H, Vandekerckhove J. J Biotechnol 78 (2000) 259–269.

Vandahl BB, Birkelund S, Demol H, Hoorelbeke B, Christiansen G, Vandekerckhove J, Gevaert K. Electrophoresis 22 (2001) 1204–1223.

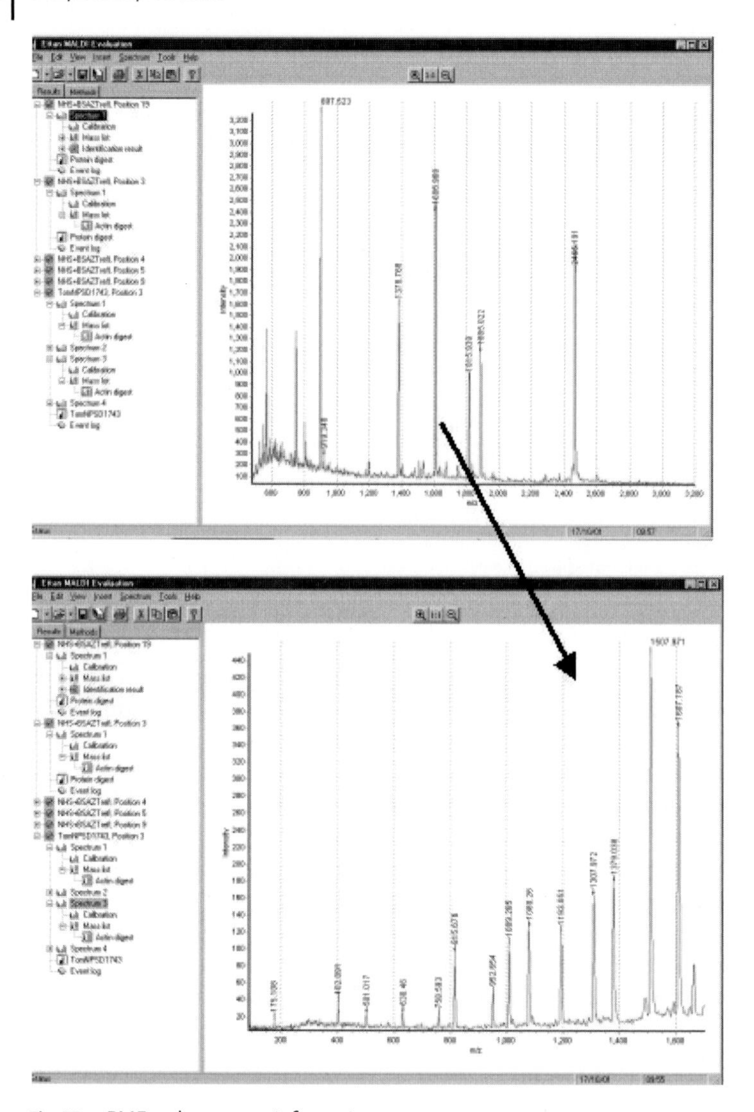

Fig. 77: PMF and sequence information acquired with a single instrument for unambiguous protein identification. Top panel – PMF, lower panel – PSD of one selected from the PMF.

2.5.2 LC-MS/MS

The combination of liquid chromatography and electrospray tandem mass spectrometry offer a powerful approach for proteome analysis. As introduced in section 2.3.3 tandem mass spectrometry enables the investigation and identification of components in complex mixtures. Resultantly methods have been developed taking advantage of this benefit and been applied successfully in two different proteomic approaches.

2.5.2.1 Multidimensional LC-MS/MS

One approach to proteomics is eliminating two-dimensional electrophoresis completely. In such cases, the protein mixture is digested and the resultant peptides eluted directly from a RP HPLC column into the mass spectrometer, where multiple peptide product ion MS/MS spectra are acquired (McCormack *et al.* 1997; Link *et al.* 1997) However, using this method with increasingly complex mixtures the likelihood that peptides will co-elute will increase and result with the MS being unable to select all the peptides for fragmentation. Hence two dimensional chromatography systems have been developed, significantly improving the resolution of the chromatographic separation and simplifying the mixture presented to the mass spectrometer (Link *et al.* 1999). In this method the total protein lysate is digested, acidified and subjected to cation-exchange chromatography. The bound peptides are eluted with increasing ionic strength onto a (reversed-phase) RP-HPLC column. The peptides are bound due to their hydrophobic properties and subsequently eluted into the mass spectrometer (figure 78). The further development of this approach, MUdPIT, has been reported for the rapid, large scale proteome analysis of *Saccharomyces cerevisiae* (Wasburn *et al.* 2001). The protein mixture is fractionated into a soluble fraction and an insoluble fraction (containing membrane proteins). The former fraction is digested with trypsin and subjected to the 2DLC-MS/MS described above and the insoluble fraction is cleaved with cyanogen bromide and secondly with trypsin before 2DLC-MS/MS. Using this approach over 1400 yeast proteins were identified, with no bias demonstrated to any particular class of protein.

McCormack AL, Schieltz DM, Goode S, Yang G, Barnes D, Drubin D, Yates JR III. Anal Chem. 69 (1997) 767–776.
Link AJ, Carmack E, Yates JR III. Int J Mass Spectrom Ion Proc 160 (1997) 303–316.
Link AJ, Eng J, Schieltz DM, Carmack E, Mize GJ, Morris DR, Garvik BM, Yates JR III. Nature Biotech 17 (1999) 676–682.
Washburn MP, Wolters D, Yates JR III. Nature Biotech 19 (2001) 242–247.

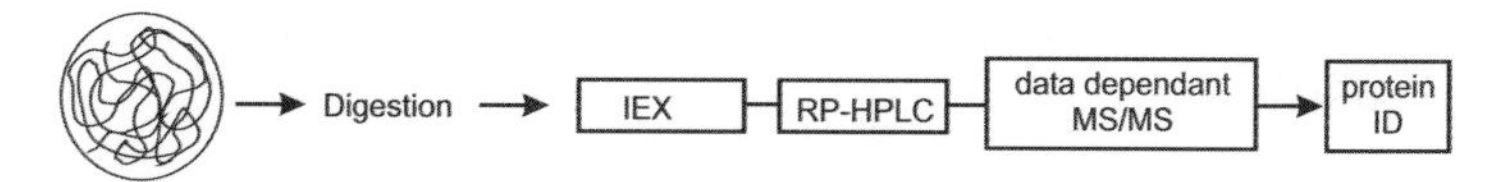

Fig. 78: Workflow for the multidimensional LC separation of protein mixtures with subsequent protein ID using tandem mass spectrometry.

Spahr CS, Susin SA, Bures EJ, Robinson JH, Davis MT, McGinley MD, Kroemer G, Patterson SD. Electrophoresis 21 (2000) 1635–1650.

Alternatively, a complex peptide mixture can be simplified using affinity capture of particular derivatised peptides. Spahr *et al.* 2000; reported the reversible biotinylation of cysteine residues, and affinity capture of cysteine-only containing peptides. The subsequent peptide mixture eluted into the mass spectrometer is significantly simplified.

The above approaches take particular advantage of the data-dependent analysis functionality provided by this type of instrumentation, which are essential for this proteomic approach. Data dependent analysis in this fashion requires the selection of a peptide precursor ion with sufficient signal intensity in MS mode, subsequent isolation and fragmentation of this ion to give a product ion MS/MS spectrum before returning back to MS mode to repeat the process. The speed of spectral acquisition and state switching in the mass spectrometer is the current limiting step in this technique.

2.5.2.2 **Affinity chromatography and LC-MS/MS**

Fields S, Song O. Nature 340 (1989) 245–246; Uetz et al. Nature 403 (200) 623–627. Ito et al. Proc Natl Acad Sci USA. 98 (2001) 459–474. Neubauer G, King A, Rappsilber J, Calvio C, Watson M, Ajuh P, Sleeman J, Lamond A, Mann M. Nature Genet 20 (1998) 46–50. Rigaut G, Shevchenko A, Rutz B, Wilm M, Mann M, Seraphin B. Nature Biotechnol 17 (1999) 1030–1032. Husi H, Ward M, Choudhary JS, Blackstock WP, Grant SGN. Nature Neurosci 3 (2000) 661–669.

Increasing numbers of studies demonstrate that proteins involved in cell mechanisms rarely act on their own, but in complexes of two or more proteins. The yeast two-hybrid technology has enabled the detection of interactions between two proteins (Fields and Song, 1989) and the technology has been applied to the comprehensive analysis of *Saccharomyces cerevisiae* (Uetz *et al.* 2000; Ito *et al.* 2001). Additional method developments have enabled the efficient affinity based isolation of multiple interactions and subsequent analysis of the protein interactions by mass spectrometry (Neubauer *et al.*1998; Rigaut *et al.* 1999; Husi *et al.* 2000).

Two recent publications have combined the affinity isolation approach with on-line LC-MS/MS for the systematic identification of protein complexes in *Saccharomyces cerevisiae* (Gavin *et al.* 2002; Ho *et al.* 2002). Both groups employed a similar approach (see figure 79). Firstly, affinity tags were bound to a wide variety of proteins. Secondly, the DNA encoding the affinity tagged proteins was introduced into yeast cells. Thirdly the affinity tag, and subsequently the protein and those complexed to it, were isolated by affinity chromatography. Finally, the eluted proteins were separated by 1D SDS PAGE, digested with trypsin and analysed by LC-MS/MS. The former publication identified 232 distinct multiprotein complexes, whilst the latter identified over 3600 associated proteins.

Gavin et al. Nature 415 (2002) 141–147.

Ho et al. Nature 415 (2002) 180–183.

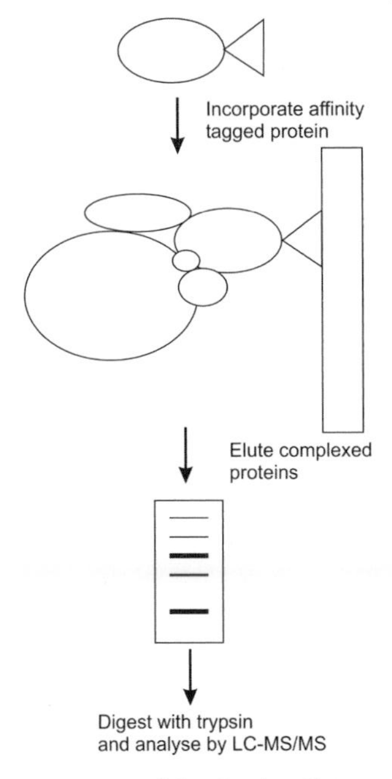

Fig: 79: Workflow for the affinity interaction, LC-MS/MS approach to proteomics.

2.5.3
Quantitative proteomics

Profiling protein expression is an important tool for understanding the molecular dynamics of a cell. The utilisation of the methods explained so far give negligible information regarding the differential expression experienced between two (or more) samples. Several groups have reported methods for quantitation of protein expression (Pasa-Tolic *et al.* 1999; Oda *et al.* 1999), but this section will briefly discribe ICAT.

Pasa-Tolic et al. J Am Chem Soc 121 (1999) 7949–7950.

Oda Y, Huang K, Cross FR, Cowburn D, Chait BT. Proc Natl Acad Sci USA 96 (1999) 6591–6596.

Gygi SP, Rist B, Gerber SA, Turecek F, Gelb MH, Aerbersold R. Nature Biotechnol 17 (1999) 994–999.

Isotope-coded affinity tags (ICAT) This technique was reported by Gygi *et al.* 1999; describing the site specific labelling of cysteine residues with isotope-coded affinity tags (figure 80). The reagent has three constituent groups; firstly biotin which provides affinity for binding with an avidin column, secondly a linker which incorporates stable isotopes and thirdly a reactive group, iodoacetamide, which labels cysteine residues. Two tags exist, one with no deuteriums in the linker (light) and one with 8 deuteriums in the linker (heavy). The strategy requires that a two protein mixture, control and test are labelled with the light and heavy tag respectively. The two derivatised fractions are then combined and digested before affinity isolated using an avidin column. The cysteine labelled peptides are separated by RP-HLPLC and eluted into the mass spectrometer. The two forms of the same peptide will differ in mass by 8 Da, and the ratio between the two determines the relative quantification. Using data dependent analysis the product ion MS/MS spectrum is acquired to identify the protein from the derivatised peptide.

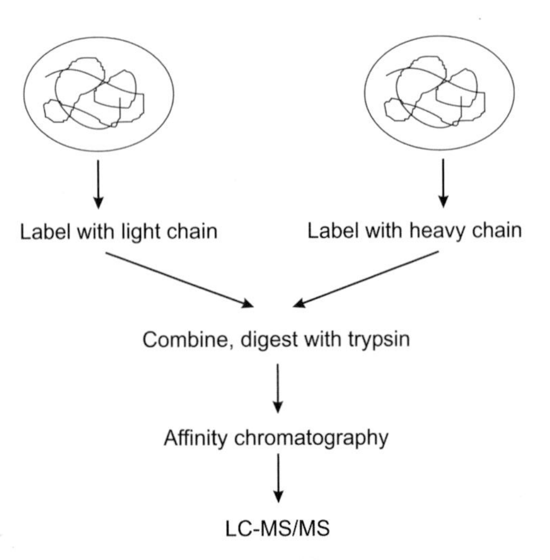

Fig. 80: ICAT reagent and workflow.

Part II
Course Manual

Equipment, Consumables, Reagents

Sample preparation

Instrumentation

Freezer	–20 °C, –80 °C
Laboratory centrifuge	for 1.5 mL tubes, at least 13,000 rpm
Heating block	for 1.5 mL tubes, at least 95 °C
Micropipettes adjustable	from 2 to 1,000 μL

Optional (not always needed):
Sonicator probe

Consumables

Molecular grinding kit,
Disposable, powder free, gloves

Reagents

Urea, CHAPS, Dithiothreitol (DTT), Bromophenol Blue, SDS, Tris, Glycerol (85 %), Pharmalytes pH 3–10 (~ 40 % w/v), IPG Buffer,
PlusOne 2D Clean-up kit
PlusOne Microdialysis kit
PlusOne 2D Quantification kit
Iodoacetamide

Optional (not always needed):
Deoxyribonuclease I (DNase I)
Ribonuclease I (RNase A and RNase B)
Ribonuclease I "A" (RNase A)
Pefabloc (AEBSF), PMSF, Pepstatin,
Thiourea, Sulfobetains

2-D electrophoresis

Instrumentation

IPGphor	IEF system for 2-D electrophoresis
Regular strip holders 18 cm, 24 cm	for standard IEF in Immobiline DryStrips
Cup-loading strip holders 24 cm	for basic gradients and very high sample loads
Reswelling Tray	for rehydrating 12 IPG strips 7–24 cm
Ettan DALT twelve	SDS PAGE system for high-throughput 2-D electrophoresis
Ettan DALT six	SDS PAGE system for medium-throughput 2-D electrophoresis
with:	
Multitemp III	thermostatic circulator
EPS 3501 XL	programmable power supply 3500 V
or	
EPS 600	power supply 600 V

Gel Caster for 14 gels
Gel Caster for 6 gels
PreCast Gel Cassette 1.0 mm
Gel Casting Cassette 1.0 mm
Gel Casting Cassette 1.5 mm
Blank Cassette Insert
Separator Sheets 0.5mm
Filler Sheets 1.0 mm
Cassette Rack
Equilibration tube set 24 cm
Thin plastic ruler (300 to 500 μm)

Heating block for 1.5 mL tubes, at least 95 °C
Magnetic stirrer
Laboratory shaker
Ring stand,
Hand roller, scissors, spatulas, forceps with bent pointed tips,
Assorted glassware: beakers, measuring cylinders, erlenmeyers, test tubes etc.
Magnetic stirrer bars in different sizes
Squeeze bottle
Graduated pipettes of 5 and 10 mL + pipetting device (e.g. Peleus ball)

Micropipettes adjustable from 2 to 1,000 μL
Waterproof pen, black, like Edding 3000

Optional:

FilmRemover	apparatus for removing gels from support films
Gradient maker	for multiple gel caster

Laboratory platform, adjustable, Pinchcock clamp

Consumables

IEF electrode strips
Disposable gloves, tissue paper, filter paper, pipette tips, microcentrifuge tubes, test tubes with screw caps 15 and 50 mL,

Reagents

IPG Dry Strips 18 or 24 cm, different pH gradients
IPG Cover Fluid
IPGphor Strip Holder Cleaning Solution (neutral detergent).

Ettan DALT II gels and buffer kit:

DALT II Gel 12.5% homogeneous 6/pk
DALT II Buffer Kit
Urea, CHAPS, Dithiothreitol (DTT), Bromophenol Blue, SDS, Tris, Pharmalytes pH 3–10 (~ 40 % w/v), IPG Buffers,
Acrylamide IEF solution (40%*T*, 3%*C*), or Acrylamide PAGE 40 % solution, N,N′-methylenebisacrylamide (Bis), Glycerol (85 %), Ammonium persulfate, TEMED, glycine, GelSeal silicone grease, Agarose M or Agarose NA
Butanol or Isopropanol,

Optional (not always needed):

Deoxyribonuclease I (DNase I)
Ribonuclease I (RNase A and RNase B)
Ribonuclease I "A" (RNase A)

Molecular Weight Markers range 14,400–94,000 and
IEF sample application pieces,

Gel labels cut from a printed paper or overhead film.
Bind-silane, Decon™ 90
Thiourea, alternative detergents, sulfobetains, isopropanol
Acid Violet 17, phosphoric acid, trichloroacetic acid, imidazole, zinc sulphate, EDTA-Na_2

Staining

Instrumentation

Processor Plus with large tray	apparatus for automated silver staining
Staining Kit with Orbital shaker	set of trays for multiple gel staining
Stainless steel staining trays	
Heating stirrer	
Kitchen foil welding apparatus	
Gel dryer	gel drying frames and loading platform

Consumables

Disposable gloves	
Transparent smooth sheet protectors	A4 or letter format

Reagents

PhastGel Blue tablets	Coomassie Brilliant Blue R-350
Acetic acid	

PlusOne Silver staining kit

Coomassie Brilliant Blue G-250, Tris, o-phosphoric acid, ammonium sulfate, methanol,

SYPRO Ruby

2-D DIGE

Instrumentation

Typhoon 9000	multifluorescence and phosphorimaging scanner

Reagents

CyDyes, Dimethylformamide, Lysine, Magnesium acetate,

Software

Ettan DeCyder

Evaluation

Instrumentation

ImageScanner	desk top scanner modified for electrophoresis gels
Typhoon 9000	multifluorescence and phosphorimaging scanner

Computer	(Windows NT, 2000 or XP)
Printer	

Software

ImageMaster 2D Elite	2D evaluation software
Imagemaster 2D Database	database add-on
Ettan Progenesis	fully automated 2-D software
Grey step tablet	

Spot picking

Instrumentation

Ettan spot picker

Consumables

Manual spot picking
Powder free gloves
500 μL Eppendorff tubes
Scalpel
Surgical needles

Reagents

Ammonium bicarbonate
Acetonitrile
MilliQ water

Digestion

Instrumentation

Ettan digester

Consumables

Reagents

DDT
Iodoacetamide
Sequence grade trypsin
Ammonium bicarbonate
Trifluoracetic acid
Acetonitrile
MilliQ water

Mass Spectrometry

Instrumentation

Ettan MALDI-ToF Pro

Consumables

Ettan MALDI-ToF Pro target slides
Gel loader pipette tips
ZipTips

Reagents

α-cyano-4-hydroxycinnamic acid
2,5 Dihydroxybenzoic acid
Heptafluorobutryic acid
RP resin
Trifluoroacetic acid
Acetonitrile
Methanol
Formic acid
MilliQ water

Step 1: Sample preparation

It is impossible to explain here all ways of sample preparation in detail. The following instructions should be seen as a starting point. More hydrophobic proteins require more sophisticated extraction procedures.

> Note:
> ***During a course at least one sample with guaranteed good performance should be selected, which can be applied without too many preparation steps. E.coli extract is a very good test and demonstration sample.***

Sample preparation is the most sensitive step in the entire procedure. The proteins, which got lost here, cannot be identified. Modifications of proteins lead to wrong conclusions in proteome analysis.

See page 15 ff for explanations

- To avoid protein losses, the treatment of the sample must be kept to a minimum.
- To avoid protein modifications, the sample should be kept as cold as possible.
- To avoid losses and modifications, the preparation time should be kept as short as possible.

It should also be noted that there are more than one possible procedures to treat a sample. A 100 % representation of the proteins contained in a cell will never be reached in practice. Usually the method, which will display the highest number of different proteins, is chosen. When there is special interest in a certain group of proteins, which are under-represented with the default method, an alternative procedure has to be applied.

Example: When – for a given sample – the highest number of proteins get extracted under alkaline conditions, a number of basic proteins will not be included, because their solubility close to their isoelectric points is low. This means: For the analysis of the basic proteins, acidic extraction conditions have to be employed.

The procedure described for basic proteins in this chapter is based on a procedure developed for optimal extraction and concentration of plant proteins.

First the default procedure and lysis buffer is described (see tables 2, 3, and 4). Alternative or additional reagents should only be used, when the results are not sufficiently good.

Pharmalytes and IPG buffers are amphoteric and acquire a net charge of zero during IEF.

The first dimension, isoelectric focusing, is relatively sensitive to salts and other ionic contaminations. Thus the use of ionic buffers must be avoided as much as possible.

The detection methods listed in table 1 describe categories of sample loads. For facts and comments on detection methods see pages 80 f and Step 4: Staining of gels.

Tab. 1: Protein loads (on large gels, using 18 cm and 24 cm strips):

Radiolabelled	***1 μg – 50 μg***	***"analytical"***
Silver stain	20 μg – 200 μg	"analytical"
SYPRO Ruby stain	100 μg – 700 μg	"analytical"
Zinc imidazol negative stain	200 μg – 500 μg	"analytical"
Coomassie Blue stain	500 μg – 2 mg	"preparative"
Coomassie Blue stain	1 mg – 10 mg	"preparative" on narrow pH intervals

The maximum applicable sample loads depend on the length of the IPG Drystrip, and the kind of pH gradient.

A sample for 2-D electrophoresis is very precious, only high purity reagents should be used.

1 Stock solutions

For the sake of reproducibility it is proposed to use exact concentrations of bromophenol blue.

1% Bromophenol blue does not go into solution without Tris

1 % (w/v) Bromophenol blue solution		
Bromophenol blue	1 % (w/v)	100 mg
Tris-base		60 mg
Water, deionised	dissolve	10 mL

Tab. 2:

Standard solubilisation cocktail "lysis buffer" (10 mL)	
9 mol/L urea	5.4 g
4 % (w/v) CHAPS	400 mg
1 % (w/v) DTT (= 65 mmol/L)	100 mg
0.8% (w/v) Pharmalyte pH 3 to 10*)	200 µL
0.002 % Bromophenol blue**)	10 µL
Water, deionised, make up to	10 mL

*) or IPG buffer of respective pH interval
**) from 1 % BPB blue solution

Important:
Prepare the solution freshly, shake to dissolve the urea, do not warm it higher than 30 °C to avoid carbamylation.

Removal of isocyanate If the purity of the urea is in doubt:

- dissolve 5.7 g urea in deionised water, fill up to 10 mL.
- Warm the tube to get the urea completely in solution: Do not exceed 37 °C .
- Add 100 mg mixed bed ion exchanger Amberlite IRN-150. Stir for 10 min.
- Filter through paper.
- Add the other additives according to table 2.

The lysis buffer can be produced in larger quantities and stored frozen in aliquots at a temperature deeper than –60 °C.

Repeated freeze thawing must be avoided.

Tab. 3:

Standard rehydration solution	
8 mol/L urea	4.8 g
0.5 % (w/v) CHAPS	50 mg
0.28 % (w/v) DTT	28 mg
10 % (v/v) glycerol	1.2 mL
0.5 % (w/v) Pharmalyte 3–10	125 µL
0.002 % Bromophenol blue	10 µL
Water, deionised, make up to	10 mL

Removal of isocyanate If the purity of the urea is in doubt:

- dissolve 4.8 g urea in deionised water containing 10 % (v/v) glycerol, fill up to 10 mL.
- Warm the tube to get the urea completely in solution: Do not exceed 37 °C.

- Add 100 mg mixed bed ion exchanger Amberlite IRN-150. Stir for 10 min.
- Filter through paper.
- Add the other additives according to table 3.

Repeated freeze thawing must be avoided.

The rehydration solution can be produced in larger quantities and stored frozen in aliquots at a temperature at –20 °C or colder.

Liquid samples

To analyse protein solutions such as serum, plasma etc. the solubilising mixture is diluted to the desired protein concentration with rehydration solution.

Very diluted samples, like cell culture supernatant:
In general, the lower additive concentration like in the rehydration solution should be adequate. However, some samples might need higher concentrations, indicated on the right half of table 4.

Tab. 4: Amounts of additives to 1 mL sample

Sample	*1 mL*	*Sample*	*1 mL*
8 mol/L urea	0.75 g	9 mol/L urea	0.94 g
0.5 % (w/v) CHAPS	8 mg	2 % (w/v) CHAPS	36 mg
0.28 % (w/v) DTT	5 mg	1 % (w/v) DTT	18 mg
0.5 % (w/v) Pharmalyte 3–10	20 μL	0.8 % (w/v) Pharmalyte 3–10	36 μL
0.002 % Bromophenol blue	2 μL	0.002 % Bromophenol blue	2 μL
Total volume	**1.6 mL**		**1.85 mL**

2
Examples

The volumes are given for 24 cm IPG strips only. For paper bridge loading 18 cm strips are recommended. The sample would be diluted in 450 μL for anodal and 350 μL for cathodal loading.

Example 1: E.coli extract (ideal as test sample)

Sample: lyophilised cells of E.coli Strain B (ATCC 11303), SIGMA EC-11303

lyophilised E. coli	30 mg	Measured protein content:
lysis buffer	1 mL	~12 mg protein / mL

Freeze (–20 °C) and thaw two times in a microcentrifuge tube. Centrifuge for 10 min with 13,000 rpm.

24 cm IPG strip	*Sample supernatant*	*Rehydration solution*
Analytical run: (~180 μg protein)		
Rehydration loading	15 μL mix with	435 μL
Cup loading	15 μL*) at anodal side	450 μL (> 6 hours pre-rehydration)
Preparative run: (~1 mg protein)		
Rehydration loading	83 μL mix with	367 μL
Cup loading	83 μL at anodal side	450 μL (> 6 hours pre-rehydration)

*) mix with 50 μL rehydration solution before you apply it.

Figure 1 shows a typical 2-D electrophoresis pattern of an E. coli extract applied with rehydration loading.

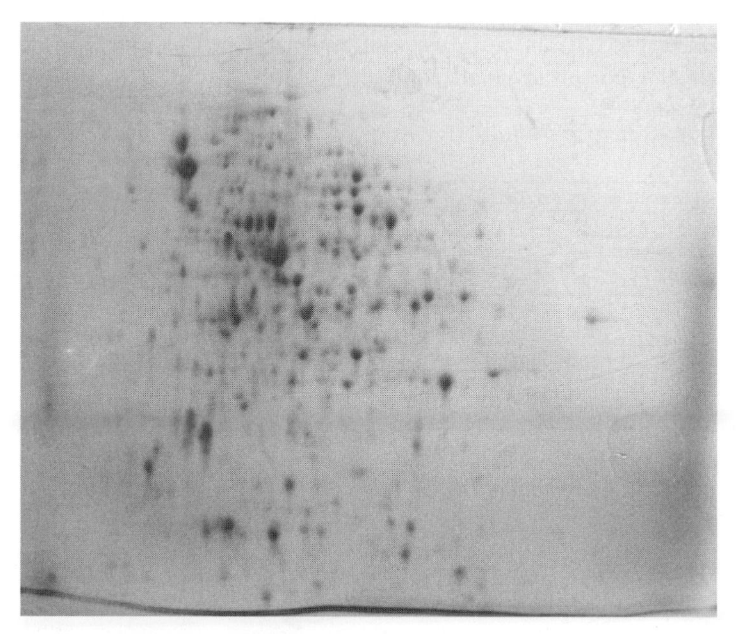

Fig. 1: E. Coli extract (~600 μg protein), IPG DryStrip pH 3–10, Ettan DALT gel, fast Coomassie staining

Example 2: Yeast cell lysate (Saccharomyces cerevisiae)

Lyophilized yeast	300 mg
lysis buffer	2.5 mL

sonicate for 10 min at 0 °C;
centrifuge for 10 min at 10 °C with 42,000 g.

24 cm IPG strip	*Sample supernatant*	*rehydration solution*
Analytical run:		
Rehydration loading	30 µL mix with	420 µL
Cup loading	30*) µL at anodal side	450 µL (> 6 hours pre-rehydration)
Preparative run:		
Rehydration loading	300 µL mix with	150 µL
Cup loading	100 µL at anodal side	450 µL (> 6 hours pre-rehydration)

*) mix with 50 µL rehydration solution before you apply it.

Tissue

The tissue of interest is sliced with a scalpel to obtain an appropriately sized piece. It can be frozen with liquid nitrogen and broken into small fragments in a mortar and pestle.

Example 3: Calf liver

Liver acetone powder, Calf, SIGMA L-7876	
Acetone powder	10 mg
lysis buffer	1 mL

For small tissue amounts a disposable grinding kit is very useful (see figure 2):

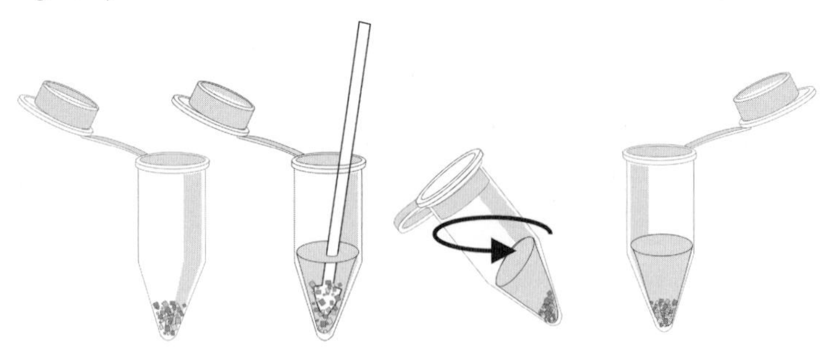

Fig. 2: Grinding of tissue material

1. Centrifuge grinding tubes briefly at maximum speed to pellet the grinding resin. Use a micropipette tip to remove as much of the liquid as possible from the grinding resin pellet.
2. Add 500 µL of lysis buffer to the grinding tube. Vortex to resuspend the grinding resin.
3. Add 5 mg of the acetone powder to the tube.

4. Or 20 to 50 mg of animal or plant tissue.
5. Use a pestle to thoroughly grind the sample.
6. Centrifuge the tube to remove resin and cellular debris. Centrifuge for 5–10 min to maximum speed.
7. Carefully transfer the supernatant to another tube.

24 cm IPG strip	*Sample supernatant*	*rehydration solution*
Analytical run:		
Rehydration loading	30 μL mix with	420 μL
Cup loading	30 μL*) at anodal side	450 μL (> 6 hours pre-rehydration)
Preparative run:		
Rehydration loading	300 μL mix with	150 μL
Cup loading	100 μL at anodal side	450 μL (> 6 hours pre-rehydration)

*) mix with 50 μL rehydration solution before you apply it.

3 Microdialysis

Dialysis of the sample is usually not recommended, because large membrane areas can cause adsorption of proteins. Less proteins are lost with specially designed minidialysis tubes (see figure 3). The fastest removal of small ions is achieved with an 8 kDa cut-off membrane.

In the Tris-glycine buffer system peptides smaller than 10 kDa anyhow co-migrate with the front, they are not displayed in the gel.

Salt and buffer ions can be removed efficiently by dialysing 2-D samples against at least 40 × sample volume of 8 mol/L urea / 1 % DTT solution for 2 hours to overnight. The 2-D sample should be prepared in lysis buffer. Other, more expensive additives such as CHAPS and carrier ampholytes do not need to be included in the dialysis solution. These components are afterwards added back, in the original concentrations.

1. Open the tube and rinse each tube briefly with water from a squeeze bottle to remove the storage solution. Place the dialysis caps in a clean beaker with the membrane sides facing downward and cover with deionised water.
2. Keep the caps in water; do not allow them to dry.
3. Directly before use remove excess water from the cap with a pipette tip.
4. Pipette the sample into the dialysis tube. For 250 μL dialysis tubes, use 10–250 μL of sample. For 2 mL dialysis tubes, use

200 μL – 2 mL of sample. Position the dialysis cap on the tube and tighten firmly.

5. Invert the dialysis tube ensuring that the entire sample rests on the dialysis membrane.

Note:
Very viscous samples frequently do not initially rest on the dialysis membrane. In this case the dialysis tube should be centrifuged briefly in the inverted position: Place the dialysis tube in a 50 mL centrifuge tube with the cap facing downward. Centrifuge at 500 rpm for 5 s. Spinning longer or faster may rupture the membrane. Check the tube to make sure that the entire sample has moved onto the dialysis membrane.

6. Place a float on each dialysis tube. Keeping the dialysis tube inverted, push the bottom of the tube through the hole in the float until the float stops against the dialysis cap.
7. Place the dialysis tube / float assemblies, cap-end down, in a beaker containing 8 mol/L urea / 1 % DTT solution. Check that the dialysis membranes fully contact the solution in the beaker and that there are no large air bubbles trapped beneath the dialysis membranes. Tilt the tube or squirt solution to remove air bubbles if necessary. Use a magnetic stirrer during dialysis.

Note:
Optimal dialysis time will depend on several factors, including the nature and volume of the sample, the M_r cut-off of the dialysis membrane and the temperature. Normally, 2 h to overnight dialysis is sufficient. Dialysis has to be conducted at room temperature, otherwise urea would crystallize.

8. During dialysis, the contents of the dialysis tube should be mixed by inverting or tapping the tube once or twice. If necessary, repeat the short centrifugation.
9. Following dialysis, remove the dialysis tube from the float and immediately collect the sample in the bottom of the tube by centrifugation. Centrifuge the tube for 5 s at 500 rpm. Do not spin longer or faster as this might rupture the membrane.
10. Remove and discard the dialysis cap. Replace with a normal cap for storage.

Re-use of the dialysis tube is not recommended.

4 Precipitation

A crude extract can contain contaminations with phospholipids and nucleic acids, which are visualized with silver staining as horizontal streaks in the acidic part of the gel. Protein precipitation is therefore often employed to remove the contaminating substances. Protein precipitation is also used to concentrate proteins from samples that are too dilute for effective 2-D analysis.

The 2-D clean-up kit procedure uses a combination of a unique precipitant and co-precipitant to quantitatively precipitate the sample proteins. The proteins are pelleted by centrifugation and the precipitate is washed to further remove non-protein contaminants. The mixture is centrifuged again and the resultant pellet is resuspended into the lysis buffer.

The sample can contain 1 μg to 1 mg protein in a volume of 1 to 100 μL (Procedure A). Protein can be processed from larger volumes by scaling up the procedure (Procedure B).

Hint:
Always position the microcentrifuge tubes in the centrifuge with the cap hinge facing outward. This way the pellet will always be on the same side of the tube so it can be left undisturbed, minimizing loss.

- Store the wash buffer in a –20 °C freezer.
- Centrifuge the sample; it must be free of solid particles.

Procedure A: for sample volumes of 1–100 μL

Process the protein samples in 1.5 mL microcentrifuge tubes. All steps should be carried out with the tubes in an ice bucket unless otherwise specified.

1. Transfer 1–100 μL protein sample (containing 1 μg to 1 mg protein) into a 1.5 mL microcentrifuge tube.
2. Add 300 μL precipitant and mix well by vortexing or inversion. Incubate on ice (4–5 °C) for 15 minutes.
3. Add 300 μL co-precipitant to the mixture of protein and precipitant. Mix by vortexing briefly.
4. Centrifuge the tubes in a microcentrifuge set at maximum speed (at least 12,000 × g) for 5 minutes.
5. Remove the tubes from the centrifuge as soon as centrifugation is complete. A small pellet should be visible. Proceed rapidly to the next step to avoid resuspension or dispersion of the pellet.

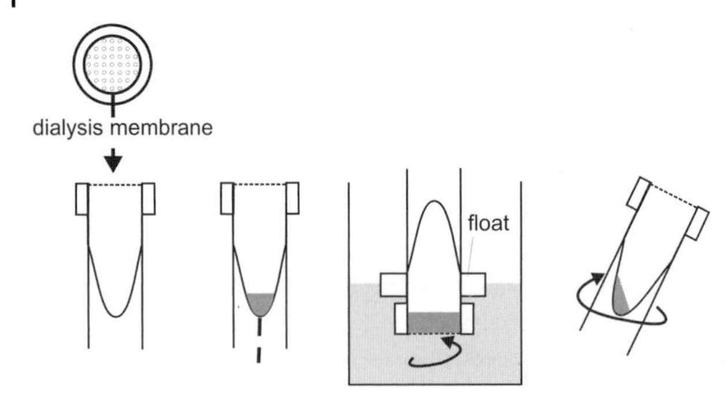

Fig. 3: Microdialysis for removal of small ionic components

6. Remove as much of the supernatant as possible by decanting or careful pipetting. Do not disturb the pellet.
7. Carefully reposition the tube in the microcentrifuge as before; with the cap-hinge and pellet facing outward. Centrifuge the tubes again to bring any remaining liquid to the bottom of the tube. A brief pulse is sufficient. Use a pipette tip to remove the remaining supernatant. There should be no visible liquid remaining in the tubes.
8. Pipette 25 µL of deionised water on top of each pellet. Vortex each tube for 5–10 seconds. The pellets should disperse, but not dissolve in the water.
9. Add 1 ml of pre-chilled wash buffer and 5 µL wash additive. Vortex until the pellets are fully dispersed.

> Note:
> ***The protein pellet will not dissolve in the wash buffer.***

10. Incubate the tubes at –20 °C for at least 30 min. Vortex for 20–30 second once every 10 minutes.

> Note:
> ***The tubes can be left at this stage at –20 °C for up to 1 week with minimal protein degradation or modification.***

11. Centrifuge the tubes in a microcentrifuge set at maximum speed (at least 12,000 × g) for 5 minutes.
12. Carefully remove and discard the supernatant. A white pellet should be visible. Allow the pellet to air dry in until it turns translucent.

> Note:
> ***Do not over dry the pellet. If it becomes too dry, it will be difficult to resuspend.***

13. Resuspend each pellet in an appropriate volume of lysis buffer or rehydration solution. Vortex the tube for 30 s. Incubate at room temperature and either vortex or work up and down in a pipette to fully dissolve.

   Note:
   If the pellet is large or too dry, it may be slow to resuspend fully. Sonication or treatment with the sample grinding kit can speed resuspension.

14. Centrifuge the tubes in a microcentrifuge set at maximum speed (at least 12,000 × g) for 5 minutes to remove any insoluble material and to reduce any foam. The supernatant may be loaded directly onto first dimension IEF or transferred to another tube and stored at –80 °C for later analysis.

Procedure B: for dilute samples of more than 100 µL
All steps should be carried out with the tubes in an ice bucket unless otherwise specified.

1. Transfer the protein sample into a tube that can be centrifuged at 8000 × g. The tube must have a capacity at least 12 × greater than the volume of the sample. Use only polypropylene, polyallomer or glass tubes.

   Note:
   The Wash Buffer used later in the procedure attacks many plastics. This limits the choice of centrifuge tube materials.

2. For each volume of sample, add 3 volumes of precipitant. Mix well by vortexing or inversion. Incubate on ice (4–5 °C) for 15 minutes.
3. For each original volume of sample, add 3 volumes of co-precipitant to the mixture of protein and Precipitant. Mix by vortexing briefly.
4. Centrifuge the tube(s) at 8000 × g for 10 minutes. Remove the tubes from the centrifuge as soon as centrifugation is complete. A small pellet should be visible. Proceed rapidly to the next step to avoid resuspension or diffusion of the pellet.
5. Remove as much of the supernatant as possible by decanting or careful pipetting. Do not disturb the pellet.
6. Carefully reposition the tubes in the centrifuge as before, with the pellet facing outward. Centrifuge the tubes again for at least 1 min to bring any remaining liquid to the bottom of the tube. Use a pipette tip to remove the remaining supernatant. There should be no visible liquid remaining in the tubes.

7. Pipette enough deionised water on top of each pellet to cover the pellet. Vortex each tube for several seconds. The pellets should disperse, but not dissolve in the water.
8. Add 1 mL of pre-chilled wash buffer for each original volume of sample. Add 5 μL wash additive (regardless of the original sample volume, use only 5 μL wash additive). Vortex until the pellets are fully dispersed.

> Note:
> ***The protein pellet will not dissolve in the wash buffer.***

9. Incubate the tubes at –20 °C for at least 30 min. Vortex for 20–30 s once every 10 minutes.

> Note:
> ***The tube(s) can be left at this stage at –20 °C for up to 1 week with minimal protein degradation or modification.***

10. Centrifuge the tubes at 8000 × g for 10 minutes.
11. Carefully remove and discard the supernatant. A white pellet should be visible. Allow the pellet to air dry in until it turns translucent.

> Note:
> ***Do not over dry the pellet. If it becomes too dry, it will be difficult to resuspend.***

12. Resuspend pellets in lysis buffer or rehydration solution. The volume of lysis buffer or rehydration solution used can be as little as 1/20 the volume of the original sample. Vortex the tube for 30 s. Incubate at room temperature and either vortex or work up and down in a pipette to fully dissolve.

> Note:
> ***If the pellet is large or too dry, it may be slow to resuspend fully. Sonication or treatment with the sample grinding kit can speed resuspension.***

13. Centrifuge the tubes at 8000 × g for 10 minutes to remove any insoluble material and to reduce any foam. The supernatant may be loaded directly onto first dimension IEF or transferred to another tube and stored at –80 °C for later analysis.

5
Basic proteins

Damerval *et al.* (1986) have developed an acidic extraction / precipitation procedure for plant samples, in order to concentrate the proteins and to get rid of polyphenols. The yield of proteins with pIs higher than pH 7 is much higher when extracted under acidic conditions than with lysis buffer. A modification of this method has been successfully employed for the extraction of basic proteins from mouse liver, myeloblasts, human heart, and yeast (Görg *et al.* 1997, 1998). The procedure takes overnight. Extraction of basic proteins can be speeded up with the following protocol, using the 2-D clean-up kit.

Damerval C, DeVienne D, Zivy M, Thiellement H. Electrophoresis 7 (1986) 53–54.
Görg A, Obermaier C, Boguth G, Csordas A, Diaz J-J, Madjar J-J. Electrophoresis 18 (1997) 328–337.
Görg A, Boguth G, Obermaier C, Weiss W. Electrophoresis 19 (1998) 1516–1519.

Extraction of basic proteins with the 2-D clean-up kit

Process the protein samples in 1.5 mL microcentrifuge tubes. All steps should be carried out with the tubes in an ice bucket unless otherwise specified.

Hint:
Always position the microcentrifuge tubes in the centrifuge with the cap hinge facing outward. This way the pellet will always be on the same side of the tube so it can be left undisturbed, minimizing loss.

1. Freeze sample in a deep freezer at –80 °C.
2. Grind sample in pre-frozen mortar with pestle.
3. Suspend 10 mg of the powder in a 1.5 mL microcentrifuge tube in 300 μL of –20 °C cold precipitant solution and mix.
4. Add 300 μL co-precipitant / 0.1 % (w/v) DTT to the mixture of protein and precipitant. Mix by vortexing briefly.
5. Incubate for 30 min at –20 °C.
6. Centrifuge the tube in a microcentrifuge set at maximum speed (at least 12,000 × g) for 5 minutes.
7. Remove the tubes from the centrifuge as soon as centrifugation is complete. A small pellet should be visible. Proceed rapidly to the next step to avoid resuspension or dispersion of the pellet.
8. Remove as much of the supernatant as possible by decanting or careful pipetting. Do not disturb the pellet.
9. Carefully reposition the tube in the microcentrifuge as before; with the cap-hinge and pellet facing outward. Centrifuge the tubes again to bring any remaining liquid to the bottom of the tube. A brief pulse is sufficient. Use a pipette tip to remove the

remaining supernatant. There should be no visible liquid remaining in the tubes.

10. Resuspend the pellet in –20 °C cold wash solution / 0.1 % (w/v) DTT.
11. Incubate the tubes at –20 °C for 30 min. Vortex for 20–30 second once every 10 minutes
12. Centrifuge the tube in a microcentrifuge set at maximum speed (at least 12,000 × g) for 5 minutes.
13. Carefully remove and discard the supernatant. A white pellet should be visible. Allow the pellet to air dry in until it turns translucent.

Note:
Do not over dry the pellet. If it becomes too dry, it will be difficult to resuspend.

14. Resuspend each pellet in an appropriate volume of lysis buffer. Vortex the tube for 30 s. Incubate at room temperature and either vortex or work up and down in a pipette to fully dissolve.

Note:
If the pellet is large or too dry, it may be slow to resuspend fully. Sonication or treatment with the sample grinding kit can speed resuspension.

15. Centrifuge the tubes in a microcentrifuge set at maximum speed (at least 12,000 × g) for 5 minutes to remove any insoluble material and to reduce any foam. The supernatant is applied directly onto first dimension IEF by cup loading on a pre-rehydrated gel strip or transferred to another tube and stored at –80 °C for later analysis.

6
Very hydrophobic proteins

Thiourea procedure

Always treat thiourea-containing solutions with mixed bed ion exchanger.

For *membrane* proteins and other very hydrophobic proteins a combination *urea/thiourea* in the solubilisation solution can be very helpful to get more proteins into solution (Rabilloud, 1998). The modified lysis buffer is then composed as follows:

7 mol/L urea	4.2 g
2 mol/L thiourea	1.5 g
with H_2O_{dist} fill up to	9.5 mL

Removal of isocyanate

- Warm the tube to get the urea and thiourea completely in solution: Do not exceed 37 °C .
- Add 100 mg mixed bed ion exchanger Amberlite IRN-150. Stir for 10 min.
- Filter through paper.
- Add the other additives:

4 % CHAPS	400 mg
1 % (w/v) DTT	100 mg
2 % (w/v) Pharmalyte pH 3 to 10*)	500 μL
0.002 % Bromophenol blue**)	10 μL

*) or IPG buffer
**) from 1 % BPB blue solution

Important:
Prepare the solution freshly, shake to dissolve the urea, do not warm it higher than 30 °C to avoid carbamylation.

Also this lysis buffer can be produced in larger quantities and stored frozen in aliquots at a temperature deeper than – 60 °C.

Repeated freeze thawing must be avoided.

SDS procedure

- Formation of oligomers can be prevented.
- Organisms with tough cell walls sometimes require boiling for 5 minutes in 1 to 2 % SDS before they are diluted with lysis buffer.
- Very hydrophobic proteins might require extraction with high percentage of SDS.

Example: Human plasma
According to Sanchez *et al.* (1995).

Sanchez J-C, Appel O, Pasquali C, Ravier F, Bairoch A, Hochstrasser DF. Electrophoresis 16 (1995) 1131–1151.

10 μL human plasma	mix with 6.25 μL 10% SDS, 2.3 % DTT
Heat for 5 min at 95 °C	
Dilute with	500 μL lysis buffer

Note:
SDS does not always completely separate from the proteins during IEF, even under high field strength. SDS can also be removed by precipitation with the clean-up kit.

7
Quantification

The procedure described is compatible with such common sample preparation reagents as 2 % SDS, 1 % DTT, 8 mol/L urea, 2 mol/L thiourea, 4 % CHAPS, 2 % Pharmalyte and 2 % IPG Buffer.

The volume range of the assay is 1–50 μL and the linear range for quantification is 0–50 μg. The assay has a sensitivity threshold of 0.5 μg:

> Hint:
> ***Always position the microcentrifuge tubes in the centrifuge with the cap hinge facing outward. This way the pellet will always be on the same side of the tube so it can be left undisturbed, minimizing loss.***

The assay should be performed at room temperature.

Preliminary preparations

- Prior to performing the assay, prepare an appropriate volume of working colour reagent by mixing 100 parts of colour reagent A with 1 part colour reagent B. Each individual assay requires 1 mL of working colour reagent.
- Working colour reagent can be stored at 4–8 °C for up to one month or as long as the optical density (OD) of the solution remains below 0.025 at 480 nm.

Procedure

1. Prepare a standard curve according using the 2 mg/mL bovine serum albumin (BSA) standard solution provided with the kit. Set up six tubes and add standard solution according to table 5. Tube 1 is the assay blank, which contains no protein.

Tab. 5: Preparation of standard curve

Tube number	*1*	*2*	*3*	*4*	*5*	*6*
Volume of 2 mg/mL BSA standard solution	0 μL	5 μL	10 μL	15 μL	20 μL	25 μL
Protein quantity	0 μg	10 μg	20 μg	30 μg	40 μg	50 μg

Note:
The accuracy of the assay is unaffected by the volume of the sample as long as the sample volume is 50 μL or less. It is therefore unnecessary to dilute standard or sample solutions to a constant volume.

2. Prepare tubes containing 1–50 μL of the sample to be assayed. Duplicates are recommended. The useful range of the assay is 0.5–50 μg and it is also recommended to assay more than one sample volume or dilution to ensure that the assay falls within this range.
3. Add 500 μL precipitant to each tube (including the standard curve tubes). Vortex briefly and incubate the tubes 2–3 min at room temperature.
4. Add 500 μL co-precipitant to each tube and mix briefly by vortexing or inversion.
5. Centrifuge the tubes at a minimum of 10,000 × g for 5 min. This sediments the protein.
6. Decant the supernatants. Remove the tubes from the centrifuge as soon as centrifugation is complete. A small pellet should be visible. Proceed rapidly to the next step to avoid resuspension or dispersion of the pellets.
7. Carefully reposition the tubes in the microcentrifuge as before, with the cap-hinge and pellet facing outward. Centrifuge the tubes again to bring any remaining liquid to the bottom of the tube. A brief pulse is sufficient. Use a pipette tip to remove the remaining supernatant. There should be no visible liquid remaining in the tubes.
8. Add 100 μL of copper solution and 400 μL of deionised water to each tube. Vortex briefly to dissolve the precipitated protein.
9. Add 1 mL of working colour reagent to each tube (See "Preliminary preparation" for preparing the working colour reagent). Ensure instantaneous mixing by introducing the reagent as rapidly as possible. Mix by inversion.
10. Incubate at room temperature for 15–20 min.
11. Read the absorbance of each sample and standard at 480 nm using water as the reference. The absorbance should be read within 40 min of the addition of working colour reagent.

Note:
Unlike most protein assays, the absorbance of the assay solution decreases with increasing protein concentration. Do not subtract the blank reading from the sample reading or use the assay blank as the reference.

12. Generate a standard curve by plotting the absorbance of the standards against the quantity of protein. Use this standard curve to determine the protein concentrations of the samples.

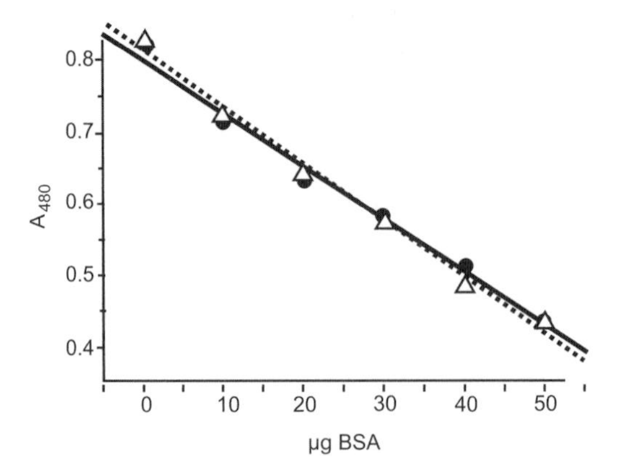

● standard curve with BSA dissolved in water

△ standard curve with BSA dissolved in first-dimension sample solution (8mol/L urea, 4 % CHAPS, 40 mmol/L DTT, 2 % Pharmalyte pH 3-10).

Fig. 4: Protein assay with the novel 2D quantification kit. The standard curves of BSA in absence and presence of usually disturbing additives are almost identical.

8
SDS samples for HMW proteins separation

SDS sample buffer

2 % (w/v) SDS		2 g
25% glycerol (v/v)	85% solution	30 mL
50 mmol/L Tris HCl pH 8.8	1.5 mol/L stock solution	3.5 mL
0.01% Bromophenol Blue		10 mg
Water, deionised	make up to	100 mL

Store at room temperature.

- Dissolve 50 mg of DTT in 100 µL deionised water.
- Dissolve 20 mg of iodoacetamide in 100 µL deionised water.
- Add 10 µL DTT solution to 100 µL SDS sample buffer.
- Mix 25 µL sample supernatant with 25µL SDS sample buffer.

Do not heat it, the sample contains urea!

- Add 10 µL iodoacetamide solution to the 50 µL SDS sample solution. Leave it at room temperature for 1 hour.
- Apply 20 µL of this solution on the SDS gel.

This is a 1-D sample

Step 2: Isoelectric focusing

In a course you can run several parallel experiments. When E.coli and/or calf liver protein extracts should be analysed, it is possible to start three IEF experiments on the first afternoon:

The same time schedule can be employed for the optimization process for a certain sample.

- Reswelling IPG strips in the reswelling tray without samples for cup loading on the second day.
- Rehydration loading of the protein extracts in the reswelling tray for running them face-up in the cup loading strip holder on the second day.
- Rehydration loading under low voltage and focusing over night in the regular strip holder on the IPGphor.

In the tables for instruments settings you find the appropriate times for starting und finishing the experiment on time.

Disposable gloves must be worn during each step to avoid contamination of the sample.

- Label the IPG strips by writing a number on the hydrophobic back before rehydration (figure 1).

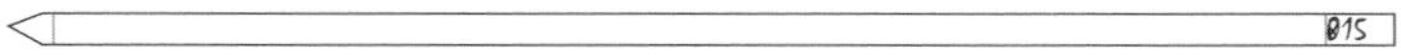

015

Fig. 1: The IPG strips should be labelled with a waterproof pen.

Choose 18 cm long IPG strips instead of 24 cm strips, when molecular weight markers and 1-D samples should be applied onto the SDS gel. As already mentioned on page 24, high molecular weight proteins are not well represented in 2-D gels. But they can be displayed as 1-D sample on the SDS gel.

1
Reswelling tray

For the course: for rehydration loading apply preparative load here.

This device is used for both, rehydration loading of samples directly into the IPG strips or pre-rehydration of IPG strips for cup loading.

Tab. 1: Standard rehydration solution

8 mol/L urea	4.8 g
0.5 % (w/v) CHAPS	50 mg
0.28 % (w/v) DTT	28 mg
10 % (v/v) glycerol	1.2 mL
0.5 % (w/v) Pharmalyte 3–10	125 µL
0.002 % Bromophenol blue	10 µL
Water, deionised, fill up to	10 mL

Repeated freeze thawing must be avoided.

see pages 190 ff

The rehydration solution can be produced in larger quantities and stored frozen in aliquots at a temperature at –20 °C or colder.

For rehydration loading dilute sample accordingly.

The grooves are numbered from 1–12 to allow sample identification. But, to be on the safe side, label the IPG strips with a waterproof pen.

- Be sure that the reswelling tray has been carefully cleaned and dried before use.
- Adjust the feet to level the tray horizontally, using the inbuilt spirit level as a control (see fig. 2).
- Pipette rehydration solution containing the sample into the first slot as a streak slightly shorter than the strip to be rehydrated.

Starting at the basic end might damage the – usually softer – basic gel surface.

- Remove the cover film from the IPG strip starting at the acidic (pointed) end.

Air bubbles are removed by lifting the strip up again with a forceps.

- Place the strip into the slot with the dried gel side down avoiding air bubbles.

When some of the sample flows onto the back of the film, it will be pushed down around the edge of the strip, when the paraffin oil is pipetted on it.

to avoid urea crystallisation and oxygen contact.

- Pipette 3 mL paraffin oil over the strip by starting on both ends of the strip, moving to the center.
- Repeat this procedure for all samples to be analysed.
- Close the tray with the sliding coverlid.

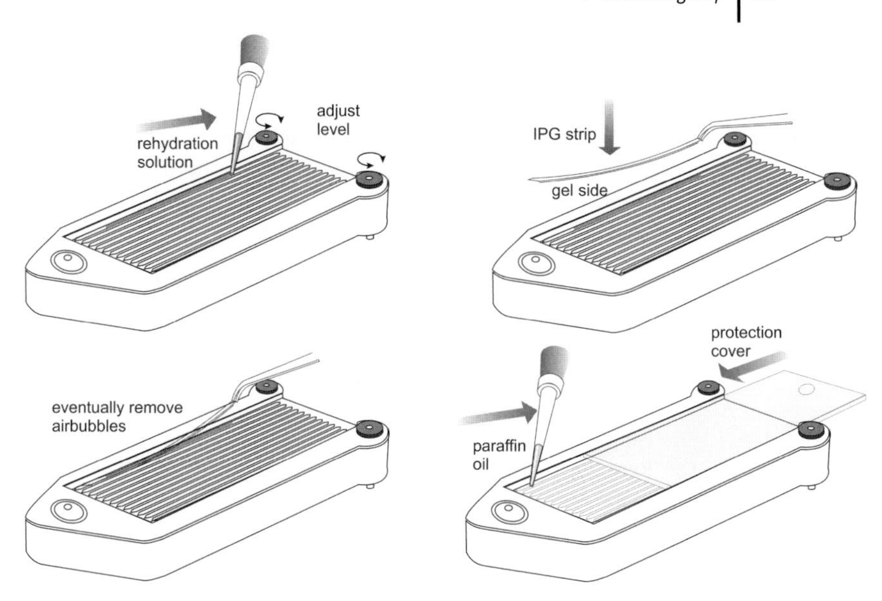

Fig. 2: Rehydration of IPG strips in individual grooves in the reswelling tray.

Rehydration time

without sample	> 6 h
including sample	> 12 h or overnight *)

*) Large protein molecules need a long time to diffuse into the strip.

After rehydration

- Rinse the surface of the strips with distilled water using a squeeze bottle and then place them for a few seconds on their edges on a damp filter paper to drain excess liquid, so that the urea on the surface does not crystallize out.

The strips can be run on the Multiphor flatbed chamber in the *Immobiline DryStrip Kit* (tray for up to 12 strips) or on the IPGphor in individual *Cup-loading strip holders.*

Cleaning of the reswelling tray The reswelling tray is thoroughly cleaned with detergent, using a toothbrush or Q-tips for the slots. After rinsing it with deionised water it must be completely dry before the next use.

2
Rehydration loading and IEF in IPGphor strip holders

For the course: use analytical protein load here.

The strip holders have numbers to allow sample identification. But to be on the safe side: label the IPG strips with the waterproof pen. The IPGphor chamber must be horizontally leveled on the bench. Rehydration and IEF separation are carried out at 20 °C.

Never use new strip holders without cleaning them before the first run.

- Be sure, that the strip holders are carefully cleaned and dried.
- Place strip holders on the cooled electrode contact areas of the power supply with the pointed end on the anodal contact area.
- Pipette rehydration solution mixed with sample into the strip holder as a streak from electrode contact to electrode contact (see figure 3).

Starting at the basic end might damage the – usually softer – basic gel surface.

- Remove the cover film from the IPG-strip starting at the acidic (pointed) end.

The distance from the platinum contact to the end of the tray is shorter at the anodal side.

- Starting at the anodal side, place the IPG strip with the pointed acidic end into the strip holder – dried gel side down. Slowly lower the basic end into the strip holder.

Should an air bubble be caught, lift the strip up with forceps and slowly lower it down again. You might need to repeat this procedure.

When some of the sample flows onto the back of the film, it will be pushed down around the edge of the strip when the paraffin oil is pipetted on it.

to avoid urea crystallisation and oxygen contact.

- Pipette 3 mL paraffin oil over the strip by starting on both ends of the strip, moving to the center.
- Place the plastic cover on the strip holder.
- Close the safety lid.
- Enter the running conditions.

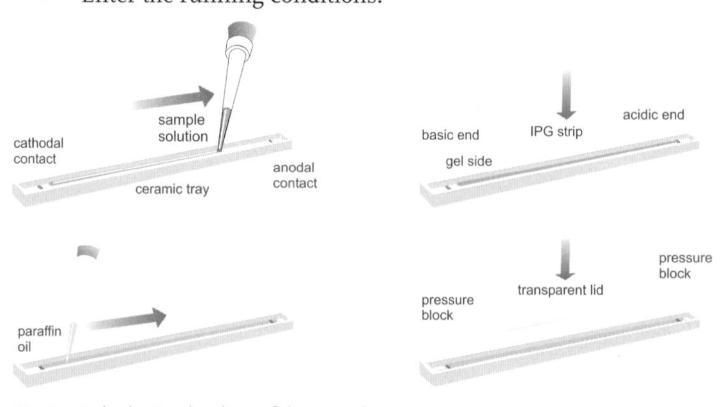

Fig. 3: Rehydration loading of the sample solution in the regular strip holder of the IPGphor.

Tab. 2: Instrument settings for rehydration loading in regular strip holders

Rehydration time			0 h			
Temperature			20 °C			
Current per strip			50 µA			
Running mode			step & hold			
Strip length			18 cm		24 cm	
pH gradient			3–10 L	3–10 NL	3–10 L	3–10 NL
Step 1	step & hold	60 V	12 h		12 h	
Step 2	step & hold	500 V	1 h		1 h	
Step 3	step & hold	1000 V	1 h		1 h	
Step 4	step & hold	8000 V	20 kVh	20 kVh	40 kVh	40 kVh
analytical			2 h 20 min	2 h 20 min	4 h	4 h
Total time			16 h 20 min	16 h 20 min	18 h	18 h
Course timing. start at [h]:			*16:40*	*16:40*	*15:00*	*15:00*
ready for2nd dim. [h]:			*9:00*	*9:00*	*9:00*	*9:00*

For preparative sample loads (> 1 mg) increase the final focusing step by 15 % of the proposed Volthours.

When an overnight run is finished early in the morning, refocus the proteins by applying 8000 V on the strips for 15 minutes before equilibration, staining, or storage of the strips.

Observations during the run

In each IPG strip the bromophenol blue tracking dye will slowly start to build a band at the cathodal end, which will migrate towards the anode. When all samples have the same salt content, they will run almost at the same level. Do not worry; when the bromophenol blue bands of a few samples migrate slower, this can always happen.

During the last few hours you will observe a strong black, sometimes also an additional yellow band at the anode: this is boromophenol blue collected here, at the very end it becomes very acidic turning the bromophenol blue into yellow.

- Before the run is finished: place a tray with detergent solution next to the instrument for the strip holders.

Cleaning of the regular strip holders

The strip holders must be carefully cleaned after each IEF. The solutions must never dry in the strip holder. Cleaning is very effective, if the strip holders are first soaked a few hours in a solution of 2–5% of the specially supplied detergent in hot water.

Clean strip holders should be handled with gloves to avoid contamination.

SDS solution in absence of buffer has a neutral pH.

The strip holder slot should be vigorously brushed with a toothbrush using a few drops of undiluted IPGphor Strip Holder Cleaning Solution. Then it is rinsed with deionised water.

Sometimes protein deposits are left on the bottom of the strip holder after IEF. This happens when highly abundant proteins have been squeezed out of the gel surface at their isoelectric point (see figure 19 on page 48). It is not always easy to remove these proteins, particularly when they are sticky like serum albumin. In this case the strip holders should be boiled in 1 % (w/w) SDS with 1 % (w/v) DTT for 30 minutes before the slot is cleaned with the toothbrush.

> Important
> ***Strip holders may be baked, boiled or autoclaved. But, because of the specially treated surface they must not be exposed to strong acids or basis, including alkaline detergents.***

> Note:
> ***The strip holder must be completely dry before use.***

3 IEF in the cup loading strip holder (rehydration loaded strips)

For the course: use preparative protein load here.

The strip holders have numbers to allow sample identification. The IPGphor chamber must be horizontally leveled on the bench. IEF separation is carried out at 20 °C.

Figure 4 shows all the parts needed and the proper placements of strips, pads, and electrodes.

Never use new strip holders without cleaning them before the first run.

- Be sure, that the cup loading strip holders are carefully cleaned and dried.
- Place strip holders on the cooled electrode contact areas of the power supply with the pointed end on the anodal contact area.
- Starting at the basic side, place the IPG strip – gel side facing up – into the strip holder with the pointed end towards the anode side. Be sure that the protruding film at the basic end touches the end of the strip holder. The aligner protrusions help to keep the strip centered.

to avoid urea crystallisation and oxygen contact.

- Pipette 4 mL paraffin oil over the strip.

The pads must be damp, not wet.

- Cut 5 mm long pads from IEF electrode strips and soak them with deionised water. Blot them on filter paper and place them on top of the ends of the strip. The pads should sit completely on the gel surface. If longer strips are required for removal of salt, there must be an overlapping of at least 5 mm.

- Place the electrodes on the pads as shown in figure 4.
- Place the plastic cover on the strip holder.
- Close the safety lid.

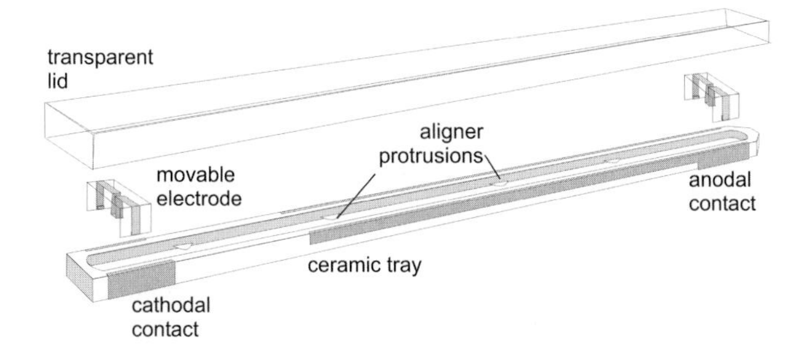

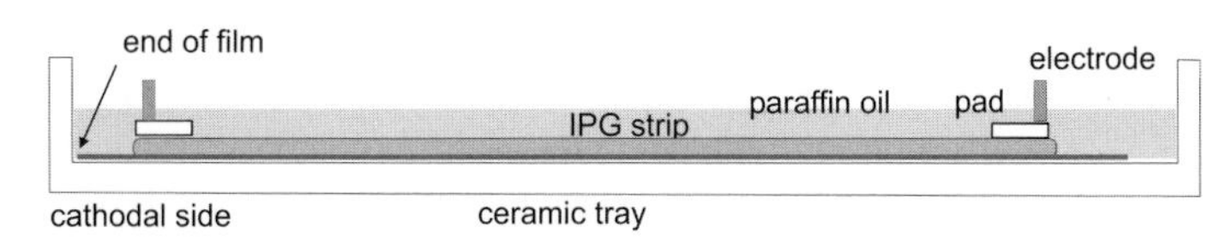

Fig. 4: Arrangement of the cup loading strip holder for running rehydration-loaded IPG strips.

Tab. 3: Instrument settings for cup loading strip holders (cup and rehydration loading)

Rehydration time			0 h			
Temperature			20 °C			
Current per strip			50 µA			
Strip length			18 cm		24 cm	
pH gradient			3–10 L	3–10 NL	3–10 L	3–10 NL
Step 1	step & hold	300 V	3 h		3 h	
Step 2	gradient	1000 V	6 h		6 h	
Step 3	gradient	8000 V	3 h		3 h	
Step 4	step & hold	8000 V	16 kVh	16 kVh	24 kVh	24 kVh
			2 h	2 h	3 h	3 h
Total time			14 h	14 h	15 h	15 h
Course timing:		*start at [h]:*	*13:00*	*13:00*	*13:00*	*13:00*
		stop at [h]:	*3:00*	*3:00*	*4:00*	*4:00*
refocus with 8000 V for:			*15 min*	*15 min*	*15 min*	*15 min*

The total volthours should be less than 50 Vh.

For preparative sample loads (> 1 mg) increase the final focusing step by 15 % of the proposed Volthours.

When an overnight run is finished early in the morning, refocus the proteins by applying 8000 V on the strips for 15 minutes before equilibration, staining, or storage of the strips.

4
Cup loading IEF

For the course: use analytical protein load here.

The strip holders have numbers to allow sample identification. The IPGphor chamber must be horizontally leveled on the bench. IEF separation is carried out at 20 °C.

Figure 5 shows all the parts needed and the proper placements of strips, pads, sample cups, and electrodes.

Never use new strip holders without cleaning them before the first run.

- Be sure, that the cup loading strip holders are carefully cleaned and dried.
- Place strip holders on the cooled electrode contact areas of the power supply with the pointed end on the anodal contact area.
- Starting at the basic side, place the IPG strip – gel side facing up – into the strip holder with the pointed end towards the anode side. Be sure that the protruding film at the basic end touches the end of the strip holder. The aligner protrusions help to keep the strip centered.
- Pipette 4 mL paraffin oil over the strip.

to avoid urea crystallisation and oxygen contact.

The pads must be damp, not wet.

- Cut 5 mm long pads from IEF electrode strips and soak them with deionised water. Blot them on filter paper and place them on top of the ends of the strip. The pads should sit completely on the gel surface. If longer strips are required for removal of salt, there must be an overlapping of at least 5 mm.
- Place the sample cup close to the electrode pad:
  - gel no.1 at the anode,
  - gel no.2 at the cathode,
  - gel no.3 two cups, one at each electrode,

  Continue with the next sample in the same way

This can be done for a course, it is also useful for optimization work for samples to be analysed.

The cups contain now paraffin oil at the bottom.

Each sample cup has four little feet, which touch the bottom of the strip holder; then the funnel sits tight on the gel surface to prevent leakage of the sample.

Note:
The cup can straddle on the alignment protrusions, if necessary.

- Dilute samples with rehydration solution to 50 to 100 μL for optimum protein entry. Divide one sample into two halves for the double cup application.
- Pipette samples into the cups by underlaying them below the paraffin oil.
- Place the electrodes on the pads as shown in figure 5.
- Place the plastic cover on the strip holder.
- Close the safety lid.
- Use the same running conditions, indicated in table 3, as for the rehydration loaded strips (containing preparative amounts).

In higher concentrated samples more proteins tend to aggregate and precipitate.

You can run both types of strips together on one chamber, because most of the separation is voltage gradient controlled.

Cleaning of the cup loading strip holders

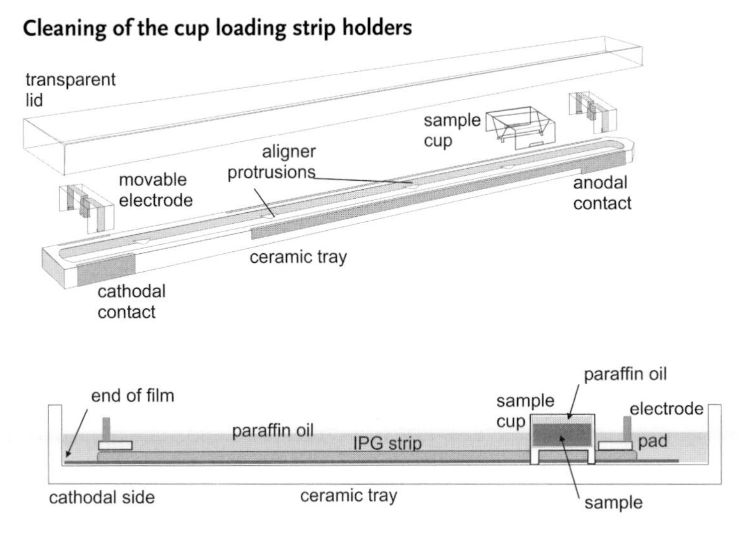

Fig. 5: Schematic drawing of cup loading.

The strip holders must be carefully cleaned after each IEF. The solutions must never dry in the strip holder. Cleaning is very effective, if the strip holders are first soaked a few hours in a solution of 2–5% of the specially supplied detergent in hot water.

Cleaning of the cup loading strip holder is easier than the regular strip holder.

The strip holder slot should be vigorously brushed with a toothbrush using a few drops of undiluted IPGphor Strip Holder Cleaning Solution. Then it is rinsed with deionised water.

Clean strip holders should be handled with gloves to avoid contamination.

> Important:
> ***Strip holders may be baked, boiled or autoclaved. But, because of the specially treated surface they must not be exposed to strong acids or basis, including alkaline detergents.***

Note:
The strip holder must be completely dry before use.

The IPG strips are directly transferred to equilibration in SDS buffer for the second dimension run or stored in a deep freezer at –60 to –80 °C.

Some of the strips can be stained for inspection, whether the IEF step was successful.

5
Staining of IPG strips

Acid violet 17 staining of IPG strips

3% phosphoric acid	21 mL	85 % H_3PO_4	make up to 1 L H_2O_{deion}
11% phosphoric acid	76.1 mL	85 % H_3PO_4	make up to 1 L H_2O_{deion}
1 % stock dye solution	1 g	Acid violet 17	in 100 mL H_2O_{deion} heat to 50 °C on magnetic stirrer
0.1 % Staining solution	1 mL	1 % stock solution	plus 9 mL of 11% phosphoric acid

Staining procedure

- fix for 20 min in 20 % TCA,
- wash for 1 min in 3 % phosphoric acid,
- stain for 10 min in 0.1 % Acid violet 17 staining solution,
- wash 3 × 1 min with $H_2O_{deion,}$
- impregnate with 5 % glycerol, air dry.

Note:
This staining is not reversible!

Reversible imidazole zinc staining of IPG strips

This is not a very good staining procedure for IPG strips, but it offers the possibility to store prerun IPG strips for second dimension runs when no deep freezer is available. Note, that for analytical runs you need double protein load because of protein losses during this procedure.

Fixing solution			
0.2 mol/L imidazole, 0.1 % SDS	2.72 g	Imidazole	
	+ 0.2 g	SDS	dissolve in 200 mL H_2O_{deion}

Staining solution			
0.2 mol/L zinc sulfate			
	5.74 g	Zn SO_4	dissolve in 200 mL H_2O_{deion}

Mobilisation solution			
50 mmol/L EDTA,			dissolve in 200 mL H_2O_{deion}
25 mmol/L Tris; pH 8.3	0.61 g	Tris	adjust to pH 8.3
	+ 3.72 g	EDTA-Na_2	with a few grains of Tris

Staining and preservation procedure

The time of the steps is critical; do not exceed the time periods!

- fix for 15 min in 0.2 mol/L imidazole / 0.1 % SDS on a shaker,
- rinse for 1 min in H_2O_{deion},
- stain (negative) for 5 min in 0.2 mol/L zinc sulfate,
- rinse for 1 min in H_2O_{deion},
- store in a few mL of new fixing solution diluted 1 : 10 with deionised water in a sealed bag.

Background becomes white, bands become visible against a dark background.

Stable in a cool place: at least 6 months!

Mobilisation

- 6 min in 10 mL per IPG strip with vigorous shaking

The background becomes completely transparent.

Now the stip can be equilibrated like a fresh or freezer-stored IPG strip.

Step 3: SDS Polyacrylamide Gel Electrophoresis

In this book only vertical gels are described. Comprehensive instructions for flatbed gels are found in the books quoted on page 63.

In a course the use of ready-made gels is highly recommended, because a good result gives much more confidence in the technique than a so called "trouble shooting gel". The gel casting procedure should, of course, be included in the course. But things can also go wrong, and then the rest of the course would be dependent on the performance of the gel casting procedure.

Wear disposable gloves during the entire procedure.

1
Casting of SDS polyacrylamide gels

Laboratory-made gels are cast in glass cassettes using relatively simply designed equipment. Casting gels on support films requires more sophisticated equipment. Film-backed gels are therefore only prepared in industrial scale. The preparation of gels with two different gel casters is described, and how to cast gradient gels without a pump. The following protocols are limited to the preparation of 1 mm thick gels, for 1.5 mm gels one third more monomer volume is used; there is space for a lower number of cassettes in the casters.

1.1
Stock solutions

Use a readymade Acrylamide, Bis stock solution containing 40%*T*, 3%*C*.

Much safer than weighing acrylamide and Bis powder.

Or prepare

Acrylamide powder could also be used, but is not recommended because of health hazards during weighing.

Acrylamide, Bis solution (40%T, 3%C)		
40 % acrylamide PAGE solution	40 % *T*	1 L
Bis	3 % *C*	12 g
Mix thoroughly in the supplied bottle		

Store in the refrigerator

> Caution!
> ***Acrylamide and Bis are toxic in the monomer form. Avoid skin contact and dispose of the remains ecologically. Polymerise the remains with an excess of ammonium persulfat.***

1.5 mol/L gel buffer Tris-Cl pH 8.8 (4 × concentrated)		
Tris-base	1.5 mol/L	181.8 g
SDS	0.4 % (w/v)	4 g
Water, deionised	dissolve	800 mL
4 mol/L HCl	titrate to	pH 8.8
Water, deionised	make up to	1 L

Store in the refrigerator

A 10 % solution is mixing more homogeneously with the monomer solution than a 40 % solution.

Ammonium persulfate solution (APS)		
Ammonium persulfate	10 % (w/v)	1 g
Water, deionized	dissolve	10 mL

Prepare fresh

TEMED is used undiluted as 100 %, because it is added to the monomer solution at the beginning.

1% Bromophenol blue does not go into solution without Tris.

1 % (w/v) Bromophenol blue solution		
Bromophenol blue	1 % (w/v)	100 mg
Tris-base		60 mg
Water, deionised	dissolve	10 mL

Displacing solution		
Tris-Cl (1.5 mol/L, pH 8.8), SDS	0.375 mol/L	30 mL
Glycerol (85 %)	50 % (v/v)	71 mL
Bromophenol blue (1 %)	0.002 %	240 μL
Water, deionised	make up to	120 mL

Prepare fresh

Overlay solution (water saturated butanol)	
n or *t*-butanol	50 ml
Water, deionised	5 mL

Shake and use top phase to overlay the monomer solutions.

Gel storage solution		
TrisCl (1.5 mol/L, pH 8.8), SDS	0.375 mol/L	500 mL
Water, deionised	make up to	2 L

Store in the refrigerator

1.2 Cassettes for laboratory made gels

Hinged cassettes

The standard cassettes are hinged on one side for easy handling. Note that one of the class plates are offset to make application of the IPG strip easier.

Inspect the cassettes carefully for any dirt or dust particles.

Separate and low fluorescence glass plates

For spot picking, the gels need to be covalently bound to one glass plate. In this case hinged cassettes are not useful, and separate glass plates are used – one with spacers and one without.

For fluorescence detection methods, special borosilicate glass plates are needed, which are usually not hinged.

Treating glass plates with Bind-Silane

The gel is bound to the glass plate without spacers; this one has to be treated.

> Note:
> ***It is important that glass plates are properly clean.***

Before re-use, the plates are placed in a 5% (v/v) Decon™ 90 solution overnight. Do not leave plates standing in this solution for a longer time, because this will cause etching due to the alkali nature of Decon.

- Thoroughly wash the plate to be treated. Any gel fragments from previous gels must be removed. The careful cleaning of the glass plates before casting is important, to ensure a uniform coating with the Bind-Silane and, to avoid keratin contamination.

- Thoroughly rinse the plates with deionised water to remove Decon.
- Dry the plate using a lint-free tissue or leave them to air dry.

Bind-Silane working solution	
Ethanol	8 mL
Acetic acid	200 μL
Bind-Silane	10 μL
Water, deionised	1.8 mL

- Pipette 4 mL of the Bind-Silane solution onto the plate and distribute equally over the plate with a lint-free tissue. Cover the plate to prevent dust contamination and leave to air dry on the bench for one hour.
- Polish the plate with a lint-free tissue, moistened with a small amount of deionised water or ethanol.
- The gels will stay attached to the glass during electrophoresis, staining procedures, scanning, spot picking, and storage.

Prepare the cassettes with gel labels cut from printed paper or overhead film placed at the corner on the opposite of the filling side (see figure 1).

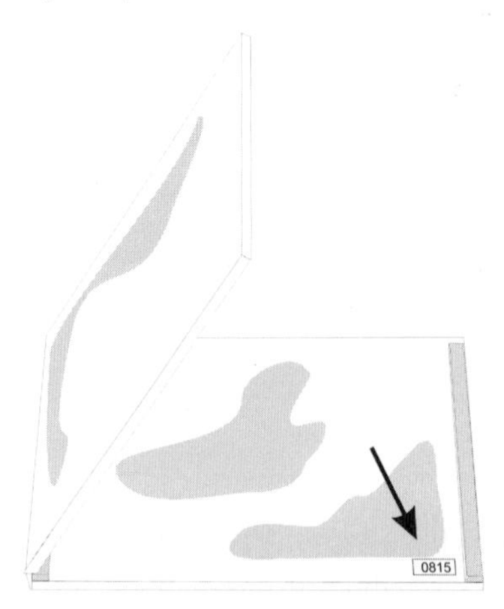

Fig. 1: Hinged cassette with gel label made from paper or overhead film.

1.3 Multiple Gel Caster (up to 14 gels)

1.3.1 Preparation of the gel caster

The gel caster should be set up in a tray, for the eventual case that liquid is overflowing. Never forget to place the plastic separator sheets between the cassettes; they would firmly stick to each other after gel polymerisation. When less than 14 gels should be prepared, block the nonused space with blank cassettes, placing separator sheets between each plate.

Be sure, that gel caster, separator sheets and cassettes are perfectly clean and dry.

Dirty cassettes cause vertical streaking in silver-stained gels.

- Place the gel caster in a tray. Tip it back, so it rests on the support legs.

According to the instruction manual the hinged cassettes should be placed into the caster with the opening edges to the filling side, some proteomics teams have experienced to get a straighter gel surface, when they are inserted with the hinged edge to the filling side.

For non-hinged cassettes the orientation does not matter.

- Place the cassettes alternating with separator sheets into the caster as shown in figure 2, the offset edges off the cassettes up.

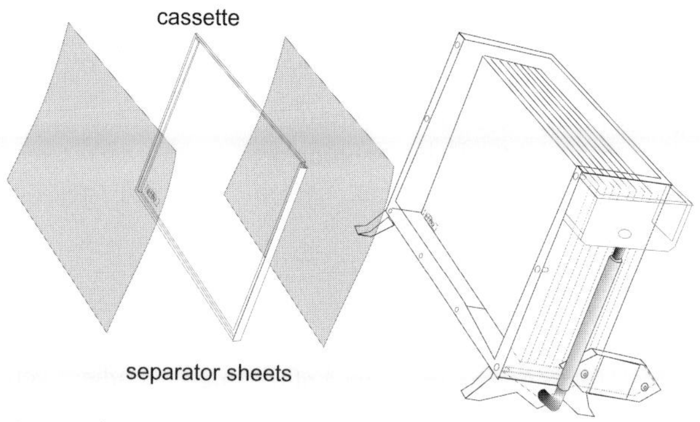

Fig. 2: Placing cassettes and separator sheets into the caster. Here the hinged edges are oriented towards the filling side.

Finally a separator sheet and a filling plate are inserted to complete the stack up to the edge of the caster. Do not overfill it. This would cause pressure on the gels during polymerisation, when the solution starts to get warm.

- Apply a thin film of GelSeal silicone grease on the foam gasket before you place it into the groove of the front plate.

- Turn the first four screws into the bottom holes and place the front plate on the caster with the bottom slots on the screws. Apply the rest of the screws (figure 3) and tighten them evenly. Do not use too much force; the sealing gasket should be compressed evenly to prevent leakage.

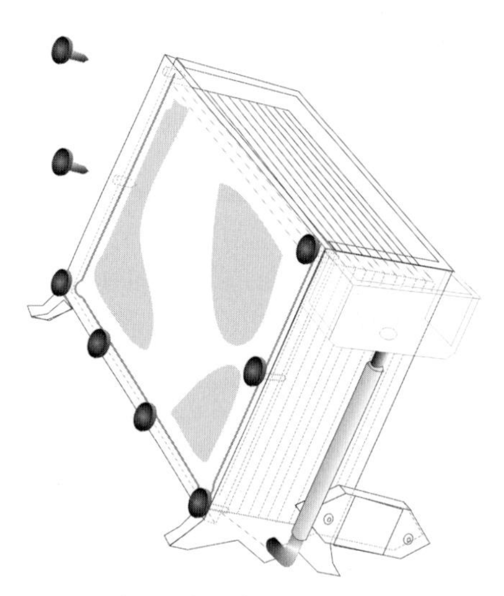

Fig. 3: Closing the gel caster.

- Tip the caster to the front.
- Level the gel caster horizontally.
- Prepare 15 mL of the overlay solution.
- Prepare 120 mL of displacing solution.

1.3.2
Homogeneous gels

Very good results are usually obtained with homogeneous gels, particularly when they run in five hours instead of overnight. Gradient gels need to be prepared in special cases.

- Insert the plastic tube into the hole in the hydrostatic balance chamber. This tube should be connected with a flexible tubing to a funnel, which sits in a ring stand. Figure 4 shows the whole assembly.
- Fill the hydrostatic balance chamber with 100 mL of the displacing solution.

Monomer solutions

- TEMED is added together with the stock solutions, it cannot catalyse the polymerisation without ammonium persulfate.
- It might be necessary to filter the monomer solution before ammonium persulfate is added.
- The ammonium persulfate should not be added before everything is in place, because there are only about 10 minutes left to pour the gel until it starts to gel.

This is not done in a course, but it should be done when highly sensitive mass spectrometry analysis of spots is performed.

Monomer solutions with selected % T, and 3 % C for 14 gels with 1 mm thickness, 900 mL

	7.5 % T	*10 % T*	*12.5 % T*	*15 % T*
Acrylamide, Bis solution (40%*T*, 3%*C*)	169 mL	225 mL	281 mL	338 mL
1.5 mol/L Tris-Cl, pH 8.8, 0.4 % SDS	225 mL	225 mL	225 mL	225 mL
TEMED (100) %	450 μL	450 μL	450 μL	450 μL
Water, deionised, make up to	900 mL	900 mL	900 mL	900 mL
Mix and degas with a water jet pump				
10 % ammonium persulfate	3.6 mL	3.6 mL	3.6 mL	3.6 mL

Add ammonium persulfate short before gel casting.

The amounts of TEMED and ammonium persulfate are based on the author's experience. In order to get reproducible gels with a straight edge, the concentration of TEMED is higher and the concentration of ammonium persulfate is lower than in the instrument instruction. Furthermore, at basic pH, ammonium persulfate can react with the Tris; this effect is minimised by adding more TEMED, and reducing the ammonium persulfate content. For summer laboratories without air conditioning or in warm countries it might be necessary to reduce the ammonium persulfate amount by 10 to 20 %.

It is important to use only high quality of TEMED and ammonium persulfate.

- Add the ammonium persulfate to the monomer solution and mix.
- Pour the gel solution into the funnel, avoiding air bubbles (see figure 4).
- Stop pouring when the level has reached 3 cm below the upper edges of the casting cassettes.
- Remove the plastic tube with the funnel from the balance chamber.

The dense displacing solution will now flow down the connecting tube, and fill the V-chamber and the sloped bottom of the caster. The gel solution level will rise to 1 cm below the cassette edges. A thin blue layer should be visible at the bottom of the cassettes.

Should the volume of the displacing solution in the balance chamber not be sufficient to produce the thin blue layer in the entire area of the caster bottom, and top more displacing solution into the bal-

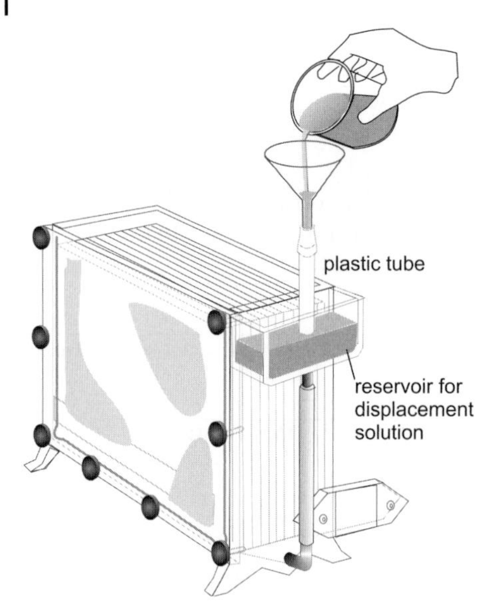

Fig. 4: Casting a homogeneous gel.

ance chamber until the blue layer has distributed over the entire bottom. This saves you from cutting off protruding gel pieces at the bottom edge of each cassette.

- Immediately pipette 1 mL overlay solution into each cassette (see figure 5).

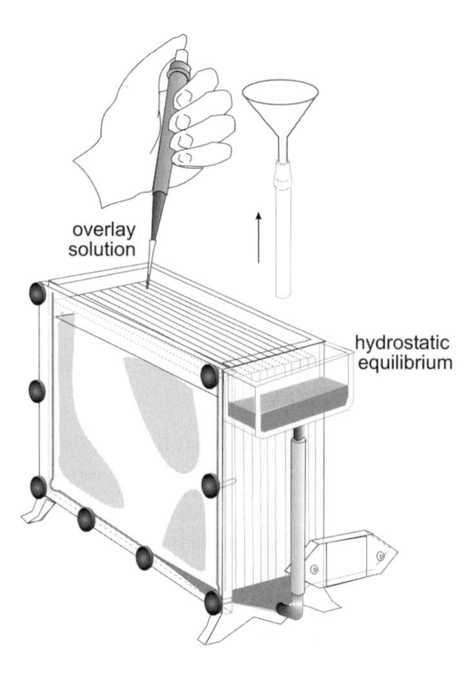

Fig. 5: Removing the tube and overlaying the gel solution.

About 50 mL monomer solution should be left over. Let it polymerise in a small beaker before you discard it.

- Let the gel polymerise for one hour.
- Take the cassettes out and wash the caster. If the gels are left in the caster overnight, the alcohol of the overlay solution will start to corrode the caster material. This does not destroy the instrument, but will look ugly.
- During unloading, rinse the top surface of each gel with distilled water to remove the butanol, and rinse the cassettes to remove excess polyacrylamide.

> Note:
> ***When enough displacing solution had been used, there is no need to cut off protruding gel pieces at the bottom edge of each cassette.***

- Inspect each gel cassette for eventual air bubbles. Gels with air bubbles should not be used.
- Store the usable gels in an airtight container with about 100 mL gel storage solution at room temperature overnight, for complete polymerisation. When the gels are not used during the next day, place the container with the gels into a cold room or a refrigerator.
- Rinse the gel caster and the separator sheet with mild detergent and then with deionised water.
- Let them dry in the air.

After the gel matrix has formed, there is still a silent polymerisation going on, which is temperature dependent. It influences the separation property of the gel.

1.3.3 Gradient gels

Gradient gels are more complicated to prepare than homogeneous gels. When homogeneous gels are run for a short time only, they achieve comparable resolution. However, for some samples, gradient gels may show a better spot distribution. Most instructions describe how to cast gradient gels with a peristaltic pump. The author has experienced that very reproducible gels are achieved without a pump, which safes a lot of work with calibrating, setting up and cleaning. Instead of a pump, gravitation is used; the flow rate is controlled by the level of the laboratory platform carrying the gradient maker and with a pinchcock clamp on the tubing. A pinchcock clamp can close the tubing completely or reduce the opening of the tubing in several steps, thus reducing the flow rate.

Pinchcock clamps are available in different sizes.

Gradient maker set up

- Place the clean and dry gradient maker with a magnetic stirrer on an adjustable laboratory platform or on a shelf board: the level must be at least 10 cm above the edge of the caster.

Note:
For reproducible gels, use always exactly the same level for the gradient maker position. Use always the same shelf board or measure and note the position height of the gradient maker.

- Insert the plastic tube into the hole in the hydrostatic balance chamber.
- Place a pinchcock clamp on the flexible connection tubing, and close it completely.
- Connect the plastic tube with the flexible tubing to the gradient maker.
- Fill the hydrostatic balance chamber with 100 mL of the displacing solution.
- Place a magnetic stir bar – 2 to 3 cm long – into the mixing chamber.
- Close the connecting valve between the reservoir and mixing chamber.

The pinchcock clamp is closed as well.

Monomer solutions for gradient gels

- The dense solution contains 20 % glycerol to stabilise the gradient.
- TEMED is added together with the stock solutions, it cannot catalyze the polymerisation without ammonium persulfate.
- It might be necessary to filter the monomer solution before ammonium persulfate is added.
- The ammonium persulfate should not be added before everything is in place, because there are only about 10 minutes left to pour the gel until it starts to gel.

This is not done in a course, but it should be done when highly sensitive mass spectrometry analysis of spots is performed.

Monomer solutions with selected % *T*, and 3 % *C* for 14 gradient gels with 1 mm thickness:

Light solution, 450 mL

	8 % T	*10 % T*	*12 % T*	*... % T*
Acrylamide, Bis solution (40%*T*, 3%*C*)	90 mL	113 mL	135 mL	. . . mL
1.5 mol/L Tris-Cl, pH 8.8, 0.4 % SDS	113 mL	113 mL	113 mL	113 mL
TEMED (100 %)	225 µL	225 µL	225 µL	225 µL
Water, deionised, make up to	450 mL	450 mL	450 mL	450 mL
Mix and degas with a water jet pump				
10 % ammonium persulfate	1.8 mL	1.8 mL	1.8 mL	1.8 mL

Add ammonium persulfate short before gel casting.

Dense solution, 450 mL

	14 % T	*16 % T*	*20 % T*	*... % T*
Acrylamide, Bis solution (40%*T*, 3%*C*)	158 mL	180 mL	225 mL	. . . mL
1.5 mol/L Tris-Cl, pH 8.8, 0.4 % SDS	113 mL	113mL	113 mL	113 mL
Glycerol (85 %)	110 mL	110 mL	110 mL	110 mL
TEMED (100 %)	225 µL	225 µL	225 µL	225 µL
Water, deionised, make up to	450 mL	450 mL	450 mL	450 mL
Mix and degas with a water jet pump				
10 % ammonium persulfate	1.4 mL	1.4 mL	1.4 mL	1.4 mL

Add ammonium persulfate short before gel casting. Note, that less APS is added to the dense solution!

The amounts of TEMED and ammonium persulfate are based on the author's experience. In order to get reproducible gels with a straight edge, the concentration of TEMED is higher and the concentration of ammonium persulfate is lower than in the instrument instruction. Furthermore, at basic pH, ammonium persulfate can react with the Tris; this effect is minimised by adding more TEMED, and reducing the ammonium persulfate content. For summer laboratories without air conditioning or in warm countries it might be necessary to reduce the ammonium persulfate amount by 10 to 20 %.

It is important to use only high quality of TEMED and ammonium persulfate.

Control of gel polymerisation for gradient gels The start of the polymerisation is controlled with the amount of ammonium persulfate. The catalyst amount must not be too high in order to avoid overheating in the caster. The dense solution must contain less catalyst, because the polymerisation should start at the top and proceed slowly down to the bottom. This is very important for two reasons:

The polymerisation is easier controlled with varying the amount of APS than with TEMED.

- Thermal convection is prevented, which would distort the gradient.
- During polymerisation the gel contracts. If the lower portion would polymerise first, it would pull the upper solutions down, resulting in a curved upper edge.

When conditions are correct, the gels will pull the displacing solution down.

Figure 6 shows the setup for casting a gradient gel.

- Add the ammonium persulfate to the monomer solutions and mix.
- Pour the dense solution into the reservoir.
- Carefully open the slide valve and let just enough solution flow to fill the connector channel. Close the valve. If too much fluid has flown into the mixing chamber, pipette it back to the reservoir.
- Pour the light solution into the mixing chamber.

Check for air bubbles.

Do not mix too fast; air bubbles must be avoided.

- Start the magnetic stirrer.
- Open the pinchcock clamp to the next setting, the one next to close the tubing completely. The light solution should start to flow down the tubing.
- Open the sliding valve of the gradient maker. Now dense solution will flow into the mixing chamber and will be mixed with the light solution.
- Adjust the flow rate to a higher speed with the pinchcock clamp. But the speed must not be too high to avoid mixing of solutions in the V-shaped chamber at the bottom. Casting time should be around 10 minutes.

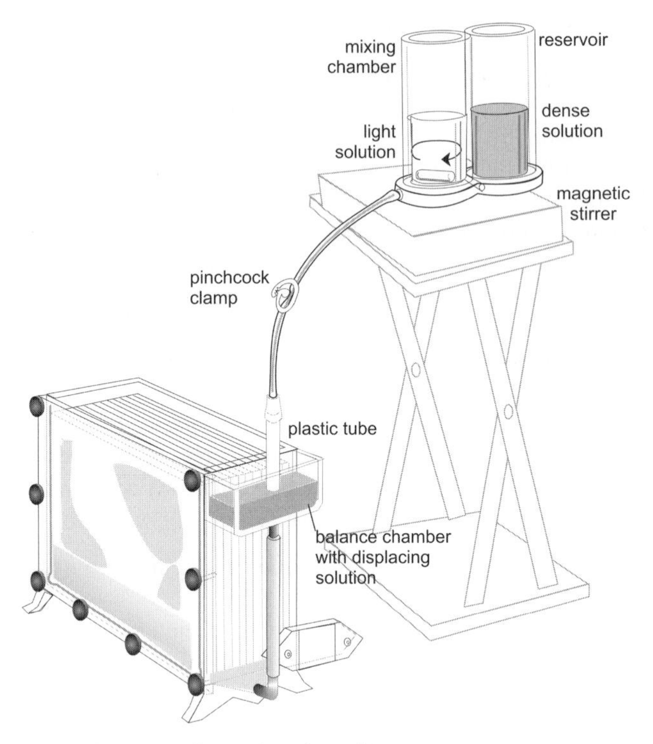

Fig. 6: Casting gradient gels in the multi caster.

After the 10 minutes the gradient maker should be empty. If you estimate, that it will take longer, open the pinchcock to the next position.

When the level has reached 3 cm below the upper edges of the casting cassettes, the gradient maker should become empty.

- Close the pinchcock clamp before the connection tubing runs empty, but not later than the liquid level has reached 3 cm below the upper edges of the cassettes.

- Remove the plastic tube with the connection tubing from the balance chamber and rinse the magnetic stirrer.

If monomer solution is left, fill it in a measuring cylinder; next time, adjust the volumes of the monomers solutions to the appropriate amount. The gels will this time have a slightly flatter gradient slope than intended, but the separation quality will not suffer much.

The dense displacing solution will now flow down the connecting tube, and fill the V-chamber and the sloped bottom of the caster. The gel solution level will rise to 1 cm below the cassette edges. A thin blue layer should be visible at the bottom of the cassettes.

Should the volume of the displacing solution in the balance chamber not be sufficient to produce the thin blue layer in the entire area of the caster bottom, and top more displacing solution into the balance chamber until the blue layer has distributed over the entire bottom. This saves you from cutting off protruding gel pieces at the bottom edge of each cassette.

- Immediately pipette 1 mL overlay solution into each cassette (see figure 5).
- Let the gel *polymerise* for two hours.
- Take the cassettes out and wash the caster. If the gels are left in the caster overnight, the alcohol of the overlay solution will start to corrode the caster material. This does not destroy the instrument, but will look ugly.
- During unloading, rinse the top surface of each gel with distilled water to remove the butanol, and rinse the cassettes to remove excess polyacrylamide.

Note:
When enough displacing solution had been used, there is no need to cut off protruding gel pieces at the bottom edge of each cassette.

- Inspect each gel cassette for eventual air bubbles. Gels with air bubbles should not be used.
- Store the usable gels in an airtight container with about 100 mL gel storage solution at room temperature overnight, for complete polymerisation. When the gels are not used during the next day, place the container with the gels into a cold room or a refrigerator.
- Rinse the gel caster and the separator sheet with mild detergent and then with deionised water.
- Let them dry in the air.

After the gel matrix has formed, there is still a silent polymerisation going on, which is temperature dependent. It influences the separation property of the gel.

1.4
Gel Caster for up to six gels

1.4.1 Preparation of the gel caster

Large gel formats require a casting box, also when only a few gels are prepared. Otherwise the gels would have an uneven thickness, they would be thicker in the center. The caster for few gels is laid flat on the bench for loading. Never forget to place the plastic separator sheets between the cassettes; they would firmly stick to each other after gel polymerisation. When less than 6 gels should be prepared, block the nonused space with blank cassettes, placing separator sheets between each plate.

Dirty cassettes cause vertical streaking in silver-stained gels.

Be sure, that gel caster, separator sheets and cassettes are perfectly clean and dry. Homogeneous gels are cast through the filling channel in the back wall of the caster, gradient gels must be poured through the filler post at the bottom of the front plate. For casting homogeneous gels, leave the V-shaped rubber insert in place; for gradient gels, remove it.

- Place the cassettes alternating with separator sheets into the caster starting with a separator sheet, the offset edges off the cassettes up.

Finally a separator sheet and – eventually – filling sheets are inserted to complete the stack up to the edge of the caster. Do not overfill it. This would cause pressure on the gels during polymerisation, when the solution starts to get warm.

- Apply a thin film of GelSeal silicone grease on the foam gasket before you place it into the groove of the front plate.
- Turn the two screws into the bottom holes and place the front plate on the caster with the bottom slots on the screws. Apply the six clamps and tighten the screws evenly. Do not use too much force; the sealing gasket should be compressed evenly to prevent leakage.
- For casting, place the gel caster upright in a tray, for the eventual case that liquid is overflowing.
- Prepare 7 mL of the overlay solution.

1.4.2 Homogeneous gels

Very good results are usually obtained with homogeneous gels, particularly when they run in five hours instead of overnight. Gradient gels need to be prepared in special cases.

The gel strip produced tin the filling channel can easily be removed.

For homogeneous gels the monomer solution is introduced through the filling channel at the back, no displacing solution is needed.

- Close the filler port at the bottom of the front plate with the cap or with a piece of tubing and a pinchcock clamp.

Monomer solutions

- TEMED is added together with the stock solutions, it cannot catalyze the polymerisation without ammonium persulfate.
- It might be necessary to filter the monomer solution before ammonium persulfate is added.
- The ammonium persulfate should not be added before everything is in place, because there are only about 10 minutes left to pour the gel until it starts to gel.

This is not done in a course, but it should be done when highly sensitive mass spectrometry analysis of spots is performed.

Monomer solutions with selected % T, and 3 % C for 6 gels with 1 mm thickness, 450 mL

	7.5 % T	*10 % T*	*12.5 % T*	*15 % T*
Acrylamide, Bis solution (40%*T*, 3%*C*)	84 mL	113 mL	141 mL	169 mL
1.5 mol/L Tris-Cl, pH 8.8, 0.4 % SDS	113 mL	113 mL	113 mL	113 mL
TEMED (100) %	225 µL	225 µL	225 µL	225 µL
Water, deionised, make up to	450 mL	450 mL	450 mL	450 mL
Mix and degas with a water jet pump				
10 % ammonium persulfate	1.8 mL	1.8 mL	1.8 mL	1.8 mL

Add ammonium persulfate short before gel casting.

The amounts of TEMED and ammonium persulfate are based on the author's experience. In order to get reproducible gels with a straight edge, the concentration of TEMED is higher and the concentration of ammonium persulfate is lower than in the instrument instruction. Furthermore, at basic pH, ammonium persulfate can react with the Tris; this effect is minimised by adding more TEMED, and by reducing the ammonium persulfate content. For summer laboratories without air conditioning or in warm countries it might be necessary to reduce the ammonium persulfate amount by 10 to 20 %.

It is important to use only high quality of TEMED and ammonium persulfate.

- Add the ammonium persulfate to the monomer solution and mix.
- Pour the gel solution into the funnel, avoiding air bubbles (see figure 7).

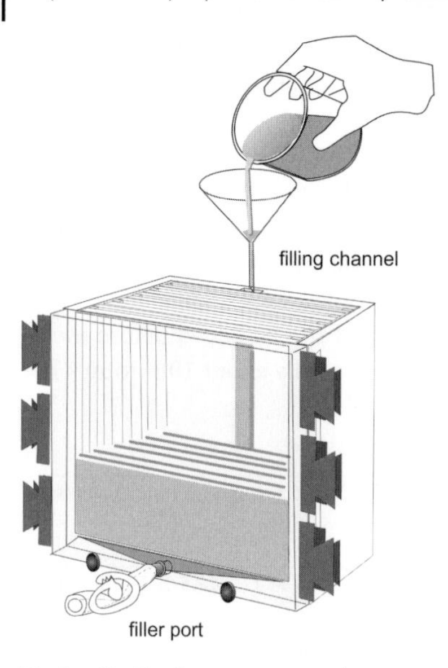

Fig. 7: Casting homogeneous gels.

- Stop pouring when the level has reached 1 cm below the upper edges of the casting cassettes.
- Immediately pipette 1 mL overlay solution into each cassette (see figure 8).

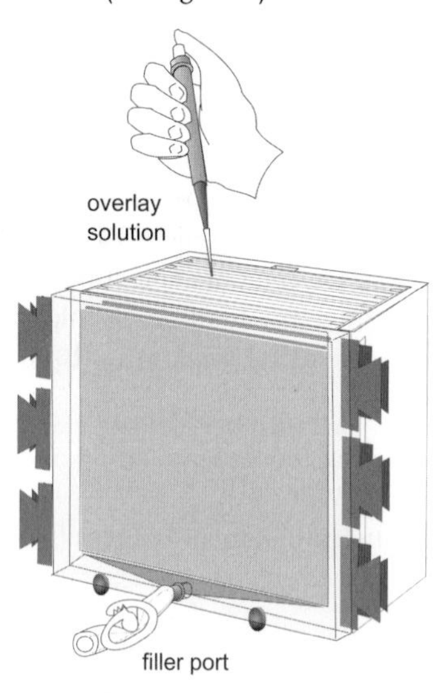

Fig. 8: Overlaying the gel solution.

Monomer solution, which has been left over, is polymerised in a small beaker before it is discarded.

- Let the gel polymerise for one hour.
- Take the cassettes out and wash the caster. If the gels are left in the caster overnight, the alcohol of the overlay solution will start to corrode the caster material. This does not destroy the instrument, but will look ugly.
- During unloading, rinse the top surface of each gel with distilled water to remove the butanol, and rinse the cassettes to remove excess polyacrylamide.
- Inspect each gel cassette for eventual air bubbles. Gels with air bubbles should not be used.
- Store the usable gels in an airtight container with about 100 mL gel storage solution at room temperature overnight, for complete polymerisation. When the gels are not used during the next day, place the container with the gels into a cold room or a refrigerator.
- Rinse the gel caster and the separator sheets with mild detergent and then with deionised water.
- Let them dry in the air.

After the gel matrix has formed, there is still a silent polymerisation going on, which is temperature dependent. It influences the separation property of the gel.

1.4.3 Gradient gels

Gradient gels are more complicated to prepare than homogeneous gels. When homogeneous gels are run for a short time only, they achieve comparable resolution. However, for some samples, gradient gels may show a better spot distribution. Most instructions describe how to cast gradient gels with a peristaltic pump. The author has experienced that very reproducible gels are achieved without a pump, which safes a lot of work with calibrating, setting up and cleaning. Instead of a pump, gravitation is used; the flow rate is controlled by the level of the laboratory platform carrying the gradient maker and with a pinchcock clamp on the tubing. A pinchcock clamp can close the tubing completely or reduce the opening of the tubing in several steps, thus reducing the flow rate.

Pinchcock clamps are available in different sizes.

For gradient gels the monomer solution has to be introduced through the filling port on the bottom of the front plate. In this case we need to fill the dead volume of the V-shaped chamber in the bottom of the caster with displacing solution.

- Remove the V-shaped rubber insert from the caster.
- Prepare 200 mL of displacing solution.

Gradient maker set up

- Place the clean and dry gradient maker with a magnetic stirrer on an adjustable laboratory platform or on a shelf board: the level must be at least 10 cm above the edge of the caster.

Note:
For reproducible gels, use always exactly the same level for the gradient maker position. Use always the same shelf board or measure and note the position height of the gradient maker.

- Place a pinchcock clamp on the flexible connection tubing, and close it completely.
- Connect the filling port with the flexible tubing to the gradient maker.
- Place a magnetic stir bar – 2 to 3 cm long – into the mixing chamber.
- Close the connecting valve between the reservoir and mixing chamber.

The pinchcock clamp is closed as well.

Monomer solutions for gradient gels

- The dense solution contains 20 % glycerol to stabilise the gradient.
- TEMED is added together with the stock solutions, it cannot catalyze the polymerisation without ammonium persulfate.
- It might be necessary to filter the monomer solution before ammonium persulfate is added.
- The ammonium persulfate should not be added before everything is in place, because there are only about 10 minutes left to pour the gel until it starts to gel.

This is not done in a course, but it should be done when highly sensitive mass spectrometry analysis of spots is performed.

Monomer solutions with selected %T, and 3 % C for 14 gradient gels with 1 mm thickness:

Light solution, 450 mL

	8 % T	*10 % T*	*12 % T*	*... % T*
Acrylamide, Bis solution (40%*T*, 3%*C*)	42 mL	53 mL	63 mL	. . . mL
1.5 mol/L Tris-Cl, pH 8.8, 0.4 % SDS	53 mL	53 mL	53 mL	53 mL
TEMED (100 %)	105 μL	105μL	105 μL	105 μL
Water, deionised, make up to	210 mL	210 mL	210 mL	210 mL
Mix and degas with a water jet pump				
10 % ammonium persulfate	840 μL	840 μL	840 μL	840 μL

Add ammonium persulfate short before gel casting.

Dense solution, 450 mL

	14 % T	*16 % T*	*20 % T*	*... % T*
Acrylamide, Bis solution (40%*T*, 3%*C*)	74 mL	84 mL	105 mL	. . . mL
1.5 mol/L Tris-Cl, pH 8.8, 0.4 % SDS	53 mL	53 mL	53 mL	53 mL
Glycerol (85 %)	50 mL	50 mL	50 mL	50 mL
TEMED (100 %)	105 μL	105 μL	105 μL	105 μL
Water, deionised, make up to	210 mL	210 mL	210 mL	210 mL
Mix and degas with a water jet pump				
10 % ammonium persulfate	670 μL	670 μL	670 μL	670 μL

Add ammonium persulfate short before gel casting. Note, that less APS is added to the dense solution!

The amount of TEMED and ammonium persulfate are based on the author's experience. In order to get reproducible gels with a straight edge, the concentration of TEMED is higher and the concentration of ammonium persulfate is lower than in the instrument instruction. Furthermore, at basic pH, ammonium persulfate can react with the Tris; this effect is minimised by adding more TEMED and reducing the ammonium persulfate content. For summer laboratories without air conditioning or in warm countries it might be necessary to reduce the ammonium persulfate amount by 10 to 20 %.

It is important to use only high quality of TEMED and ammonium persulfate.

Control of gel polymerisation for gradient gels The start of the polymerisation is controlled with the amount of ammonium persulfate. The catalyst amount must not be too high in order to avoid overheating in the caster. The dense solution must contain less catalyst, because the polymerisation should start at the top and proceed slowly down to the bottom. This is very important for two reasons:

The polymerisation is easier controlled with varying the amount of APS than with TEMED.

- Thermal convection is prevented, which would distort the gradient.
- During polymerisation the gel contracts. If the lower portion would polymerise first, it would pull the upper solutions down, resulting in a curved upper edge.

When conditions are correct, the gels will pull the displacing solution down.

Figure 9 shows the setup for casting a gradient gel in the caster for six gels.

- Add the ammonium persulfate to the monomer solutions and mix.
- Pour the dense solution into the reservoir.
- Carefully open the slide valve and let just enough solution flow to fill the connector channel. Close the valve. If too much fluid has flown into the mixing chamber, pipette it back to the reservoir.
- Pour the light solution into the mixing chamber.

Check for air bubbles.

Do not mix too fast; air bubbles must be avoided.

- Start the magnetic stirrer.
- Open the pinchcock clamp to the next setting, the one next to close the tubing completely. The light solution should start to flow down the tubing.
- Open the sliding valve of the gradient maker. Now dense solution will flow into the mixing chamber and will be mixed with the light solution.
- Adjust the flow rate to a higher speed with the pinchcock clamp. But the speed must not be too high to avoid mixing of solutions in the V-shaped chamber at the bottom. Casting time should be around 10 minutes.

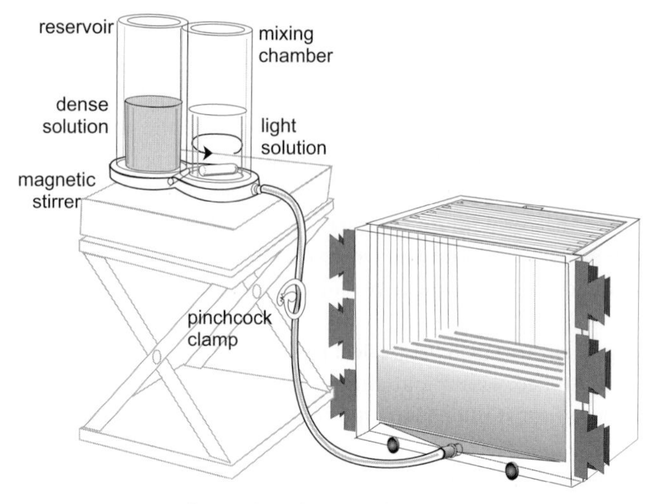

Fig. 9: Casting gradient gels in the caster for six gels.

After the 10 minutes the gradient maker should be empty. If you estimate, that it will take longer, open the pinchcock to the next position.

- Close the pinchcock clamp before the connection tubing runs empty.
- Close the sliding valve between the mixing chamber and the reservoir.
- Pour the 200 mL displacing solution into the mixing chamber of the gradient maker.
- Open the pinchcock clamp to the same position like casting the gradient.
- When the monomer solutions have reached the desired level, which is 1 cm below the upper edges of the cassettes, close the pinchcock clamp.

The displacing solution – visible because of the added bromophenol blue – should fill the V-shaped chamber completely. This saves you from cutting off protruding gel pieces at the bottom edge of each cassette.

- Immediately pipette 1 mL overlay solution into each cassette (see figure 8).
- Let the gel polymerise for two hours.
- Take the cassettes out and wash the caster. If the gels are left in the caster overnight, the alcohol of the overlay solution will start to corrode the caster material. This does not destroy the instrument, but will look ugly.
- During unloading, rinse the top surface of each gel with distilled water to remove the butanol, and rinse the cassettes to remove excess polyacrylamide.

Note:
When enough displacing solution had been used, there is no need to cut off protruding gel pieces at the bottom edge of each cassette.

- Inspect each gel cassette for eventual air bubbles. Gels with air bubbles should not be used.
- Store the usable gels in an airtight container with about 100 mL gel storage solution at room temperature overnight, for complete polymerisation. When the gels are not used during the next day, place the container with the gels into a cold room or a refrigerator.
- Rinse the gel caster and the separator sheet with mild detergent and then with deionised water.
- Let them dry in the air.

After the gel matrix has formed, there is still a silent polymerisation going on, which is temperature dependent. It influences the separation property of the gel.

2 Inserting ready-made gels into cassettes

Ready-made gels on a film support are supplied in airtight aluminum bags and need to be inserted into specially designed reusable cassettes. They are run in a vertical mode in the vertical systems for large gels.

- Place the cassette for ready-made gels on the bench top with the hinge down, plastic frame to the left Clean the inner side of the glass plate thoroughly with a 2 % (w/v) SDS water solution, rinse it with water and dry it completely with a lint-free tissue paper. Pipette 1 ml gel buffer onto the glass plate as a streak along the spacer of the closing side (see figure 10).

Do not use more than 1 mL, because excess liquid between gel surface and glass plate has to be removed completely with (surface and glass plate has to be removed completely with a roller.

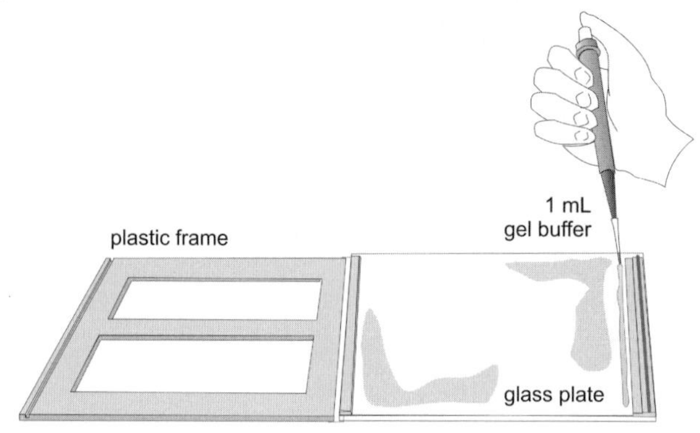

Fig. 10: Application of 1 mL gl buffer on the glass plate along the spacer of the closing side.

- Open the gel package by cutting around the gel on three sides at about 1 cm from the edge to avoid cutting the gel or the support film. Remove the gel from the package.

The gel is cast onto a plastic support film and does not cover the film entirely. The support film protrudes approximately 15 mm beyond the top (– or cathodal) edge of the gel and approximately 5 mm at the lateral sides.

- Remove the protective plastic sheet from the gel. Handling the gel only by the side support film margins, hold it, gel-side down, over the glass plate. Align the right edge of the gel with the inner edge of the side spacer next to the opening side, flex the gel downward slightly and lower it slowly toward the glass plate from right to left (see figure 11). Take care that the bottom edge of the gel is flush of the bottom edge of the glass plate. The protruding side support film margins (not the gel) should rest on top of the side spacers.

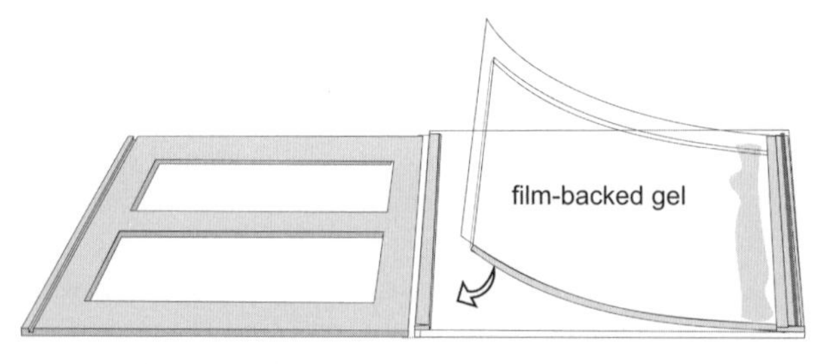

Fig. 11: Placing the film-backed gel into the cassette.

With this procedure almost no air bubbles are caught between glass plate and gel surface. The gel edge near to the opening side will take up a few microliters of gel buffer; this prevents eventual current leakage at the closing side, when gel edge should be slightly thinner than the cassette spacer.

- Press out any bubbles or liquid from between the gel and the glass with a roller (figure 12). Press firmly against the plastic support film with the roller and roll over the entire gel.
- Snap the plastic frame to the glass plate and press the edges tightly together. Ensure that the cassette is closed completely: an incompletely closed cassette causes a strongly curved front because of current leakage.

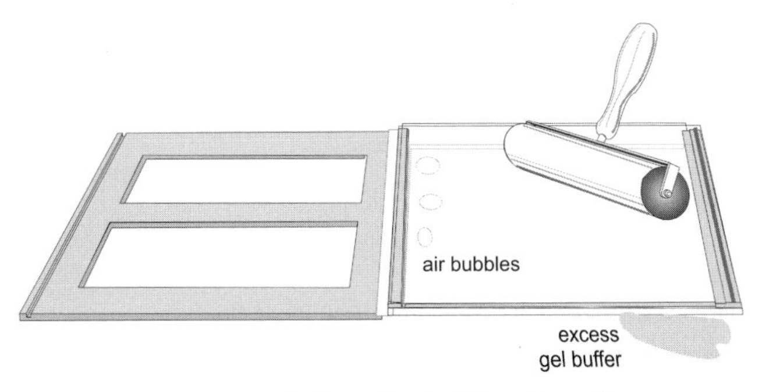

Fig. 12: Removing excess buffer and air bubbles with a roller.

- Keep them in the cassette rack upside down to prevent drying of the upper edge, to which the IPG strip will be applied.

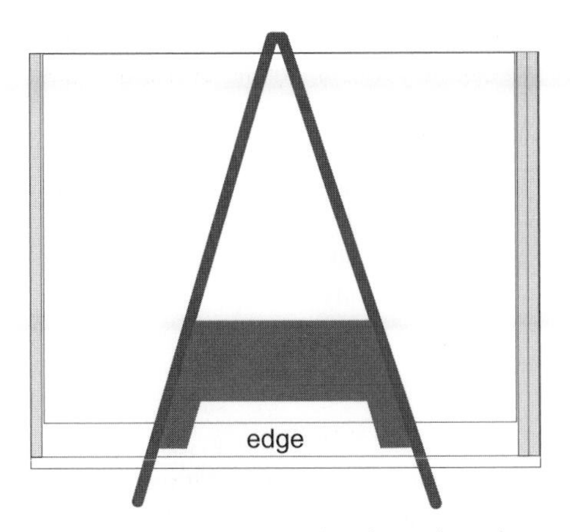

Fig. 13: The cassettes are placed into the rack upside down to prevent drying of the surface.

3
Preparation of the SDS electrophoresis equipment

3.1
Stock solutions for the running buffers

The cathodal buffer is identical for laboratory-made gels containing Tris-Cl pH 8.8 and the readymade gels containing PPA-Cl pH 7.0:

Cathodal buffer (5 to 10 × conc) = "Laemmli buffer":

Tris- base	0.25 mol/L	30.4 g
Glycine	1.92 mol/L	144.0 g
SDS	1 % (w/v)	10 g
Water, deionised	make up to	1L

Do not titrate the Tris-glycine buffer!

Store at room temperature.

> Note:
> ***When the instruments are loaded with more than half of the possible gel number, use 2 × cathodal buffer instead of 1 × buffer to prevent buffer depletion.***

In the standard procedure the same running buffer for the anode and cathode tank is used.

This saves costs, particularly when large volumes of anodal buffer are used.

1 × cathodal buffer is usually used as anodal buffer for Tris-Cl pH 8.8 gels. When care is taken, that anodal and cathodal buffers do not mix during electrophoresis, the anodal buffer concentrate can be composed as follows:

Anodal buffer (10 × conc) for Tris-Cl buffered gels

Tris- base	0.25 mol/L	30.4 g
SDS	1 % (w/v)	10 g
Water, deionised	make up to	800 mL
4 mol/L HCl	titrate to	pH 8.4
Water, deionised	make up to	1L

Store at room temperature.

Also in this case, care must be taken, that anodal and cathodal buffers do not mix during the run.

When the anodal buffer is mixed with the used cathodal buffer after the run, there will be enough Tris in the buffer for repeated use as anodal buffer. This saves work and reagent costs. Of course, the cathodal buffer must be new for each electrophoretic run.

Anodal buffer for ready-made gels: The ready-made gels contain PPA-Cl. For this buffer system, the anodal buffer must be completely free of Tris ions. In a buffer kit 75 mL of the 100 × concentrated buffer are supplied, which is enough for one run in the high-through-put instrument and two runs in the instrument for up to six gels.

Note, this is a highly viscous fluid.

Anodal buffer (100 × conc) for PPA-Cl buffered gels		
Diethanolamine	5 mol/L	167 mL
Acetic acid	5 mol/L	100 mL
Water, deionised	make up to	300 mL

Store at room temperature.

3.2
Setting up the integrated high-throughput instrument

- Place the separation unit on a levelled bench. Check with a spirit level. If necessary, adjust the level with inserting plastic sheets below some feet of the separation unit.
- Turn the pump valve at the back of the separation unit to "circulate".
- Pour 750 mL Tris-containing anodal buffer or – when running readymade gels is intended – 75 mL DEA-containing buffer concentrate into the lower buffer tank.
- Fill the tank up to the mark 7.5 L with deionised water; in this way rinsing the buffer concentrate off the buffer seal tubings.

When identical buffers are used for the anode and the cathode, you can pour 950 mL buffer concentrate into the tank and fill up to the mark 9.5 L with deionised water.

- Switch the control unit on, set the temperature to 25 °C, and set the pump to "ON". The pump starts to circulate the liquid, mixes the concentrate with the water, and the buffer temperature will be adjusted to 25 °C.

3.3
Setting up the six gel instrument

- Take the cassette carrier and the upper buffer chamber out of the instrument.
- Connect the tubings to a circulating thermostat, which has been set to 10 °C.
- Pour 450 mL Tris-containing anodal buffer or – when running ready-made gels is intended – 45 mL DEA-containing buffer concentrate into the lower buffer tank.
- Fill 4 L deionised water into the tank.
- Plug the cable for the pump in. The pump starts to circulate the liquid, mixes the concentrate with the water.

4
Equilibration of the IPG strips and transfer to the SDS gels

4.1
Equilibration

Equilibration stock solution		
6 mol/L urea		180 g
30% glycerol (v/v)	85% solution	176 mL
50 mmol/L Tris HCl pH 8.8	1.5 mol/L stock solution	17 mL
0.01% Bromophenol Blue		50 mg
Water, deionised	make up to	500 mL
Dissolve completely		
Then add		
2 % (w/v) SDS		10.0 g

Do not add the SDS from the beginning; urea and SDS together would need a long time to become dissolved.

Store in 50 mL aliquots in a freezer at –20 °C or deeper.

- Place each IPG strip into an individual equilibration tube and equilibrate with 10 mL buffer.

For the course it does not matter, when two strips containing the same sample are equilibrated together in the same tube.

Equilibration		
15 min	10 mL equilibration stock solution	plus 100 mg DTT
15 min	10 mL equilibration stock solution	plus 250 mg iodoacetamide

4.2
Application of the IPG strips onto the SDS gels

Agarose sealing solution	
0.5 % Agarose M or NA	0.5 g
0.01 % Bromophenol Blue	10 mg
SDS cathode buffer (1 × conc)	100 mL

- Heat on a heating stirrer until agarose is completely dissolved. Pipet 2 mL aliquots into reaction cups and store them at room temperature. Alternatively use a microwave oven. For ready-made gels aliquoted sealing solutions are supplied with the buffer kit.

Both types of cassettes (those for lab-cast and for pre-cast gels) have a “longer” glass plate (see figure 14).

- Lay the cassette on the bench with the longer glass plate down, the protruding edge oriented towards the operator.
- Place the strip with the acidic end to the left, gel surface up onto the protruding edge of the longer glass plate as shown in figure 14. When the strip is applied gel surface up, with the acidic end to the left side, the gels are assembled correctly in the standard orientation as indicated on page 76.

For film-backed gels it is important to slide the IPG strip into the cassette with its film-side towards the glass and its gel surface towards the film backing. This does not matter for the laboratory-made gel cassettes, however, a standardised orientation should always be applied.

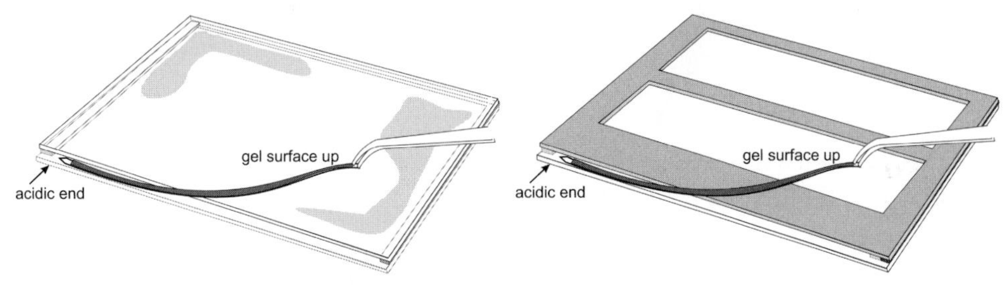

Fig. 14: Application of the equilibrated IPG strip into the SDS gel cassette.

- With the forceps move the strip into the cassette slot.
- Place the cassette into the rack, now with the IPG strip-supporting edge upside.
- With a thin plastic ruler, gently push the IPG strip down so that the entire lower edge of the IPG strip is in contact with the top surface of the SDS gel (see figure 15). Ensure that no air bubbles are trapped between the IPG strip and the slab gel surfaces. Do not push the IPG strip down too hard, it would force a gap between gel surface and glass plate and damage the gel.

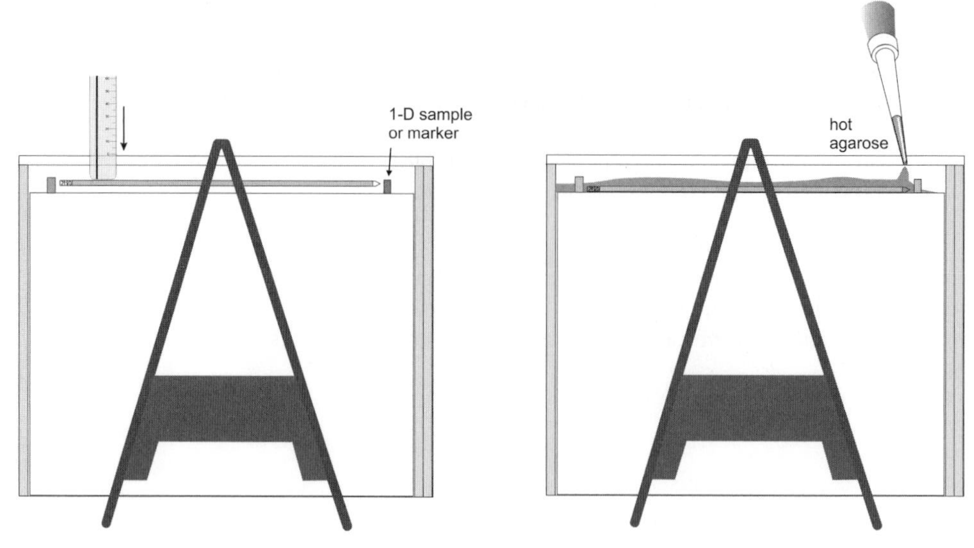

Fig. 15: Pushing the IPG strip onto the SDS gel with a ruler and sealing the strip and the 1-D sample pads in place with hot agarose.

4.3
Application of molecular weight marker proteins and 1-D samples.

Note:
Close to the spacers there are always some edge effects causing bent zones.

If molecular weight markers and 1-D samples should be applied, choose 18 cm long IPG strips instead of 24 cm strips. They should be placed at least 1.5 cm apart from the spacers to achieve straight bands.

Best results are obtained when the molecular weight marker protein solution is mixed with an equal volume of a hot 1% agarose solution prior to application to the IEF sample application piece. The resultant 0.5% agarose will gel and prevent the marker proteins from diffusing laterally.

Other alternatives are to apply the markers to a paper IEF sample application piece in a volume of 15 to 20 μL. Place the IEF application piece on a glass plate and pipette the marker solution onto it, then pick up the application piece with forceps and apply to the top surface of the gel next to one end of the IPG strip.

4.4
Seal the IPG strip and the SDS gel

- Melt each aliquot as needed in a 100 °C heat block (each gel will require 1 to 1.5 mL). Allow the agarose to cool until the tube can be hold by fingers (60 °C) and then slowly pipette the amount required to seal the IPG strip in place. Pipetting slowly avoids introducing bubbles.

For ready-made gels the agarose will also seal the narrow gap between the spacer near the hinge and the gel edge. The agarose must not be too hot, other wise it will run down through the gap before it gels. Also, in general, carbamylation of proteins must be avoided.

Carbamylated peptides would cause problems for protein identification in mass spectrometry.

5
The SDS electrophoresis run

5.1
The integrated high-throughput instrument

Note:
Never switch the pump on without liquid in the tank.

- Wet the tubings of the buffer seal with deionised water.
- Insert the cassettes between the tubings of the buffer seal, starting at the back. Slide them down to the bottom.

Do not force the cassettes down; this could damage the buffer seal. Take care, that the silicone tubings are not bent and stretched down; this would cause current leakage and buffer mixing. If you detected a bent tubing, move the upper edges of the two neighbouring cassettes slightly back and forward, to release the tubing.

- When less than 12 gels are run, insert blank cassettes into the free positions. In the front, however, a gel cassette should be inserted: this makes it easier to watch the migration of the bromophenol blue front during the run.

Turn readymade gel cassettes with the glass plates to the front, because the plastic frames are not transparent.

When all cassettes are in place, the level of the anodal buffer should be 3 to 5 mm below the lower buffer seal edge. If the level is lower, remove on cassette, add some deionised water, and slide the cassette in again.

If the level is higher, and you use the identical running buffer, it does not matter. If you use an anodal buffer different from the cathodal buffer: open the draining valve for a very short time and close it again immediately to pump the excess liquid out.

- Make up 2.5 L cathodal buffer by diluting and mixing the concentrate:
  - 1 × cathodal buffer for 1 to 6 gels (250 mL concentrate plus 2.25 L deionised water).
  - 2 × cathodal buffer for more than 6 gels, to prevent buffer depletion (500 mL concentrate plus 2 L deionised water).
- Pour the 2.5 L cathodal buffer into the upper tank.
- Close the safety lid, which contains the cathodal electrodes.

The running conditions are the same for both ready-made and laboratory-made SDS gels:

Overnight runs (set temperature to 25 °C		
Step 1	2 W per gel	Over night

Fast runs		
Setting of the programmable power supply (set temperature to 25 °C):		
Step 1	2 W per gel	45 minutes
Step 2	19 to 30 W per gel, maximum 180 W	3.5 hours to 5 hours

In a warm laboratory the buffer temperature can rise to 30 °C. This does not create a problem. When the temperature gets higher, the power setting should be reduced.

Note:
The running time is also dependent on the water quality. Water with more ions causes slower runs.

After the run

- Switch off the electric field.
- Open the safety lid.
- Remove the cassettes.
- Open the cassettes only with a piece of plastic, like a Wonderwedge™.

If you use metal spatulas or knives, you can damage the glass plates.

- Remove the gels from the cassette; be very careful, when gels are not attached to a glass plate or a support film. Place the gels immediately into fixing solution or the hot Coomassie Blue solution (see next chapter).
- Rinse the cassettes with mild detergent, tap water and deionised water and let them dry in the air.

Or dry them with lint-free tissue paper.

- Pump the used buffer out by opening the valve: Be sure that the end of the tubing is placed into a bucket or a sink. Switch the pump off immediately when the tank is empty.

Do not walk away during this procedure!

When the anodal buffer should be re-used (only possible for Tris-Cl gels), pump only 2.5 L out. When the PPA-Cl gels are run, the buffer cannot be re-used. The Tris ions have to be completely removed from the lower buffer tank by filling several times a few liters of water in and pumping it out.

- Rinse the cathode panels on the hinged safety lid with deionised water.
- Wash the lower buffer tank several times with tap water and then with deionised water to remove any disturbing ions.

5.2
The six-gel instrument with standard power supply

Note:
Never plug the pump cable in without liquid in the tank.

- Insert the cassettes into the cassette carrier and place it into the tank.
- When less than 6 gels are run, insert blank cassettes into the free positions. In the front, however, a gel cassette should be inserted: this makes it easier to watch the migration of the bromophenol blue front during the run.

Turn readymade gel cassettes with the glass plates to the front, because the plastic frames are not transparent.

When all cassettes are in place, the level of the anodal buffer should have reached the mark "LBC start fill". If necessary, add deionised water to reach the mark. The filling procedure is shown in detail in figure 16.

- Make up 800 mL cathodal buffer by diluting and mixing the concentrate:
  - 1 × cathodal buffer for running 1 to 3 gels (80 mL concentrate plus 720 mL deionised water).
  - 2 × cathodal buffer for more than 3 gels, to prevent buffer depletion (160 mL concentrate plus 640 mL deionised water).

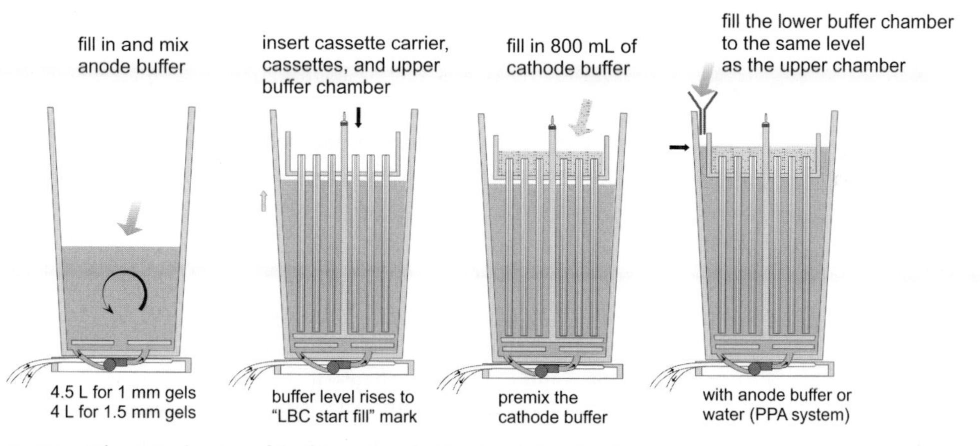

Fig. 16: Schematic drawing of the filling steps for the six-gel chamber. It is important to have a hydrostatic balance between upper and lower buffer.

- Pour the 800 mL cathodal buffer into the upper tank.
- Immediately fill more water or buffer into the lower buffer chamber to adjust the liquid level to the level in the upper

buffer. The hydrostatic balance is necessary to prevent leakage of the upper buffer tank and mixing of the buffers (see figure 16).

- Put the safety lid containing the cathodal electrodes in place and connect the cables to the power supply.

The running conditions are the same for both ready-made gels and laboratory-made SDS gels:

External thermostat is set to 10 °C

Overnight runs:		
Step 1	2 W per gel	Over night

Fast runs			
Setting of a standard power supply:			
Step 1	2 W per gel	45 min	V and mA to max
Step 2	100 W (max 30 W per gel)	3.5 hours to 5 hours	V and mA to max

Note:
The running time is also dependent on the water quality. Water with more ions causes slower runs.

After the run

- Switch off the power supply and unplug the connecting cables.
- Remove the safety lid.
- Remove the entire cassette carrier and carry it to a sink. Then remove the upper buffer tank, and the cassettes from the carrier.

When the anodal buffer should be re-used (only possible for Tris-Cl gels), the upper buffer must be collected and poured into the anodal buffer in order to bring the Tris ions back.

If you use metal spatulas or knives, you can damage the glass plates.

- Open the cassettes only with a piece of plastic, like a Wonderwedge™.
- Remove the gels from the cassette; be very careful, when gels are not attached to a glass plate or a support film. Place the gels immediately into fixing solution or the hot Coomassie Blue solution (see next chapter).

Or dry them with lint-free tissue paper.

- Rinse the cassettes with mild detergent, tap water and deionised water and let them dry in the air.
- Unplug the pump cable.
- Pour the buffer out and wash the lower buffer tank several times with tap water and then with deionised water to remove any disturbing ions.

When the PPA-Cl gels are run, the buffer cannot be re-used. The Tris ions have to be completely removed from the lower buffer tank by filling several times a few liters of water in and pumping it out.

- Rinse the cathode panels on the safety lid with deionised water.

Step 4: Staining of the gels

Only a selection of staining techniques will be described, those which are most useful for fast results, good sensitivity, image analysis, quantification, and mass spectrometry compatibility.

1
Colloidal Coomassie Brilliant Blue staining

This method has the highest sensitivity of all Coomassie Blue staining methods (ca. 30 ng per band), but takes at least overnight (Neuhoff *et al.* 1988).

Neuhoff V, Arold N, Taube D and Ehrhardt W. Electrophoresis 9 (1988) 255–262.

1.1
Solutions

Fixing solution		
o-Phosphoric acid (85 %)	1.3 % (w/v)	5 mL
Methanol	20 % (v/v)	100 mL
Water, deionised	make up to	500 mL

Prepare fresh

Staining stock solution A		
o-Phosphoric acid (85 %)	2 % (w/v)	9.5 mL
Ammonium sulfate	10 % (w/v)	40 g
Water, deionised	make up to	400 mL

Stock staining solution B		
Coomassie Brilliant Blue G-250	5 % (w/v)	2.5 g
Water, deionised	make up to	50 mL

Staining solution, freshly prepared:

- Mix 10 mL of stock staining solution B with 400 mL of stock staining solution A.
- Add 100 mL methanol.

The staining solution should never be filtered because the colloidal dye particles formed are retained on the filter.

Neutralization solution		
Tris-base	0.1 M	6 g
Water, deionised	make up to	500 mL
o-Phosphoric acid	titrate to	pH 6.5

Washing solution		
Methanol		125 mL
Water, deionised	make up to	500 mL

Stabilizing solution		
Ammonium sulfate		100 g
Water, deionised	make up to	500 mL

1.2
Staining

1. Fix for 60 minutes in fixing solution.
2. Stain overnight with staining solution.
3. Transfer gel into neutralization buffer for 1 – 3 min
4. Wash with 25 % methanol for less 1 min.
5. Transfer gel into stabilizing solution.

It is said that sensitivity close to a good silver staining procedure can be achieved with repeated staining.

For further staining, the gel has to stay in stabilizing solution for one day, and then repeat steps 2 to 5. Three staining should be enough, which takes about one week. Gel can stay in staining or stabilizing solution over the weekend without any effect on the results.

2
Hot Coomassie Brilliant Blue staining

0.025 % (w/v) Coomassie R-350 staining solution: Dissolve 1 PhastGel Blue tablet in 1.6 L of 10% acetic acid.

Staining

- Heat the solution to 90 °C and pour it over the gel, which is in a stainless steel tray.
- Place the tray on a laboratory shaker for 10 min.

Destaining

- On a rocking table in 10% acetic acid for at least 2 h at room temperature. Change solution several times; recycle it by pouring it through a filter filled with activated charcoal.

Placing a paper towel into the destaining solution adsorbs the Coomassie dye.

Staining and destaining solutions can be used repeatedly.

The gels can now be scanned and evaluated. The selected spots can be cut out with a spot picker.

Subsequent silver staining will reveal more protein spots. Silver staining with prior Coomassie Brilliant Blue staining is more sensitive than without prestaining, and detects proteins, which are otherwise missing (see page 83).

3
Silver staining

Quality of reagents

Water For silver staining the water quality must be adequate. Sometimes it is thought, that MilliQ quality is a guarantee for good results. It happens, however, frequently, that the cartridges are not exchanged for a long period of time. Sometimes, deionised water is better for silver staining than MilliQ water.

There is a simple test: Mix a few droplets of the water with a few droplets of the silver nitrate solution (from a commercial silver staining kit or laboratory-made). When the water contains chloride, the mixture becomes immediately turbid: This water cannot be used. In case of emergency: buy deionised water from the super market.

Ethanol If pure ethanol is not available or too expensive, there are two possibilities:

- Buy a commercial 40 % stock solution in a supermarket. For fixing just add the acetic acid.
- Use MEK denatured ethanol (methylethylketon).

Also known under the name Vodka.

Do not use otherwise denatured ethanol, do not use methanol.

Other reagents The reagent quality is very important for good results. The quality of silver nitrate and formaldehyde is most critical.

Never store the 37 % formaldehyde in the refrigerator or in the cold room.

Double staining

The background of the Coomassie Blue stained gel must be completely clear. Because the spots are fixed and the buffer components are already removed, the fixing steps can be omitted, and the procedure is started with the sensitizing step.

Heukeshoven J, Dernick R. Electrophoresis 6 (1985) 103–112.

The following modification of the silver nitrate technique, based on the method by Heukeshoven und Dernick (1986) is probably the most sensitive and reproducible one, when the quality of the chemicals and the water is high, and when the timing of the steps is exactly kept. The sensitivity of this method is of about 0.05 to 0.1 ng/mm^2.

If only a few gels have to be stained, an automated stainer (see figure 1) is very useful. Automatic staining is more reproducible than manual staining.

Silver staining

Step	***Solution***	***Volume***	***Gel type***	
Manual or in automated stainer	Laboratory-made solutions or silver staining kit for proteins	[mL]	1 mm unbacked [min]	1 mm on film or glass support or 1.5 mm unbacked [min]
Fixing	10 % acetic acid; 40 % Ethanol; with deionised water to	2 × 250	2 × 15	2 × 60
Sensitising	75 mL Ethanol; 10 mL Na-thiosulphate (5 % w/v); 17 g Na-acetate; 1.25 mL glutardialdehyde; with deionised water to	250	30	60
Washing	deionised water	3 to 5 × 250	3 × 15	5 × 8
Silver	25 mL silver nitrate (2.5 %); 100 µL formaldehyde; with deionised water to	250	20	60
Washing	deionised water	2 to 4 × 250	2 × 1	4 × 1
Developing	6.25 g Na-carbonate; 100 µL formaldehyde; 7 µL Na-thiosulphate (5 % w/v); with deionised water to	250	4	6
Stopping	3.65 g EDTA; with deionised water to	250	10	40
Washing	deionised water	2 to 3 × 250	3 × 5	2 × 30

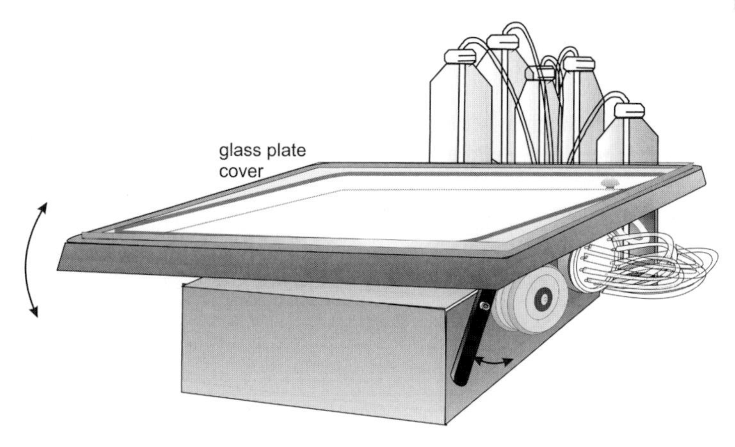

Fig. 1: Automated gel stainer

The addition of a tiny amount of Na-thiosulphate is particularly helpful to reduce the background when thick gels and film-supported 1 mm thick gels are stained.

Note:
When silver stained spots should be further analysed with mass spectrometry, some modifications of this protocol have to be made:
- ***No glutardialdehyde in the sensitiser and no formaldehyde in the silver solution.***
- ***No Na-thiosulphate in the developer.***

The staining tray must be covered with a lid to prevent contamination with keratin.

Staining of multiple gels

Up to four gels can be stained at a time with a special set of staining trays on an orbital shaker, as shown in figure 2. One tray is perforated to allow easy changing of staining liquids without breaking the gels. 1.5 L solution is needed for each step.

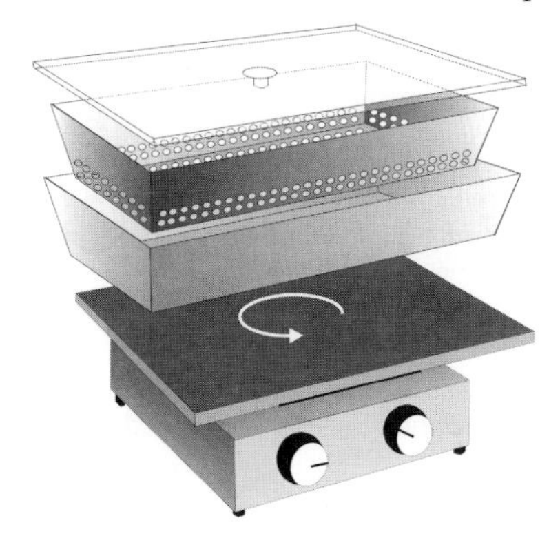

Fig. 2: Staining of multiple large format gels on an orbital shaker. Up to four gels, with or without film-backing can be stained together in one set.

4
Fluorescence staining

Rabilloud T, Strub J-M, Luche S, van Dorsselaer A, Lunardi J. Proteomics 1 (2001) 699–704.

For highest sensitivity with Sypro Ruby or RuBPS the following procedure is recommended by Rabilloud *et al*:

Fixation	20% ethanol, 7% acetic acid	250 ml	1 hour
Washing	20% ethanol	250 ml	4 × 30 minutes
Staining	SYPRO Ruby or RuBPS	700 ml	overnight
Washing	deionised water	250 ml	2 × 10 minutes

Stain and store the gels in dark plastic containers.
Do not use steel or glass trays.

5
Preserving and drying of gels

Film-backed gels can be stored in a sealed sheet protector. The open side is simply closed by sealing it with a kitchen foil welding apparatus.

Gels without film support can be dried between two sheets of cellophane, which are clamped into specially designed frames, as shown in figure 3. The cellophane must be prewetted with 10 % glycerol / water without alcohol.

During staining gels tend to swell. When a gel does not fit into such a frame anymore, it has to be re-shrunk to its original size with 30 % ethanol / 10 % glycerol.

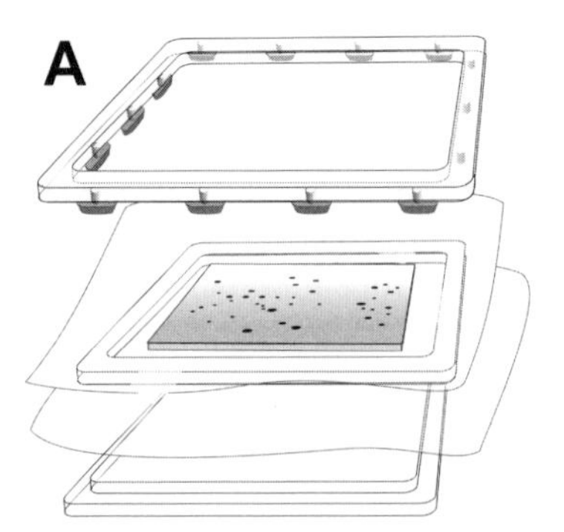

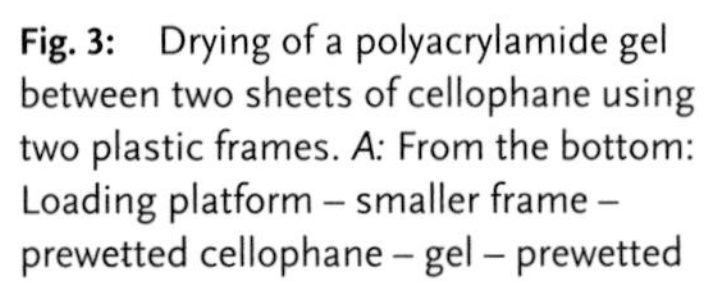

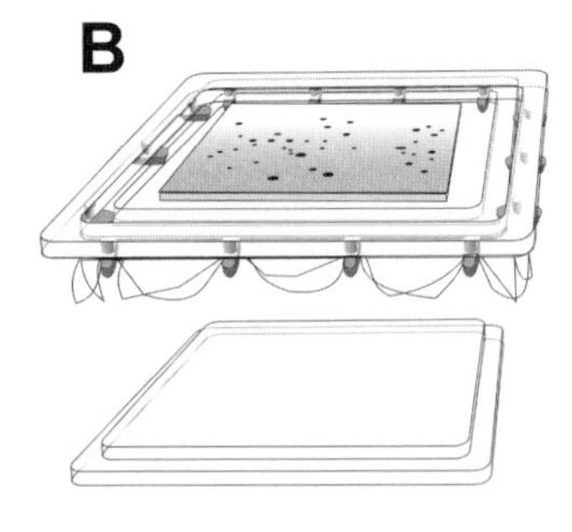

Fig. 3: Drying of a polyacrylamide gel between two sheets of cellophane using two plastic frames. *A:* From the bottom: Loading platform – smaller frame – prewetted cellophane – gel – prewetted cellophane – larger frame. *B:* The two clamped frames with the sandwich are removed from the platform; the screws are turned by 90°.

This drying procedure can also be applied on film-supported gels.

Step 5: Scanning of gels and image analysis

For scanning and image analysis the instructions and manuals of the used hardware and software must be followed. The descriptions in the following chapter should be understood as a generic way to evaluate gels, but they are based on specific instruments and programs.

1 Scanning

1.1 Gels stained with visible dyes

- Activate in the *Scanner Options*: "Always use TWAIN controls."
- Press "Scan", set the scanning parameters in the submenu in the *Scanner Control Window.*
  - If necessary, choose light source color,
    For instance: Yellow-brown silver stained spots are best scanned with blue color.
  - Transmission mode,
    Otherwise you cannot perform proper quantification.
  - Grey scale,
  - Bits per pixel as high as possible, like 16 bits,
    Important for quantification.
  - Resolution 300 dpi or lower.
    To limit the size of the resulting file.
  - Gamma correction.
    A tool for intensity adjustment at selected level.

If the value is set above "1", the intensity resolution will be higher for the strong signals, if it is set below "1", the weaker signals are displayed with a higher intensity resolution.

- These parameters are saved under "Settings".

Calibration of the scanner

Note: It is much easier to calibrate the scanner than calibrating each gel after scanning.

At first the measured dimensionless intensity has to be converted into optical density values (O.D.) by calibration of the scanner with a grey step tablet.

- Choose "Recalibrate" in the program.
- Set the units, usually: "Optical Density"
- Scan a grey step tablet, available for instance from Kodak, shown in figure 1.
- Enter the step wedge intensities into the step wedge list.
- Identify the step wedge points, when a value is used, it is indicated in the image window.

The software will produce a string of points, which can be stretched to fit the steps.

- Select "Cubic spline". This curve will be saved as the default calibration, which can be loaded for each scan.

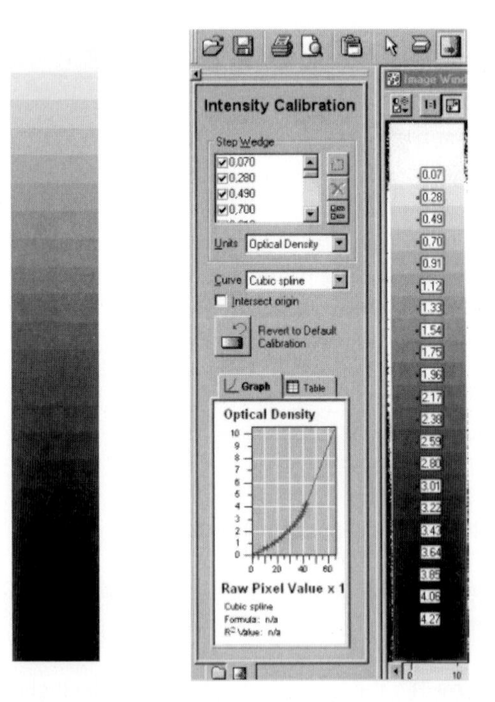

Fig. 1: Calibration of the scanner with the help of a grey step tablet.

Scanning a gel

- Pipette a few milliliters of water on the scanning plate and place the gel on it, avoiding air bubbles. Film- and glass-backed gels are placed on the scanning plate with the gel surface down. Excess water is removed with tissue paper.

1.2
Scanning fluorescent dyes

The calibration procedure is not necessary for densitometers and dedicated scientific scanners. The gels are scanned with the surface down.

2
Spot detection and background parameters

> Note:
> ***All gels to be evaluated in one experiment have to be located in one folder.***

All image files must be on the hard disk.

- In *the File menu* choose "New experiment".

Usually the buttons are similar like in most windows menus.

2.1
Automated spot detection

- Select a gel to be analysed. The gel will be displayed in an image window.
  Automatically a *Spot Detection Menu* will open up in the navigator on the left hand side.
- Select "Spot Detection Wizard".
- With the mouse select a typical spot area by drawing a box around it.
- This area will be displayed with different parameter settings in a 9-windows matrix, shown in figure 44 on page 93.
- First adjust the two parameters "sensitivity" and "operator size":

The operator size is the size of the pixel matrix around a pixel.

Those spots, which are *detected,* are marked with a blue line around the spot. The number of spots detected varies with the selected parameters. The two parameters are also displayed in the scroll bars at the bottom and the right hand side.

- Select the image, which shows the best representation of the pattern by clicking on it: it will move into the center, marked with a green frame. The two parameters are again closely varied around the set parameters.

Include small spots, but prevent too much spot merging.

- Repeat this until you are satisfied with the result.

This one eliminates false signals in the areas empty of spots.

- Secondly, adjust the third parameter "Background factor" with the help of the slider at the bottom.

The smaller the figure, the more sensitive is the spot detection.

- Adjust now the forth parameter "Noise level" in the Advanced mode.
- Click "Finish".

After the parameters are set, the spot detection occurs automatically, and the result will be displayed on the Main Image Window. The selected parameters can be applied on a series of gels with batch processing.

Hint:
It is important, that the small and weak spots are well detected. Editing of the large/saturated spots is considerably easier and more simple to perform than vice versa.

2.2 Spot filtering

It is highly recommended to perform a filtering process before spot editing, in order to reduce time and human influence in the spot editing step.

There ar many possibilities of filtering, the following three measures are proposed:

- Circularity.
- Minimum area.
- Minimum volume.

There is also the possibility of "spot selected" filtering, where the filtering parameters can be defined with the help of a selected spot.

2.3 Spot editing

Particularly in the areas of large, saturated spots you can observe inadequate spot detection. Those need to be manually edited with the toolbox "Edit". The tools are:

- Draw,
- Erase spots,
- Peak growth,

- Edge growth,
- Split.

> **Hint**
> ***Manual editing of a large gel should not take substantially longer than roughly one hour. Should you need longer, automated spot detection has not been sufficiently optimised and filtering was not carried out properly.***

2.4 Background correction

Background subtraction should be done automatically as well. Several different algorithms are offered.

- Select 'Lowest on Boundary' or 'Mode of Non Spot':
  - *Lowest on Boundary:* One pixel outside of each individual spot boundary will be chosen as the respective lowest value.
  - *Mode of Non Spot:* Each spot will be framed by a rectangular box. Around this box a frame is built, which is *n Pixel* wide in all directions. Those Pixels of the frame, which do not belong to another spot, will be chosen as the respective lowest values. The value of *n* must be chosen depending on the resolution, a large value is recommended (30–45).
- Display the background without spots in the *Main Image Window* as a control.

3 Evaluation of 2-D patterns

3.1 Reference gel

- Select in the "Edit" menu the "Reference Gel Wizard" and click "Create a new reference gel".
- Click "Next" to get to the list of gels in the experiment and choose a gel to make a reference gel from it.

You can also use a shared reference gel from another experiment.

3.2
Automatic gel matching

- Click the matching icon.
- Adjust the vector box size and the search box size.
  The software will connect "reference" spots and "slave" spots with a *vector.*
  - *Vector Box Size:* Edge length of a box at the end of the vector in pixels.
    8 is a good starting point, larger values allow a wider match tolerance, (the danger for mismatches increases)
  - *Search Box Size:* Edge length of the initial search box in pixels.
    The iterative algorithm starts with the most central reference spot, and searches for the best pattern fit. 64 as starting value is sufficient in almost all cases. The influence of this box size is smaller than the one of the *Vector Box Size.*
    Automatic matching does not always work for all spots to be matched (see figure 2).
- Add seed matches

The gel will be warped to the seed matches. In the gel overlay image, the matched and unmatched spots are indicated as open or solid circles in different colours (see figures 2 and 3).

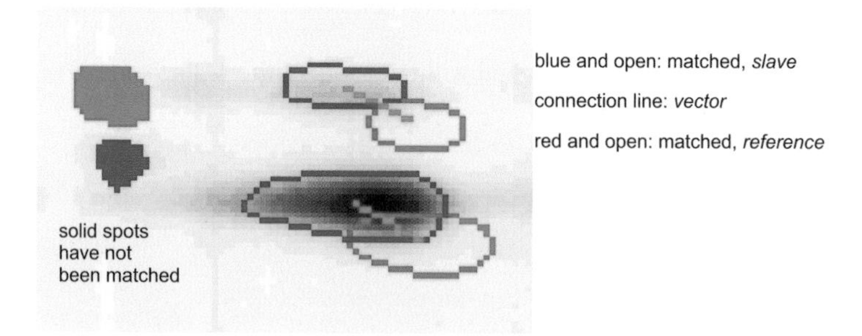

Fig. 2: Spot matching: Indication of matched and not matched spots.

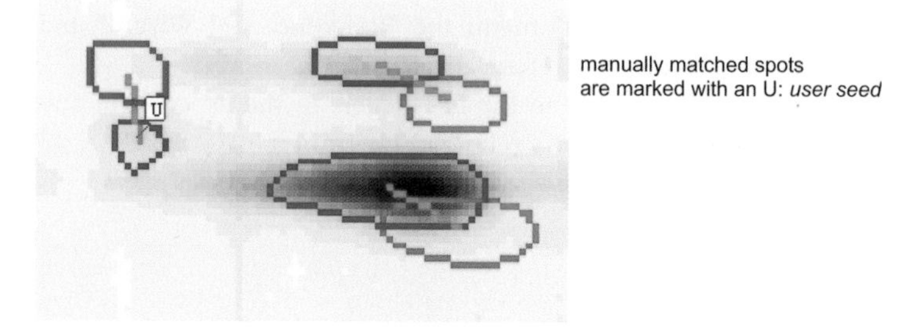

Fig. 3: Spot matching: Indication of manually entered seed matches.

The example in figures 2 and 3 shows clearly why automated matching was not performed: The resulting vector shows into a markedly different direction than the other vectors in the vicinity. In difficult cases, it is recommended to start placing seeds matches in the center of the image

- Check the matching process by displaying the gel overlay unwarped: vectors connecting the matched spots are shown (see figure 45).

Incorrect matching can easily be recognized, because vectors will point into different directions and vary in lengths in a non-plausible way.

- Update the reference gel with the non matched spots.

3.3 Normalisation

When gels are compared, the differences in protein load and staining effectiveness must be compensated with normalisation. Two modes are possible. The normalisation mode must be the same for all gels within an experiment.

This is an adjustment for quantitative comparison.

Total spot volume normalisation

- Multiply the value by the total spot area or by 100 (for percentage).

Single spot normalisation

- Choose a spot to base the normalisation: *base spot.*
- Enter normalized volume for base spot.
- Press *Spot Normalisation.*

> Note particularly for silver staining:
> ***If staining is too different, do not normalise. Repeat experiment with more similar sample loads!***

Silver staining is not linear. Employ quantification procedure as described on pages 184ff.

3.4 pI and M_r calibration

- Create *Protein Lists* for selected proteins: containing name, pI, M_r, accession number, etc.

1-D calibration

pI and M_r are calibrated separately.

- Click *Edit Ladders.*

You can also import ladders.

A "*.lad*" file will be created in a subdirectory. You create a pI and a M_r ladder for the reference gel (see also pages 53 and 79). For IPG you transfer the graphs into ladders.

- Click *Import Ladder*, it will be displayed on the gel. The ladders can be moved around with the mouse and adjusted to the related band locations on the gel image.
- Click *Calibrate.*
- The program will interpolate the values for the not assigned proteins in the respective dimension in the reference gel.

2-D calibration

Identify several spots in the reference gel with know pI and M_r:

- Click at a spot of the reference gel in the Image Window, which should be used as a calibration point.
  At least one spot in each quadrant of the gel has to be assigned.
- Click *Assign* in the Navigator window; and so on.
- Click *Calibrate.*

Now the program will interpolate the values for the not assigned proteins in the reference gel. Then the spots in the matched gels are calibrated.

3.5
Difference maps

- Click *Difference Map* in the *Main Image Window* and set the parameters in *Further Display Options.*
  The slave gels can then easily be compared with each other. For instance:
- Set those spots are in green, which have a volume increase of 50 % or more over the comparable spot.
- Set those spots are in yellow, which show a volume decrease of 50 % or more.

Within the selected boundaries matched spots are displayed with a blue line color. Not matched spots are displayed as solid blue.

In this way, up- and down-regulated spots are visible, as shown in figure 46.

3.6 Averaged gels

- From the *Edit menu* select the *Averaged Gel* wizard.
- Create a name for the averaged gel and you can add and remove gels from it.
- Set tolerance parameters in order to exclude erroneous spots, for instance from how many sub-gels a spot can be maximally absent.

Those are usually replicate gels where the same sample has been applied.

Note:
For good, statistically reliable results more than 4 replicate gels should be averaged.

3.7 Quantification

- Check the *Saturation Map* first before you draw any quantitative conclusions. Saturated spots or areas will be displayed in a colour overlay.
- Click on *Measurements Window* to have spot numbers and volumes displayed across all gels of an experiment.
- Quantitative data are graphically represented in the *Histogram Window.*

Sample load, staining method and scanner features have a strong influence.

3.8 Creating reports

Spot report
- Click on the selected spot.
- On the *Report Window* tool bar click on *Generate Spot Report.* All information on this spot across all experiment gels will be displayed.

Gel report
- Click on *Generate Gel Report* All information on this gel will be displayed.

Export to other programs
- Use copy and paste or export as an ".txt" file.

3.9 Spot picking

The spot picking list is created automatically.

Step 6: Fluorescence difference gel electrophoresis

In the standard labelling protocol, proteins are first denatured in a lysis buffer that will eliminate all secondary and tertiary structures of proteins. The protein concentration should then be determined using the quantification procedure described in Step 1: Sample preparation. CyDye probes are then added to the protein lysate and incubated on ice in the dark for thirty minutes so that 50 μg of protein is labelled with 400 ρmoles of CyDye.

When handling proteins it is important to keep them on ice at all times to reduce the effect of proteases, and not to use glassware, as many proteins will adhere to it.

The fluorescent properties of Cy2, Cy3 and Cy5 can be adversely affected by exposure to light: perform all labelling reactions in the dark in microcentrifuge tubes, and keep the exposure of protein labelled with CyDye to all light sources to a minimum.

Reconstitution of CyDye

99.8% anhydrous Dimethylformamide (DMF) less than 3 months old from day of opening.

Labelling

Microcentrifuge tubes 1.5 mL

Standard Cell wash Buffer

10 mmol/L Tris (pH 8.0), 5 mmol/L Magnesium acetate. Aliquot and store at –20 °C.

Lysis buffer

30 mmol/L Tris, 7 mol/L urea, 2 mol/L thiourea, 4% (w/v) CHAPS. Adjust to pH 8.5 with dilute HCl. Small aliquots can be stored at –20 °C.

Lysine

10 mmol/L L-Lysine

Dividing the sample

- Mix 1/3 of each of the Control and Diseased samples together to create a Standard sample. Label the standard sample with Cy2.
- Label the remaining 2/3 of the Control sample with Cy3.
- Label the remaining 2/3 of the Diseased sample with Cy5.

More complex experimental designs can be generated using the Standard sample on all gels.

> Note:
> ***CyDye probes when coupled to the protein add approximately 500 Da to the protein's mass.***

1
Preparing a cell lysate compatible with CyDye labelling

The example given here was used with the *Escherichia coli* model system, but other wash buffers might be more appropriate for different cell types. Approximately 4×10^{10} *E.coli* cells will result in a 5 to 10 mg/mL of protein in 1 mL of lysis buffer.

1. Pellet the cells in a suitable centrifuge at +4 °C.
2. Pour off all growth media, taking care not to disturb the cell pellet.
3. Resuspend cell pellet in 1 mL of Standard Cell Wash buffer in a microcentrifuge tube.
4. Pellet the cells in a bench-top microcentrifuge at 12,000 g for 4 minutes at +4 °C.
5. Remove and discard the supernatant.
6. Resuspend cell pellet in 1 mL of Standard Cell Wash buffer in a microcentrifuge tube.
7. Repeat steps 4 to 6 at least three times.
8. Ensure all the wash buffer has been removed.
9. Resuspend the washed cell pellet in 1 mL of Lysis buffer (30m mol/L Tris, 7 mol/L urea, 2 mol/L thiourea, 4% [w/v] CHAPS, pH 8.5) and leave on ice for 10 minutes.

   > Note:
   > ***If your protein concentration is less than 5 mg/mL after protein quantification, resuspend cells in a smaller volume of lysis buffer in subsequent experiments.***

10. Keep the cells on ice and sonicate intermittently until the cells are lysed.

11. Centrifuge the cell lysate at +4 °C for 10 minutes at 12,000 g in a microcentrifuge.
12. Transfer supernatant to a labelled tube. This is the cell lysate. Discard the pellet.
13. Check the pH of the cell lysate is still at pH 8.5 by spotting 5 µL on a pH indicator strip. If the pH of the cell lysate has fallen below pH 8.0 then the pH of the lysate will need to be adjusted before labelling. See "How to adjust the pH of your protein lysate".
14. Store cell lysates in aliquots at –70 °C until protein yield is to be determined.

2
Reconstituting the stock CyDye in Dimethylformamide (DMF)

CyDye is first delivered in a solid format that must be reconstituted in DMF. Each vial of CyDye will only have to be reconstituted in the DMF once. It might seem that there is little solid in each of the tubes, however, after reconstitution in DMF the CyDye will give a deep colour; Cy2-yellow, Cy3-red, Cy5-blue.

Note:
The quality of the DMF used in all experiments is critical to ensure that the protein labelling is successful. The DMF must be anhydrous and every effort should be used to ensure it is not contaminated with water. DMF after opening, over a period of time, will degrade with amine compounds being produced. Amines will react with the NHS ester CyDye reducing the concentration of dye available for protein labelling.

1. Take a small volume of DMF from its original container and dispense into a microcentrifuge tube.
2. Take the CyDye from the –20 °C freezer and leave to warm for 5 minutes at room temperature
3. After 5 minutes add 25 µL of the DMF to each new vial of CyDye.
4. Replace the cap on the dye microcentrifuge tube and vortex vigorously for 30 seconds.
5. Centrifuge the microcentrifuge tube for 30 seconds at 12,000 g in a benchtop microcentrifuge.
6. The dye can now be used.

Note:
When dyes are not being used they should be returned to the –20 °C freezer as soon as possible and stored in the dark.

Note:
After reconstitution CyDye is only stable and useable until the expiry date listed on the tube.

3
Preparing CyDye solution used to label proteins

1. Briefly spin down CyDye stock in a microcentrifuge.
2. Dilute 1 volume of the stock CyDye in 1.5 × volumes of high grade DMF. To make 400 ρmol of CyDye in 1 μL; take 2 μL of the stock dye and add 3 μL of DMF (i.e. 2000 ρmoles CyDye in 5 μL; therefore 1 μL contains 400 ρmoles.

Note:
Add the DMF first to the sterile microcentrifuge tube, followed by CyDye. 1 μL of the diluted dye now contains 400 ρmoles.

Amount of CyDye label for a protein lysate

Routinely we label 50 μg of protein with 400 ρmoles of CyDye.

In each tube of CyDye received there will be 25ηmoles of dye that has been resuspended in 25 μL of DMF to create a CyDye stock solution.

Although we recommend 400 ρmoles of CyDye per 50 μg of protein to be labelled, between 100 ρmoles and 1000 ρmoles per 50 μg of protein can be used. If labelling more than 50 μg of protein then you must keep the dye to protein ratio the same for all samples on the same gel.

Some examples of CyDye dilutions that are often used (we recommend using the highlighted example) are shown in Table 1.

Tab. 1: Examples of some widely used CyDye dilutions

Volume of stock CyDye (μL)	*Volume of added DMF (μL)*	*Total volume (μL)*	*Concentration of CyDye (ρmoles/μL)*
1	4	5	200
2	**3**	**5**	**400**
2	2	4	500
1	–	1	1 000

Note:
CyDye in the diluted form is only stable for 2 weeks at –20 °C.

Protein sample labelling with the CyDye

Note:
The recommended concentration of the protein sample is between 5 and 10 mg/ml. Samples containing as little as 1 mg/ml have been successfully labelled using the protocol below.

The amount of CyDye used in the labelling reaction will have to be determined individually for the type of protein sample being analysed. Here is one example where we labelled an *E. coli* lysate.

1. Add a volume of protein sample equivalent to 50 μg to a sterile microcentrifuge tube.
2. Add 1 μL of diluted CyDye to the microcentrifuge tube containing the protein sample (i.e. 50 μg of protein is labelled with 400 pmoles of dye for the labelling reaction).
3. Mix and centrifuge briefly in a microcentrifuge. Leave on ice for 30 minutes in the dark.
4. Add 1 μL of 10 mmol/L lysine to stop the reaction. Mix and spin briefly in a microcentrifuge. Leave for 10 minutes on ice in the dark.
5. Labelling is now finished.
6. Samples can now be stored for at least three months at –70 °C in the dark.

Now load onto the IPG strips.

Loading the samples onto the IPG strips

After the protein samples have been CyDye labelled, add an equal volume of 2 × sample buffer and leave on ice for 10 minutes.

Rehydration loading as well as cup loading can be used.

Requirements for 2D-DIGE protein lysis buffer

It is essential that the pH of the protein solution you use with a CyDye is between pH 8.0–9.0.

1. To ensure that the pH remains between pH 8.0–9.0 a buffer such as Tris, Hepes or Bicarbonate should be included in your protein solution at a concentration of approximately 30mM. Failure to include a suitable buffer will mean that the pH of the solution will fall below pH 8.0 resulting in little or no pro-

tein labelling. The Lysis buffer is required to work at +4 °C so the pH should be checked when the solution is chilled.

2. The protein solution should not contain any added primary amine compounds BEFORE labelling.

> Note:
> ***Primary amines, such as ampholytes, will compete with the proteins for CyDye. The result will be fewer CyDye labelled proteins, which might affect your data after scanning and spot detection.***

Requirements for a cell wash buffer

A cell wash buffer should not lyse the cells, but it should dilute and remove any growth media or reagents that might affect a CyDye labelling process.

> Note:
> ***A cell wash buffer should not contain any primary amines.***

A range of cell wash solutions such as 75 mmol/L Phosphate Buffered Saline (PBS) can be used in conjunction with the DIGE technology as long as their compatibility with the 2D-DIGE labelling technology is evaluated in controlled experiments.

Step 7: Spot excision

Following the selection of the spots of interest, the protein spots have to be excised from the gel. This step is performed manually or automatically with commercially available spot pickers.

The procedure must be performed in clean environment such as a Laminar flow hood in order to minimise or most preferably exclude contaminating proteins such as keratin entering the protein identification workflow at this stage. The keratin(s) will be digested in the same way as the target protein(s), with keratin peptides included in the PMF spectrum, complicating the subsequent dataset. This is particularly problematic for low abundant samples. See tables 1–4 for lists of keratin peptides commonly observed in MALDI peptide mass fingerprinting (courtesy of www.matrixscience.com)

Keratin contamination can be introduced into the workflow by poor sample handling at any stage between conclusion of the electrophoresis step and digestion of the target protein. Furthermore, keratin contamination is almost certainly present in the starting acrylamide solutions and solvents; all of which need to be filtered before use. Even the staining unit should be used solely for applications leading to mass spectrometric analysis. All eppendorf tubes to be used should be rinsed with MilliQ water (or at least doubly distilled water), especially if the tubes have been in lying around in drawers before use. This step will remove any residual dust from the tubes, which may contain keratins.

All possible measures need to be taken to avoid keratin contamination.

Peptides from four common keratins have been reported to contaminate many mass spectra. The commonly observed keratin tryptic peptides from these four keratin proteins are tabulated below:

Tab. 1: K2C1_HUMAN: KERATIN, TYPE II CYTOSKELETAL 1 (K1 skin)
Nominal mass (M_r): 65847; Calculated pI value: 8.16

Position in sequence	*Experimental mass [M+H]⁺*	*Theoretical mass [M]⁺*	*Sequence*
185–196	1382.68	1383.69	SLNNQFASFIDK
211–222	1474.75	1473.74	WELLQQVDTSTR
417–431	1716.85	1715.84	QISNLQQSISDAEQR
223–238	1993.98	1992.97	THNLEPYFESFINNLR
518–548	2383.95	2382.94	GGGGGGYGSGGSSYGSGGGSYGSG GGGGGGR
549–587	3312.31	3311.30	GSYGSGGSSYGSGGGSYGSGGGGG GHGSYGSGSSSGGYR

Tab. 2: K2E keratin (Dandruff), 67K type II epidermal – human
Nominal mass (M_r): 65825; Calculated pI value: 8.07

Position in sequence	*Experimental mass [M+H]⁺*	*Theoretical Mass [M]⁺*	*Sequence*
245–253	1037.50	1036.49	YLDGLTAER
381–390	1193.60	1192.59	YEELQVTVGR
46–61	1320.60	1319.59	HGGGGGGFGGGGFGSR
71–92	1838.90	1837.89	GGGFGGGSGFGGGSGFGGGSGFSG GGFGGGGFGGGR
93–128	2831.20	2830.19	SISISVAGGGGGFGAAGGFGGR

Tab. 3: K.9. keratin 9 (Skin), type I, cytoskeletal – human
Nominal mass (M_r): 61950; Calculated pI value: 5.14

Position in sequence	*Experimental mass [M+H]⁺*	*Theoretical mass [M]⁺*	*Sequence*
233–239	897.40	896.39	MTLDDFR
224–232	1060.60	1059.59	TLLDIDNTR
242–249	1066.50	1065.49	FEMEQNLR
449–471	2510.10	2509.09 0	EIETYHNLLEGGQEDFESSGAGK
63–94	2705.20	2704.19	GGGSFGYSYGGGSGGGFSASSLGGG FGGGSR

Tab. 4: K10 (dandruff) keratin 10, type I, cytoskeletal – human
Nominal mass: 59492; Calculated pI value: 5.17

Position in sequence	*Experimental mass [M+H]⁺*	*Theoretical mass [M]⁺*	*Sequence*
442–450	1165.60	1164.59	LENEIQTYR
323–333	1365.60	1364.59	SQYEQLAEQNR
166–177	1381.64	1380.63	ALEESNYELEGK
41–59	1707.80	1706.79	GSLGGGFSSGGFSGGSFSR
423–439	2024.94	2023.93	AETECQNTEYQQLLDIK
208–228	2366.26	2365.25	NQILNLTTDNANILLQIDNAR

Manual procedure

Wearing powder-free gloves, use a clean scalpel blade to cut around the protein spot of interest. Attempt to take as little of the surrounding gel as possible. Using a surgical needle transfer the excised gel piece to a pre-rinsed 500 μL eppendorf tube.

Step 8: Sample destaining

Generally three methods are used for gel staining of 2-D gels. Coomassie Brilliant Blue, silver and Sypro Ruby.

Coomassie Blue-stained protein spots

The CBB will heavily suppress signal in both MALDI and ESI analysis. CBB will be easier to remove before digestion rather than after, because after digestion the CBB will be concentrated along with the peptide digest during the extraction stage. The CBB will be very difficult to remove by a microscale purification step (see step 9)

Add 25 μL of 75 mmol/L ammonium bicarbonate (40% ethanol) to the excised Coomassie stained gel plug. Vortex and leave to stand. The supernatant will rapidly turn blue, remove supernatant after ten minutes and replace with a further aliquot of the destain solution until the excised spot is destained. The gel plug is now ready for digestion.

The length of time required to destain will be dependant on the intensity of the stain.

Silver-stained protein spots

The silver staining method has to be compatible with MS analysis. Sensitive silver stain methods, in the order of 1–10 ng per protein spot (Rabilloud, 1999), typically required the use of glutardialdehyde as part of the sensitisation process. However, glutardialdehyde reacts with the amino groups of proteins, both ε-amino (lysine side chain) and the α-amino; cross-linking the protein to the gel. This step needs to be eliminated for optimal MS analysis (Shevchenko *et al.* 1996; Yan *et al.* 2000; Sinha *et al.* 2001). Further optimisation to the procedure was reported by Gharahdaghi *et al.* 1999; by removing the silver ions prior to digestion.

A. Prepare fresh solutions of potassium ferricyanide (30mmol/L) and sodium thiosulphate (100 mmol/L).
B. Prepare a 1:1 solution of the above reagents and immerse the excised spots in this newly prepared solution.

Volume used depends on the size of the excised gel plug.

C. Once the dark stain has been removed, wash the spot with water.
D. Equilibrate the excised spot in ammonium bicarbonate (200 mmol/L) for 15 minutes, remove supernatant and replace with a second aliquot of ammonium bicarbonate and leave for 15 minutes.
E. The gel plug is now ready for digestion.

Step 9: In-gel digestion

Perform the reduction and alkylation step prior to digestion. Remember to use the same alkylation reagent as was used between the first and second dimensions of 2-D electrophoresis.

A. Reduction. Add 10 µL of dithiothreitol solution (5 mmol/L in 25 mmol/L ammonium bicarbonate) to the gel plug (or sufficient to cover the gel piece) and incubate for 30 minutes at 60 °C.
B. Alkylation. Add 10 µL of iodoacetamide (55 mmol/L in 25mmol/L ammonium bicarbonate) to the gel plug and stand at room temperature for 30 minutes in the dark.
C. Remove the supernatant, wash with ammonium bicarbonate (25 mmol/L), remove supernatant and wash with acetonitrile.
D. Take the excised gel plug and cut into smaller pieces with the scalpel (2–4 mm^2). Press down on the gel plug with the surgical needle holding the gel plug in place; allowing the cutting of the plug. Transfer the gel plug pieces to the rinsed eppendorf tube.
E. Dehydration step. Dehydrate gel plug with acetonitrile (3 × 25 µL, 10 minutes each). The gel plug will begin to look white, after 3 washes it will be completely white.
F. Dry plug further in a speed vac or drying chamber, until gel plug appears "dust like."
G. Rehydration step. Apply enzyme in buffer (10 µL at 40 ng µLin 50 mmol/L ammonium bicarbonate) to the dried gel plug and incubate on ice for 45 minutes.
H. Digestion. After 45 minutes remove supernatant and add enough buffer solution (without enzyme) to cover the hydrated gel piece.
I. Incubate at 30 °C, for between 1 hour and overnight (a preliminary peptide mass fingerprint can be obtained after 1 hour by sampling from the digestion mixture).

J. Extraction step. Add 20 μL of 50 mmol/L ammonium bicarbonate and sonicate for 10 minutes. Add 20 μL of Acetonitrile : 5% TFA (1:1) and sonicate for 10 minutes. Remove supernatant and dispense into a separate tube.

K. Add 20 μL of Acetonitrile : 5% TFA (1:1) and sonicate for 10 minutes. Combine supernatant extracts. Repeat one further time.

L. Add 10 μL acetonitrile and sonicate. Remove and combine with earlier supernatant extracts.

Step 10:
Microscale desalting and concentration of sample

The sample preparation for MALDI (and ESI) is a crucial step. The analyte must be incorporated into the matrix crystals, a process that is significantly upset by the presence of contaminating salts and buffers (see table 1 for MALDI salt/buffer compatibility). Sample preparation can be tailored to a particular matrix. For instance α-cyano-4-hydroxycinnamic acid is insoluble in water. Sampling directly from the digestion mixture, the analyte solution can be spotted onto a preformed thin layer of matrix on the MALDI target (thin film method). Once the dried spot has formed, the spot can undergo significant washing with 0.1%TFA solution on the target surface, removing the salt contamination. As the analyte has been incorporated into the matrix crystal, it is preferably bound during the washing step (Vorm *et al.* 1994). Addition of nitrocellulose to the matrix solution allows for improved desalting and improved binding on the MALDI target (Shevchenko *et al.* 1996).

On the contrary 2,5-dihydroxybenzoic acid is a water soluble matrix. It is possible to sample directly from the digest mixture without any sample clean up (on target washing is not applicable) as the matrix excludes the contaminants from the crystallisation process

However if sampling from the digest mixture is not applicable (or if this approach is not attractive) then the peptides are extracted as detailed in the section above. However, the combined extract volume (~100 µL) may be too dilute for successful analysis of the peptide mixture; in these instances the sample must be concentrated prior to analysis. Discussed earlier, simply concentrating the sample will concentrate all the contaminants as well, hence a clean up step is recommended. Kussman *et al.* 1997 described the use of microscale purification columns using RP resin.

Kussman et al. J Mass Spectrom 32 (1997) 593–601.

Tab. 1: Salt/buffer compatibility with MALDI analysis

Type of impurity	*Concentration*
Phosphate buffers	< 20 mmol/L
Tris buffer	< 50 mmol/L
Detergents	< 0.1%
Alkali metal salts	< 1 mol/L
Guanidine	< 1 mol/L
Ammonium bicarbonate	< 30 mmol/L
Glycerol	To be avoided
SDS	To be avoided
Sodium azide	To be avoided

Microscale desalting and concentration

Prevents the packing from being eluted during preparation and use.

A. Take a gel-loader pipette tip and very carefully pinch the tapered lower end of the tip with a pair of flat armed tweezers.
B. Prepare a suspension of reversed phase resin in methanol.
C. Add 50 μl of methanol to the pipette tip, followed by 2–3 μL of the suspension
D. A pipette can then be used to gently push the methanol though the column. The RP resin forms a small column at the end of the tip.
E. Equilibrate the column with 0.1% TFA (20 μL)

Complete drying or lyophilisation not recommended, extensive sample loss.

F. At this stage ensure the acetonitrile concentration of the peptide extracts is sufficiently low to allow good retention of the peptides on the column. Hence dry the peptide extracts to a volume of approximately 10 μL.

Heptafluorobutyric acid is a very hydrophobic ion-pairing reagent and will make the digested peptides "very sticky" improving there retention on a reversed phase C18 column.

G. Further improve the retention of the peptide mixture by acidifying the mixture with heptafluorobutyric acid (HFBA). Add 9 μL of water and 1 μL of HFBA to the semi-dried peptide extracts, creating a 55 % concentration (v:v) of HFBA.
H. Load the acidified peptide extracts onto the column using a pipette. Gently push through the solution with a pipette. Apply the eluate back to the column; repeat 5 times to improve retention of the peptides.
I. Wash the column with 0.1%TFA (20 μL)
J. Elute peptides with 3–5 μL of MeCN: 0.5% Formic acid (1:1; v:v).

Alternatively, these columns are available commercially (ZipTip™). Follow the manufacturers instructions, though they are similar to above.

Step 11: Chemical derivatisation of the peptide digest

Guanidation and sulphonation of the peptide digest

(Keough et al. 1999; 2000; 2000a)

This method describes the procedure for the guanidation of lysine residues and the subsequent sulphonation of the α-amino group at the n-terminus of peptides in the digest mixture. The sulphonation reagent, 2-chlorosulphonyl acetylchloride is highly hygroscopic and strenuous precautions must be made to ensure that water (including moisture in the atmosphere) is absent from the reagents and reaction vessel.

A. Take combined peptide extracts and concentrate in a speed vac to a volume of 20 μL.
B. Guanidate the lysine side chain. Mix 2 mL *O*-methylisoureahydrogen sulphate solution (86 mg/mL MilliQ H_2O) with 8 mL 0.25 mol/L $NaHCO_3$, pH 11.5 and add to digest. Place it in the oven at 37 °C for 2h. (The reaction can also be performed at room temperature overnight or at 70 °C for 10 minutes).
C. Remove reagents using a reversed phase microscale desalting and concentration column as described in step 9 above, elute peptide digest and evaporate solvent to dryness.
D. Reconstitute in 10 μL THF:DIEA 19:1.
E. Add 2 μL sulphonation reagent (2 μL neat material in 1 mL THF)
F. React 1 to 2 min. at room temperature.
G. Dry completely to remove organics and excess base.
H. Reconstitute in 5–10 μL of 0.1% TFA.
I. Apply to MS target as described in step 11. 2,5 DHB is the preferred matrix of choice, though α-cyano 4 hydroxy cinnamic acid can be used.

Sulphonation of the peptide digest using the CAF chemistry

This method describes sulphonation of the α-amino group at the n-terminus of peptides in the digest using a novel, modified water stable reagent. Steps B and C can be omitted if guanidation of the lysine groups in the peptide digest is not favoured.

A. Apply peptide digest extracts to a reversed phase microscale desalting and concentration column as described in step 9 and dry down extract to a minimal volume. Alternatively the peptide can be dried down to a minimal volume and proceed from step B (or D if the guanidation step is to be omitted).
B. Reconstitute peptide digest in the guanidation reagent [mix 2 μL *O*-methylisoureahydrogen sulphate solution (86 mg/mL MilliQ H_2O) with 8 μL 0.25 mol/L $NaHCO_3$, pH 11.5].
C. Place it in 37 °C for 2 h. (The reaction can also be performed at room temperature overnight or at 70 °C for 10 minutes)
D. Prepare fresh reagent. Dissolve the CAF-reagent in 0.25 M $NaHCO_3$, pH 9.4 (10 mg/100 μL).
E. Apply 10 μL of the CAF-reagent solution to the sample and leave for 15 minutes.
F. Add 1 μL 50% hydroxylamine solution. Reverses tyrosine modifications at the hydroxyl group.
G. Add 8 μL water and 1 μL HFBA to the digest solution and apply to a reversed phase C18 column as described in section step 9.
H. Elute derivatised peptides in aceonitrile: 0.5% formic acid (1:1;v:v) 1–5 μL.
I. Apply to the MALDI target as described in section 2.2.1.2.

Derivatisation of peptide digest with SMA

Unlike the sulphonation reactions described above, this reaction involves the derivatisation of both the α-amino group at the n-terminus of each peptide in the digest and the ε-amino group of the lysine side chain.

A. Carefully dry down the peptide extracts from the digestion process.
B. Prepare the SMA reagent on ice. Prepare a 2 % solution of the SMA reagent in 200 m mol/L MOPS buffer pH 7.8 (2 mg/100 μL).
C. Add 7 μL of the NHS reagent to the peptide digest.
D. Leave at room temperature for 15 minutes.
E. Add 7 μL of water, acidify 1 μL of HFBA and apply to a reversed phase microscale desalting and concentration column to remove buffer and reagent and concentrate the sample as in section 2.2.1.1. Alternatively the buffer and reagent can be removed using RP HPLC. Inject the sample into a loop using a Rheodyne valve. The derivatised peptides were bound to a RP C18 column (15 cm × 300 μm). A 5–95% acetonitrile gradi-

ent was performed, and the peptides eluted into three consecutive fractions (4–5 μL each fraction). An aliquot is then use for MALDI peptide mass fingerprinting and an aliquot loaded into a nanospray needle for subsequent MS/MS analysis.

F. Alternatively, a microscale desalting and concentration step can be performed.

Step 12: MS analysis

Sample preparation for MALDI

Three matrices are generally used for peptide and protein analysis. A variety of methods have been described, particularly for the use of α-cyano-4-hydroxy-cinnamic acid for peptide analysis. Notably, Vorm *et al.* described the thin film method and Jensen *et al.* 1996 incorporated nitrocellulose into a similar method for improved performance. This section will describe the dried droplet method, probably first choice for most applications for each of the three matrices concerned. It is important in each case to keep the size of the target spot as small as possible. Remember the width of the laser beam is very narrow (~20 microns) and as such the majority of the sample will not be ablated with the laser from a large sample spot.

Vorm O, Roepstorff P, Mann M. Anal Chem 66 (1994) 3281–3287.

Jensen ON, Podtelejnikov A, Mann M. Rapid Commun Mass Spectrom 10 (1996) 1371–1378.

Do not use non-volatile solvents (e.g. glycerol, polyethyleneglycol, DMSO, Triton X or 2-mercaptoethanol) with this method.

Loading the sample in high organic content (high methanol or acetonitrile causes the droplet to spread uncontrollably across the target surface.

α-cyano-4-hydroxy-cinnamic acid. Analysis of peptides and peptide digests

A. Prepare a fresh solution of α-cyano-4-hydroxy-cinnamic acid in acetonitrile:0.1 %TFA (1:1; v:v) (10 mg/mL). Prepare a fresh solution of matrix wherever possible.

It is important to keep the pH of the matrix solution below pH 4, using 0.1% TFA maintains ensures this.

B. Take 0.2–0.5 μL matrix and mix with 0.2–0.5 μL of sample. Mix thoroughly by uptaking and displacing the volume from a pipette several times. For reproducible analysis, thorough mixing of the matrix and analyte is necessary.

If the analyte concentration is sufficiently large then 1–2 μL can be mixed with several μL of matrix and subsequently vortexed before applying to the target.

C. Apply to the target and leave to dry at room temperature.

Do not heat the sample target to speed up drying process (place on a hot plate).

D. Insert target into mass spectrometer.

2,5 Dihydroxybenzoic acid. Analysis of peptide digests

the concentration of acetonitrile can be increased up to 30%.

A. Prepare a solution of 2,5-dihydroxybenzoic acid in 0.1% TFA (10mg/mL)
B. Take 0.2–0.5 μL matrix and mix with 0.2–0.5 μL of sample. Mix thoroughly by uptaking and displacing the volume from a pipette several times. For reproducible analysis, thorough mixing of the matrix and analyte is necessary.
C. Apply to the target and leave to dry
D. Insert target into mass spectrometer.

2,5-dihydroxybenzoic acid is commonly used for the analysis of oligosaccharides released from glycoproteins, the above method can be followed by the analysis of oligosaccharides.

Tip – oligosaccharide analysis
Since 2,5-DHB forms long needle-like crystals, homogeneity of the sample can be increased by redissolving the dried matrix-sample with ethanol. Apply 0.1 μL of ethanol to the dried spot.

Sinapinic acid. Analysis of proteins

A. Prepare a solution of sinapinic acid in acetonitrile and 0.1% TFA (60:40; v:v; 10mg/mL).
B. Take 0.2–0.5 μL matrix and mix with 0.2–0.5 μL of sample. Mix thoroughly by uptaking and displacing the volume from a pipette several times. For reproducible analysis, thorough mixing of the matrix and analyte is necessary.
C. Apply to the target and leave to dry
D. Insert target into mass spectrometer.

Dried droplets are quite stable; they can be kept in a drawer or in vacuum for days.

Sample preparation for ESI MS

Nanospray

In electrospray mass spectrometry the presence of salts, involatile buffers and polymeric material in a sample is often detrimental to the analysis. High amount of salts often result in partial or complete blocking of the needle orifice resulting with, at best, an intermittent and instable spray. Subsequently, sensitivity is reduced and the spectrum can be complicated with salt adducts. The ingredients in a protein digest reaction, salts and possibly detergents, are present in millimolar amounts and they have to be removed before analysis. As with MALDI analysis, a microscale desalting and concentration step often eliminates this problem and at the same time allows a concentration of the analyte(s).

A microscale desalting and purification column is prepared as explained in step 9. Peptides are eluted into the nanospray needle using 50% Methanol:0.5% Formic acid, or eluted into a sample tube and pipetted into a nanospray needle.

A microscale desalting and concentration may be less successful for removing detergents.

LC-MS/MS

Following the in-gel digestion step redissolve the digest extract in 5 μL water. (Optional, 0.5 μL can be used for MALDI MS analysis). The remaining 4.5 μL is mixed with 4.5 μL of a 0.1% aqueous formic acid solution for subsequent ESI LC-MS/MS analysis. The remainder of the digest extract solution can be loaded on an autosampler, from which 5 μL are injected into a nano-HPLC (or alternatively manually injected). Separation is typically performed using a 75 mm × 150 mm silica C18 (3 mm particle size) PepMapä column (LCPackings, Amsterdam, The Netherlands) with 0.1% formic acid in water as solvent A and 0.08% formic acid in 80% acetonitrile / 20% water as solvent B by starting after equilibration with 5% solvent B linearly increasing it to 40% within 32 min using a flow rate of 200 nL/min. The nano-HPLC is directly coupled to a tandem mass spectrometer equipped with electrospray ionisation. This method was supplied by Dr Rainer Cramer, Ludwig Institute for Cancer Research.

Step 13: Calibration of the MALDI-ToF MS

The MALDI peptide mass fingerprint can be calibrated internally or externally.

Internal calibration for peptide analysis

Trypsin autolysis during the digestion of the target protein(s) can be very useful for calibration purposes. However, it is important to know which species the trypsin is derived from, porcine or bovine. It is fortunate that the two major porcine trypsin autolysis peaks present in the peptide mass fingerprint actually bracket the majority of the mass range of interest. The internal calibration procedure will be dependent on the MS in use, but the suggested two peptides to use are

Trypsin II	842.509 $[M+H]^+$
Trypsin IV	2211.104 $[M+H]^+$

Both of these molecular weights are the monoisotopic C12 isotope, protonated form of each peptide. Other porcine trypsin autolysis peaks observed in a MALDI PMF include 515.32, 1045.56, 2283.17, and 2299.17 (2283.17 with oxidised Met)

Using an internal calibration, high-mass accuracy is achievable (sub 25 ppm).

The expected peptides from a theoretical trypsin digest of porcine and bovine trypsin are listed in the tables 1 and 2.

Tab. 1: Theoretical tryptic peptides of porcine trypsin, TRYP_PIG, (protonated monoisotopic ^{12}C mass, unprotonated average mass)

Obtained with permission from the www.matrixscience.com

Position in sequence	*Monoisotopic mass $[M+H]^+$*	*Average mass $[M]^+$*	*Sequence*
52–53	262.14	261.28	SR
54–57	515.32	514.63	IQVR
108–115	842.50	842.01	VATVSLPR
209–216	906.50	906.05	NKPGVYTK

Position in sequence	*Monoisotopic mass [M+H]⁺*	*Average mass [M]⁺*	*Sequence*
148–157	1006.48	1006.15	APVLSDSSCK
98–107	1045.56	1045.16	LSSPATLNSR
134–147	1469.72	1469.68	SSGSSYPSLLQCLK
217–231	1736.84	1736.97	VCNYVNWIQQTIAAN
116–133	1768.79	1768.99	SCAAAGTECLISGWGNTK
158–178	2158.02	2158.48	SSYPGQITGNMICVGFLEGGK
58–77	2211.10	2211.42	LGEHNIDVLEGNEQFINAAK
78–97	2283.17	2283.63	IITHPNFNGNTLDNDIMLIK
179–208	3013.32	3014.33	DSCQGDSGG...SWGYGCAQK
9–51	4475.09	4477.04	IVGGYTCAA...VVSAAHCYK
9–51	4489.11	4491.07	IVGGYTCAA...VVSAAHCYK

Obtained with permission from the www.matrixscience.com

Tab. 2: Theoretical tryptic peptides of bovine trypsin, TRY1_BOVIN, (protonated monoisotopic ^{12}C mass, unprotonated average mass).

Position in sequence	*Monoisotopic mass [M+H]⁺*	*Average mass [M]⁺*	*Sequence*
110–111	260.19	259.35	LK
157–159	363.20	362.49	CLK
238–243	633.31	632.67	QTIASN
64–69	659.38	658.76	SGIQVR
112–119	805.41	804.86	SAASLNSR
221–228	906.50	906.05	NKPGVYTK
160–169	1020.50	1020.17	APILSDSSCK
229–228	1111.55	1111.33	VCNYVSWIK
146–156	1153.57	1153.25	SSGTSYPDVLK
207–220	1433.71	1433.65	LQGIVSWGSGCAQK
191–206	1495.61	1495.61	DSCQGDSGGPVVCSGK
70–89	2163.05	2163.33	LGEDNINVVEGNEQFISASK
170–190	2193.99	2194.47	SAYPGQITSNMFCAGYLEGGK
90–109	2273.15	2273.60	SIVHPSYNSNTLNNDIMLIK
120–145	2552.24	2552.91	VASISLPTS...LISGWGNTK
21–63	4551.12	4553.14	IVGGYTCGA...VVSAAHCYK

The practice of adding synthetic peptides to the peptide digest to perform an internal calibration is not recommended for two reasons. Firstly, it is difficult to estimate the concentration of calibrants to add; too high concentration may suppress the ionisation of the peptides of interest. Secondly, the calibrants may co-incide with peptides of interest, therefore precluding these peptides from the subsequent database search.

External calibration for peptide analysis

When one or both of the autolysis peaks are absent from the spectrum, an external calibration has to be performed. This is normally achieved by calibrating a spectrum acquired from the sample spot adjacent to the sample spot whose spectrum needs to be calibrated. Hence it is often useful to have a calibration sample alongside each sample, or at least every two samples. Hence an internal calibration is performed on the calibration mixture, the calibration file saved and then applied to the spectrum of interest (the procedure will depend upon the MS instrument used). This is a useful fallback method in case successive spectra cannot be calibrated internally from the trypsin autolysis peaks.

Obviously a calibration file obtained from the internal calibration using both trypsin autolysis peaks can be applied to the spectrum acquired from an adjacent spot, without having to use a calibration mixture such as this.

As opposed to the internal calibration where only two autolysis peaks act are used as the calibrants, the external calibration mixture can contain several calibrants defining the calibration curve more efficiently. A table of potential calibrants is listed below.

Tab. 3: Suggested peptides for use in an external calibration mixture. (All molecular weights are the ^{12}C monoisotopic value).

Peptide	*Monoisotopic mass $[M+H]^+$*
Angiotensin I	1296.678
Angiotensin II	1046.535
Substance P	1347.728
[Glu]-fibrinogen b	1570.670
Renin substrate	1759.932
Apamin	2026.887
HACTH clip 18–39	2465.188
HACTH clip 7–38	3657.922

Each combination of instrument mode (detection mode (linear/reflectron), accelerating potential) need individual calibrations.

Step 14: Preparing for a database search

This section will describe the function of each category observed on a commonly used search engine web browser for peptide mass fingerprinting (figure 1).

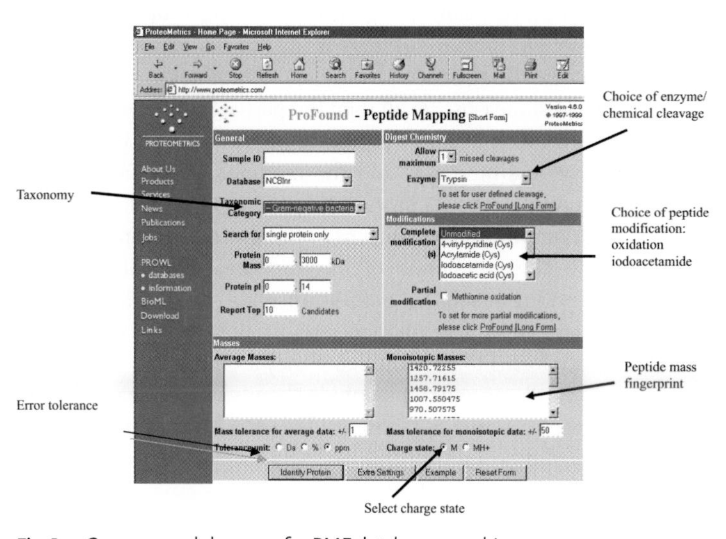

Fig. 1: Common web browser for PMF database searching

Number of missed cleavages

The number inputted will be dependent on the efficiency of the digestion. If the digestion is ideal then the enzyme will cleave at the correct site every time, affording the highest discrimination. However, enzymes rarely act in such an ideal fashion and partial peptides can be observed in the peptide mass fingerprint. Typically, one missed cleavage is selected as this. Selecting a higher number than this can reduce discrimination, as the experimental peptide mass list will be searched against more theoretical peptides.

Choice of cleavage agent

The table 8 on page 103 lists a range of enzymes available for proteolytic digestion. However, not all of the listed enzymes are suitable for subsequent protein ID. Enzymes with low specificity, particularly those generating small peptides should be avoided. Trypsin is the most commonly used enzyme as it yields peptides across a suitable mass range, relatively high specificity and locates the basic residues at the c-terminus of the peptide. This is useful for subsequent MS and MS/MS analysis.

Choice of amino acid modification

Proteins can be modified as a result of cellular processes, specifically post-translational modications; a sample preparation, including methionine oxidation and/or directly as part of an analytical method such as the reduction and alkylation of cysteine residues.

These modifications can be regarded as complete or partial for a database search. For instance, if there is uncertainty about the quantitative modification of a particular residue, or whether a modification of a particular residue exists at all then the partial modification option can be used. In this instance, the search engine will use the masses of both the native amino acid and the modified amino acid in the search. For example, when a protein is applied to the second dimension the methionine residues present in the protein are prone to oxidation due to the conditions of electrophoresis. Oxidation of a methionine residue results in a mass increase of 16 Da (nominal mass). Therefore, selecting a partial modification of the methionine will enable both forms to be identified if they are present.

Secondly, if all cysteine disulphide bonds were quantitatively reduced and the resulting cysteine residues alkylated with iodoacetamide prior to the second dimension, and the procedure repeated prior to in gel-digestion then theoretically all cysteine amino acids will now have a mass of 160 and not 103 Da (nominal mass). For a database search, the search engine requires this information if it is to correctly identify any cysteine containing peptides.

It is very difficult to predict the presence of post-translational modifications. A particular example is phosphorylation. It would be unwise to simply use a partial phosphorylation calculation on all the potential residues.

Peptide mass fingerprint

This is where the masses from the PMF are manually inputted. A trypsin digestion yields useful peptides in the 800–3500 Da mass range. Select peptides that fall within this mass range. Using masses much below this range is unadvisable as they be matrix related or too small to be significant in the search. Further Fenyo *et al.* 1998 demonstrated that a high mass peptide was useful for constraining the database. The number of peptides submitted to the search should also be considered. Too many peptides can be as problematic as too few peptides (see section 2.4.3, figure 68). Peptide masses which cannot be matched with the correct protein will contribute to the numbers of random matches, and it is possible that proteins with a number of random matches may achieve a significantly high score.

Charge state

In the case of MALDI all the peptides will be singly charged. The choice is to input the protonated form, the measured mass given in the spectrum, or the deprotonated form of the peptide. It is important that whatever option is chosen, it matches the masses used in the mass list. If these two pieces of information are different then there will be a 1Da discrepancy, and this will trivialise the performance of the instrument and significantly reducing the discrimination of the database search

Monoisotopic mass or average mass

It is important to select whether the peptide masses are the monoisotopic or average masses. If the incorrect option is selected then the peptide masses will have an error of 1Da, again trivialising the performance of the instrument and significantly reducing the discrimination of the database search.

Error tolerance

This option will determine the stringency of the database search and will be dependant on the mass accuracy of masses measured within the peptide mass fingerprint. Choose an error tolerance (ppm value) that will allow all the peptides in your peptide mass fingerprint to be included in the search.

If the error tolerance is too stringent then a peptide mass measure to a greater error than the tolerance limit will contribute nothing towards the score, even though it may be a peptide from the protein of interest. However, if the error tolerance is too lenient then the number of spurious matches will also increase, reducing specificity.

> Hint
>
> ***If a 100 ppm error tolerance is selected then a window of 100 ppm will be applied to all the theoretical masses in the database, only the experimental masses which have an error 100 ppm or less will have be accepted by the search.***

Extra information

Any additional information may help to constrain the search, enhancing the discrimination. Thus molecular weight and pI may be important. It may be useful to remember though that many proteins may be post-translationally modified or alternatively spliced, altering the mass and/or the pI.

Taxonomy

If the origin of the sample is known then the search can be limited to that particular species or groups of species. This is useful for fully sequenced and well characterised genomes. However if the species is not fully sequenced then it may well be useful not to restrict the search engine to just search against this species. For instance if the species of origin is mouse, it may also be advantageous to also search against rat and human; a homologous protein may be matched.

Step 15: PMF database search unsuccessful

Combined database with PMF and peptide sequence

If the PMF database search has been unsuccessful then some partial or complete sequence in combination with the PMF can be a useful approach, particularly if the sequence can be acquired with the same instrument as the peptide mass fingerprint. This approach can be performed using the MASCOT sequence query at www.matrixscience.com (see section 2.4.3 for a worked example). In this database search the same principles apply as in step 12.

The peptide mass fingerprint is inputted, and any partial sequence that has been inferred from sequence experiment can be inputted alongside the peptide mass from which it was acquired using the (seq) command:

1200.0 seq(*–abcd)

where abcd is the hypothetical amino acid sequence derived from the experimental data (obtained from a dPSD spectrum for example), *– indicates that the direction of the sequence is unknown (c-terminus – n-terminus or vice versa)

If designation of one of the residues is ambiguous then all the residues under suspicion can be inputted in the search as below:

1200.0 seq(*–uv[xy]yz)

where [xy] = amino acid residues which could not be differentiated from the experimental data.

Similarly a single peptide sequence can be searched in the same fashion, without the remainder of the peptide masses from the PMF.

A Trouble shooting

1
Two-dimensional Electrophoresis

Additional trouble shooting guides for 2-D electrophoresis are found in:

- Berkelman T, Stenstedt T. Handbook: 2-D electrophoresis using immobilized pH gradients. Principles & methods. Amersham Biosciences 80-6429-60 (1998).
- Görg A, Weiss W. Two-dimensional electrophoresis with immobilized pH gradients. In Rabilloud T, Ed. Proteome research: Two-dimensional gel electrophoresis and identification methods. Springer Berlin Heidelberg New York (2000) 107–126.

Including images of bad gels.

In the following part only a selection of cases are described, which typically occur now and then.

1.1
Isoelectric focusing in IPG strips

Symptom	*Cause*	*Remedy*
Rehydration solution is distributed unevenly within the gel strip.	Some coating in the strip holder or reswelling tray.	Wash strip holder and reswelling tray with detergent, rinse with deionised water.
	Uneven pipetting of the rehydration solution.	Pipette the solution as a streak.
	Reswelling tray or IPGphor not levelled.	Adjust the level of the reswelling tray or the IPGphor on the bench.
Rehydration liquid is left in the reswelling tray or strip holder.	Rehydration time too short.	Rehydrate at least for 6 hours without and 12 hours with sample.

Symptom	***Cause***	***Remedy***
	Liquid volume too high.	Follow the recommendations on the package.
	IPG strips improperly stored.	Always store IPG strips in the freezer, do not leave them on the bench at room temperature for too long time.
Basic part of the gel comes off during rehydration.	Surface has been damaged during removal of cover film	Always start at the acidic, pointed side to remove the cover film.
Voltage too low (8 kV not reached)	Short strips (7 cm and 11 cm) are used, 8 kV is reached in those strips only with some samples under exceptional conditions.	Nothing to worry about.
	Poor quality of urea and/or thiourea.	Use high quality urea and thiourea, remove ions with a mixed bed ion exchanger.
	Too much salt in the sample.	Remove salts by microdialysis or precipitation, replace PBS for cell washing with something non-ionic.
	TCA left in the sample from precipitation.	Add more washing steps.
Bromophenol blue band stops and does not migrate completely into the anode.	Too much salt in the sample.	See above.
Strip starts to burn at a certain position.	Too much salt in the sample.	See above.
Visible brown band develops.	Tris-chloride in the sample and rehydration solution.	Run the sample in a cup-loading strip holder.
Cover fluid (paraffin oil) leaks out of the cup loading strip holder during IEF.	High protein and salt load cause water transport that carries the oil with it.	Reduce the initial voltage and prolong the first low-voltage steps.

Symptom	Cause	Remedy
The basic part of the strip swells during IEF and becomes mechanically instable.	Many cations in the sample, for instance Tris.	Avoid adding too much Tris-base, replace it by adding 25 mmol/L spermine base, treat the sample with microdialysis, apply IEF strips soaked with deionised water between gel and electrode to accommodate ions.
Urea crystallized and IPG strip dried during IEF.	Not enough cover fluid used.	Use 3 mL cover fluid for 18 and 24 cm regular and 4 mL for cup loading strip holder.
	Cover fluid has been moved around in the strip and leaked out.	Reduce the initial voltage and prolong the first low-voltage steps.
	Running temperature was incorrect.	Set the rehydration and separation temperature to 20 °C.

1.2 SDS PAGE

Symptom	Cause	Remedy
Casting homogeneous gels: Uneven upper edge.	Improper overlay.	Carefully pipette water-saturated butanol over each monomer solution before polymerization starts.
Casting gradient gels: Curved upper edge.	Polymerization started at the bottom, and caused thermal convection, and uneven contraction of the gel.	Make sure that the polymerization starts at the top by adding less catalyst to the dense solution.
Upper buffer tank is leaking.	Hydrostatic balance between the buffers is not achieved.	Fill enough lower buffer into the tank according to the instruction.
Migration of dye front is too slow.	Poor quality of reagents and water.	Make sure, that you use only high quality of reagents, check the water.
	Upper buffer contains chloride ions.	Do not titrate the running buffer.

Symptom	*Cause*	*Remedy*
	PPA buffer system: Upper and lower buffer have mixed because of leakage of upper buffer tank.	Fill enough lower buffer into the tank according to the instruction. Do not overfill lower and upper tank.
The dye front is not straight in the beginning.	Uneven conductivity across the IPG strips originating from the fixed pH gradient.	Nothing to worry about. The dye front will become straight when it has reach about the middle of the cassette.
The dye front becomes curved during the run.	Current leakage in the ready-made gel cassette.	Remove all excess buffer with the roller. Close the cassette properly.
Irregular dye front.	Poorly polymerised gels.	Optimize catalyst amounts and use high quality reagents.
	Smiling effect because of overheating.	Use cooling during the run and/or reduce the power setting.

1.3 Staining

Symptom	*Cause*	*Remedy*
Background formed like a "sail" in the basic part of the gel.	Complexes between SDS and basic carrier ampholyte.	Increase fixing time or destaining time. Try alternative IPG buffer or carrier ampholytes.
Silver staining: negative spots.	Dependent on structure of protein.	Stain briefly with Coomassie blue first.
Silver staining: "Donut"-shaped spots.	Happens with highly abundant proteins.	Stain briefly with Coomassie blue first.
Silver staining: Dark background.	Silver nitrate reduced by sulphuric compounds, like softeners.	Use glass or stainless steel trays only.
	Bad water quality.	Check water with a few drops of silver nitrate solution before you use it for silver staining.
Sypro Ruby or RuBPS staining: Dye precipitates.	Not sufficiently washed gel or wrong tray material.	Wash the gels after staining, use only dark polypropylene containers.

1.4
DIGE fluorescence labelling

Symptom	*Cause*	*Remedy*
The pH of the protein lysate is less than pH 8 prior to labelling.	The lysis of the cells has caused a drop in the pH.	1. Increase the buffering capacity of your lysis buffer to 40 mmol/L Tris.
	The cell wash buffer was not completely removed prior to addition of the lysis buffer.	2. Increase the pH of the lysis buffer by the addition of a small volume of 50 mmol/L NaOH. Or add an equal volume of the lysis buffer that is at pH 9.5.
The fluorescent signal is weak when scanned on a 2D gel.	The dyes after reconstitution have a fixed lifetime in DMF that may have been exceeded.	Check the expiry date on CyDye.
	The DMF used to reconstitute CyDye was of poor quality or has been opened for longer than 3 months.	Always use the 99.8% anhydrous DMF to reconstitute your CyDye. Breakdown products of DMF include amines which compete with the protein for the CyDye labelling
	CyDye has been exposed to light for long periods of time.	Always store CyDye in the dark.
	CyDye has been left out of the –20 °C freezer for a long period of time.	Always store CyDye at –20 °C and only remove them for short periods to remove a small aliquot.
	The pH of the protein lysate is less than pH 8.	Increase the pH of the lysis buffer by the addition of a small volume of 50 mmol/L NaOH. Or add an equal volume of the lysis buffer that is at pH 9.5.
	Primary amines such as Pharmalytes or ampholytes are present in the labelling reaction competing with the protein for CyDye.	Omit all exogenous primary amines from the labelling reaction.

Symptom	*Cause*	*Remedy*
	DTT or other substances such as SDS are present in the labelling reaction at too high a concentration.	Remove the substances from the labelling reaction if not essential. If they are essential test if the reduction in labelling efficiency can be counterbalanced by increasing CyDye concentration.
	The protein lysate concentration is too low; i.e. less than 2 mg/mL.	Make a new batch of protein lysate reducing the volume of lysis buffer to increase the protein concentration. Or, increase the ratio of CyDye to protein. Or, precipitate the proteins and resuspend them in a smaller volume of lysis buffer; check the pH and concentration of the new sample before labelling.

1.5
Results in 2-D electrophoresis

Symptom	*Cause*	*Remedy*
No or very few spots.	Low protein content.	Check with reliable quantification method; check your entire procedure with a standard, like E. coli lyophylisate; try alternative sample extraction procedure.
	Problem with silver staining.	Check, whether all the solutions have been made correctly and no step has been forgotten.
Missing protein spots in the high molecular weight area.	High molecular weight proteins did not enter the IPG strip.	Perform rehydration with 50 V applied on the IPG strip for 12 hours.
	Equilibration of IPG strip was too short.	Increase equilibration steps to 2 × 20 minutes
Missing proteins.	Sample application on IPG strips was not optimized.	Try all alternatives, also cup loading on different sides.

Symptom	*Cause*	*Remedy*
Vertical streaking.	Dirty glass cassettes.	Clean glass plates thoroughly, touch them only with gloves.
	Inefficient equilibration of the IPG strip.	Equilibrate at least 2 × 15 minutes. Use iodoacetamide in the second equilibration step.
	Poor transfer of proteins from the IPG strips to the SDS gel.	Prolong the protein transfer phase at low power setting before you switch to the separation conditions.
Spots elongated in the vertical direction.	Depletion of the upper buffer.	Use 2 × Tris-glycine buffer in the upper buffer tank.
	Poor quality of chemicals.	Use only high quality chemicals. In case of doubt try reagents from an alternative supplier.
	High glycoprotein content.	Use gradient gels and Tris-borate in the cathodal tank instead of Tris-glycine.
Vertical streak in the acidic area, when thiourea was used in the first dimension.	Contaminated reagent.	Try another batch of thiourea.
Vertical gap(s).	Air bubbles between first and second dimension gel.	Place the IPG strip carefully on the SDS gel and seal it carefully with agarose.
	Amphoteric buffer in the cell culture, for instance HEPES, occupies a part of the pH gradient.	Avoid amphoteric buffers.
Double spots in vertical direction.	IPG strip not applied in the correct way.	Plastic film has to be in contact with the glass plate.
Multiple spots in vertical direction in preparative gels.	Insufficient DTT amount	Increase amount of DTT during the first equilibration step.
Clouds of spots in the low M_r area.	Peptides from protein digestion.	Add protease inhibitors to the sample; use cup loading or rehydration loading under voltage.
Horizontal streaking	Under focusing.	Prolong the volthours in the last IEF step.

Symptom	*Cause*	*Remedy*
	Over focusing: labile proteins are degrading at their pI. Happens particularly in the basic range.	Shorten the volthours in the last IEF step.
	Urea and/or detergent concentration too low.	Increase concentrations of urea to 9 mol/L and CHAPS to 2 or 4 %.
	Salt content is too high.	Treat the sample with microdialysis or precipitation.
	Incomplete rehydration of the IPG strip.	Check the rehydration conditions.
	Instability of some proteins because of wrong sample application.	Try alternatives, like cup loading at the acidic or basic end of the IPG strip.
	DTT depletion in the basic part of the gradient.	Apply a filter paper pad soaked in 1 % DTT solution between cathode and gel.
	Electroendosmosis effects at the ends of narrow interval IPG strips because of accumulated proteins of higher and lower pIs.	Place filter paper pads soaked in deionised water between electrodes and gel during the last IEF step.
Protein spots arranged like a string of beads in horizontal direction.	Differential carbamylation of some proteins because of presence of isocyanate.	Do not heat sample. Do not store urea and thiourea solutions at room temperature. Use only high quality urea and thiourea. Remove ionic compounds with mixed bed ion exchanger.
	Storage proteins in plant seed extracts, differentially glycosylated.	No artifact. Nothing to worry about.
Cloudy background.	Micelles between SDS and the nonionic or zwitterionic detergent have formed.	Reduce content of detergent in the rehydration solution for the IPG strip.
Blurred and streaky spot pattern in preparative gels.	Insufficient IEF conditions	Prolong the volthours in the last IEF step by 15 %.

Symptom	*Cause*	*Remedy*
	DTT depletion in preparative basic gels.	Add 2.5 % DTT, 20 % isopropanol, 5 % glycerol to the rehydration solution, apply paper wick soaked in rehydration solution plus 3.5 % DTT at the cathode.
Blurred spot pattern in ready-made gels.	Liquid and/or bubbles between surface of ready-made gel and glass plate.	Use only low amount of gel buffer and roll out excess buffer and air bubbles completely.
Very basic proteins are lost.	Too many volthours applied on basic gradients, autohydrolysis of basic buffering groups in the gel.	Shorten the volthours in the last IEF step.
"Wrong" proteins from other organisms, which should not be present, have been identified in the gel.	Contamination with previous sample or proteins from the laboratory environment.	Use only highly pure reagents; clean equipment thoroughly, particularly reswelling tray and strip holders for IPG strips; filter solutions through a membrane filter.

2 Mass spectrometry

Symptom	*Cause*	*Remedy*
No signal in reflectron mode MALDI MS	No accelerating voltage	Check read out from electronics gate
	Laser not firing	Contact supplier
	Laser not firing in correct place	Re-alignment of laser required
	Laser power far too low	Increase laser power
	Potentially a detector or amplifier problem	Check signal in linear mode.
Poor analyte signal	Insufficient protein present in gel plug for digestion	Run a new gel with a higher protein load.
	Trypsin inactivity	Prepare trypsin fresh on ice. Perform rehydration step on ice Ensure correct buffer for trypsin activity.

Symptom	***Cause***	***Remedy***
	Sample concentration too low	Use 0.1–10 pmol/μL final conc
	Sample concentration too high	Sample signal may be suppressed. Dilute sample to 0.1–10 pmol/μL final conc
	Poor crystallisation, presence of salt/buffer	Use a higher concentration on TFA (up to 1%) to improve ionisation. Avoid phosphorylated/sulphated buffers
	Poor crystallisation, presence of salt/buffer	Microscale purification to remove contaminants
	Poor crystallisation, presence of non-volatile contaminants	Eliminate non-volatile components from sample before analysis. Do not use non-volatile components to solubilise sample.
	Poor crystallisation, presence of visualisation agent; i.e. CBB	Destain gel plug prior to digestion
	Laser power below required threshold	Increase laser power
	Old matrix	Prepare fresh matrix
	Unsuitable matrix	See table 10 for correct matrix selection
	Known contaminants, keratin, dominate and suppress analyte signal	Address sample handling prior to digestion. See page 255 ff
	Poor digestion and extraction efficiency	Improve digestion and extraction procedure.
Poor resolution across mass range of interest	Pulse time and voltage not optimised for mass range of interest	See manufacturers recommendations
	Laser poor far above threshold level	Reduce laser power to optimal point
	Incorrect reflectron voltage	See manufacturers recommendations
Poor mass accuracy across mass range of interest	Poor resolution, unable to select ^{12}C peak of the isotopic envelope.	Improve resolution, see above.
	Unable to calibrate, one or both internal calibrants absent	Perform external calibration. In worst case scenario, respot sample with calibrants added to the mixture.

Symptom	*Cause*	*Remedy*
	Incorrectly calibrated	Check internal calibrants and their masses.
Unsuccessful or ambiguous protein ID from PMF	Insufficient peptides	Acquire more specific information; combine PMF with composition or partial sequence. Acquire actual sequence
	Protein does not exist in the database or low homology with other proteins	Acquire sequence data from product ion MS/MS or PSD experiments
	Sample is a mixture	Acquire more specific information.
	Error tolerance too high due to poor mass accuracy	Constrain the database search by improving mass accuracy
	Keratin contamination	Remove known keratin peaks from mass list (see step 6) and repeat search
	Incorrect database search settings	Check settings: enzyme, error tolerance, fixed and potential modifications, taxonomy, charge state and mono-isotopic or average masses.
	C12 isotope not selected from mass list	
Non quantitative derivatisation-CAF	Reagent not freshly prepared	Always prepare reagent freshly and use immediately
	Residual ammonium bicarbonate	Perform a micro-scale purification
	Incorrect ratio of reagent to sample	Ensure reaction conditions.
	Reaction pH not optimal	Ensure correct reaction conditions.
	Reaction pH not optimal	Ensure correct reaction conditions.

References

Altland K. IPGMAKER: A program for IBM-compatible personal computers to create and test recipes for immobilized pH gradients. Electrophoresis 11 (1990) 140–147.

Amersham Biosciences Datafile: Immobiline DryStrip visualization of pH gradients (2000) 18-1140-60.

Amersham Biosciences Handbook: Fluorescence imaging: principles and methods (2000) 63-0035-28.

Anderson BL, Berry RW, Telser A. A sodium dodecyl sulfate-polyacrylamide gel electrophoresis system that separates peptides and proteins in the molecular weight range of 2,500 to 90,000. Anal Biochem 132 (1983) 365–275.

Anderson L, Anderson NG. High resolution two-dimensional electrophoresis of human plasma proteins. Proc Nat Acad Sci USA 74 (1977) 5421–5425.

Anderson NG, Anderson NL. Analytical techniques for cell fractions XXI. Two-dimensional analysis of serum and tissue proteins: multiple isoelectric focusing. Anal Biochem 85 (1978) 331–340.

Anderson NG, Anderson NL. Analytical techniques for cell fractions XXII. Two-dimensional analysis of serum and tissue proteins: multiple gradient-slab electrophoresis. Anal Biochem 85 (1978) 341–354.

Anderson NG, Anderson NL, Tollaksen SL. Proteins of human urine. I. Concentration and analysis by two-dimensional electrophoresis. Clin Chem 25 (1979) 1199–1210.

Anderson NG, Anderson NL. The human protein index. Clin Chem 28 (1982) 739–748–1210.

Anderson L, Seilhamer J.A Comparison of selected mRNA and protein abundances in human liver. Electrophoresis 18 (1997) 533–537.

Anderson NL, Anderson NG. Proteome and proteomics: new technologies, new concepts and new words. Electrophoresis 19 (1998) 1853–1861.

Anderson UN, Colburn AW, Makarov AA, Raptakis EN, Reynolds DJ, Derrick PJ, Davis SC, Hoffman AD, Thomson S. In-series combination of a magnetic-sector mass spectrometer with a time-of-flight quadratic-field ion mirror. Rev Sci Instrum 69 (1998) 1650–1660.

Annan RS, Huddleston MJ, Verna R, Deshaies RJ, Carr SA; A multidimensional electrospray MS-based approach to phosphopeptide mapping. Anal Chem 73 (2001) 393–404.

Axelsson J, Boren M, Naven TJP, Fenyö D. Stringency in database searches for protein identification: determining the level of mass error at which search results are incorrect. Proceedings of the 49th ASMS conference on mass spectrometry and allied topics, Chicago (2001).

Baldwin MA, Medzihradszky KF, Lock CM, Fisher B, Settineri TA, Burlingame AL. Matrix-assisted laser desorption/ionisation coupled with quadrupole/orthogonal acceleration time-of-flight mass spectrometry for protein discovery, identification and structural analysis. Anal Chem 73 (2001) 1707–1720.

Banks R. Dunn MJ, Forbes MA, Stanly A, Pappin DJ, Naven T, Gough M, Harnden P, Selby PJ.The potential use of laser capture microdissection to selctively obtain distinct populations of cells for proteomic analysis. – preliminary findings. Electrophoresis 20 (1999) 689–700.

Barber M, Bordoli RS, Sedgwick RD, Tyler AN. Fast atom bombardment of solids as an ion source in mass spectrometry. Nature 293 (1981) 270–271.

Barrett T, Gould HJ. Tissue and species specifity on non-histone chromatin proteins. Biochim Biophys Acta 294 (1973) 165–170.

Beavis RC, Chait BT, Cinnamic acid derivatives as matrices for ultraviolet laser desorption mass spectrometry of proteins; Rapid Commun Mass Spectrom 3 (1989) 432–435.

Beavis RC, Chaudhary T, Chait BT; α-cyano-4-hydroxycinnamic acid as a matrix for matrix assisted laser desorption mass spectrometry. Org Mass Spectrom 27 (1992) 156–158.

Bell AW, Ward MA, Blackstock WP, Freeman HNM, Choudhary JS, Lewis AP, Chotai D, Fazel A, Gushue JN, Paiement J. Proteomics characterisation of abundant Golgi membrane proteins. J Biol. Chem 276 (2001) 5152–5165.

Berkelman T, Stenstedt T. Handbook: 2-D electrophoresis using immobilized pH gradients. Principles & methods. Amersham Biosciences 80–6429–60 (1998).

Bienvenut WV, Sanchez JC, Karmime A, Rouge V, Rose K, Binz PA, Hochstrasser DF. Toward a clinical molecular scanner for proteome research: parallel protein chemical processing before and during western blot. Anal Chem 71 (1999) 4800–4807.

Bjellqvist B, Ek K, Righetti PG, Gianazza E, Görg A, Westermeier R, Postel W. Isoelectric focusing in immobilized pH gradients: principle, methodology, and some applications. J Biochem Biophys Methods 6 (1982) 317–339.

Bjellqvist B, Hughes GJ, Pasquali C, Paquet N, Ravier F, Sanchez J-C, Frutiger S, Hochstrasser D. The focusing positions of polypeptides in immobilized pH gradients can be predicted from their amino acid sequences. Electrophoresis 14 (1993) 1023–1031.

Bjellqvist B, Sanchez J-C, Pasquali C, Ravier F, Paquet N, Frutiger S, Hughes GJ, Hochstrasser D. Micropreparative two-dimensional electrophoresis allowing the separations of samples containing milligram amounts of proteins. Electrophoresis 14 (1993) 1375–1378.

Bjellqvist B, Basse B, Olsen E, Celis JE. Reference points for comparisons of two-dimensional maps of proteins from different human cell types defined in a pH scale where isoelectric points correlate with polypeptide compositions. Electrophoresis 15 (1994) 529–539.

Blomberg A, Blomberg L, Norbeck J, Fey SJ, Larsen PM, Roepstorff P, Degand H, Boutry M, Posch A, Görg A. Interlaboratory reproducibility of yeast protein patterns analized by immobilized pH gradient two-dimensional gel electrophoresis. Electrophoresis 16 (1995) 1935–1945.

Boutry M, Posch A, Görg A. Interlaboratory reproducibility of yeast protein patterns analized by immobilized pH gradient two-dimensional gel electrophoresis. Electrophoresis 16 (1995) 1935–1945.

Boyle JG and Whitehouse CM. Time-of-flight mass spectrometry with an electrospray ion beam. Anal Chem 64 (1992) 2084–2089.

Brancia FL, Oliver SG, Gaskell SJ. Improved matrix-assisted laser desorption mass spectrometric analysis of tryptic hydrolysates of proteins following guanidation of lysine-containing peptides. Rapid Commun Mass Spectrom 14 (2000) 2070–2073.

Brown RS and Lennon JJ. Mass resolution improvement by incorporation of pulsed ion extraction/ionisation linear time-of-flight mass spectrometry. Anal Chem 67 (1995) 1998–2003.

Burkhard PR, Rodrigo N, May D, Sztajzel R, Sanchez J-C, Hochstrasser DF, Shiffer E, Reverdin A, Lacroix JS. Assessing cerebrospinal fluid rhinorrhea: A two-dimensional electrophoresis approach. Electrophoresis 22 (2001) 1826–1833.

Carr SA, Huddleston MJ, Annan RS; Selective detection and sequencing of phophopeptides at the femtomole level by mass spectrometry. Anal Biochem 239 (1996) 180–192.

Celentano F, Gianazza E, Dossi G, Righetti PG. Buffer systems and pH gradient simulation.Chemometr Intel Lab Systems 1 (1987) 349–358.

Chaurand P, Luetzenkirchen F, Spengler B. Peptide and protein identification by matrix assisted laser desorption and MALDI post-source decay time-of-flight mass spectrometry. J Am Soc Mass Spectrom 10 (1999) 91–103.

Chevallet C, Santoni V. Poinas A, Rouquie D, Fuchs A, Kieffer S, Rossignol M, Lunardi J, Gerin J, Rabilloud T. New zwitterionic detergents improve the analysis of membrane proteins by two-dimensional electrophoresis. Electrophoresis 19 (1998) 1901–1909.

Clauser KR, Baker P, Burlingame AL. Role of accurate mass measurement (± 10 ppm) in protein identification strategies employing MS or MS/MS and database searching. Anal Chem 71 (1999) 2871–2882.

Cleland WW. Dithiothreitol, a new protective reagent for SH groups. Biochemistry 3 (1964) 480–482.

Cornish TJ and Cotter RJ. A curved field reflectron time-of-flight mass spectrometer for the simultaneous focusing of metastable ions. Rapid Commun Mass Spectrom 8 (1994) 781–785.

Covey TR, Huang EC, Henion JD. Structural characterisation of protein tryptic peptides by liquid chromatography, mass spectrometry and collision induced dissociation of their doubly charged molecular ions. Anal Chem 63 (1991) 1193–1200.

Damerval C, DeVienne D, Zivy M, Thiellement H. Technical improvements in two-dimensional electrophoresis increase the level of genetic variation detected in wheat-seedling protein. Electrophoresis 7 (1986) 53–54.

Dunbar BS. Two-dimensional electrophoresis and immunological techniques. Plenum Press, New York (1987).

Dunn MJ, Burghes AHM. High-resolution two-dimensional polyacrylamide gel electrophoresis. I. Methodological procedures. Electrophoresis 4 (1983a) 97–116.

Dunn MJ, Burghes AHM. High-resolution two-dimensional polyacrylamide gel electrophoresis. II. Analysis and applications. Electrophoresis 4 (1983b) 173–189.

Dunn MJ. Gel electrophoresis of proteins. Bios Scientific Publishers Alden Press, Oxford (1993).

Dunn MJ. Detection of total proteins on western blots of 2-D polyacrylamide gels. In. Link AJ. Ed. 2-D Proteome Analysis Protocols. Methods in Molecular Biology 112. Humana Press, Totowa, NJ (1999) 319–329.

Dunn MJ, Ed. From genome to proteome. Advances in the practice and application of proteomics. WILEY-VCH, Weinheim (1999).

Emmett MR and Caprioli RM. Micro-electrospray mass spectrometery: ultra high sensitivity analysis of peptides and proteins. J Am Soc Mass Spectrom 5 (1994) 605–613.

Eng JK, McCormack AL, Yates JR III. An approach to correlate tandem mass spectra data of peptides with amino acid sequences in protein databases. J Am Soc Mass Spectrom 5 (1994) 976–989.

Fenn JB, Mann M, Meng CK, Wong SK, Whitehouse CM. Electrospray ionisation for mass spectrometry of large biomolecules. Science 246 (1989) 64–71.

Fenyö D, Qin J, Chait BT. Protein identification using mass spectrometric information. Electrophoresis 19 (1998) 998–1005.

Fernandez-Patron C, Castellanos-Serra L, Hardy E, Guerra M, Estevez E, Mehl E, Frank RW. Understanding the mechanism of the zinc-ion stains of biomacromolecules in electrophoresis gels: Generalization of the reverse-staining technique. Electrophoresis 19 (1998) 2398–2406.

Field S, Song O. A novel genetic system to detect protein-protein interactions. Nature 340 (1989) 245–246.

Fountoulakis M, Takacs B, Langen H. Two-dimensional map of basic proteins of Haemophilus influenza. Electrophoresis 19 (1998) 761–766.

Gavin AC, Bösche M, Krause R, Grandi P, Marzioch M, Bauer A, Schultz J, Rick JM, Michon AM, Cruciat CM, Remor M, Höfert C, Schelder M, Brajenovic M, Ruffner H, Merino A, Klein K, Hudak M, Dickson D, Rudi T, Gnau V, Bauch A, Bastuck S, Huhse B, Leutwein C, Heurtier MA, Copley RR, Edelmann A, Querfurth E, Rybin V, Drewes G, Raida M, Bouwmeester T, Bork P, Seraphin B, Kuster B, Neubauer G, Superti-Furga G. Functional organisation of the yeast proteome by systematic analysis of protein complexes. Nature 415 (2002) 141–147.

Gevaert K, De Mol H, Sklyarova T, Houthaeye T, Vandekerckhove J. A peptide concentration and purification method for protein characterisation in the subpicomole range using matrix assisted laser desorption/ionisation-postsource decay (MALDI-PSD) sequencing. Electrophoresis 19 (1998) 909–917.

Gevaert K, Eggermont L, Demol H, Vandekerckhove J. A fast and convenient MALDI-MS based proteomic approach; identification of components scaffolded by the actin cytoskeleton of activated human thrombocytes. J Biotechnol 78 (2000) 259–269.

Gevaert K, Demol H, Martens L, Hoorelbeke B, Puype M, Goethals M, Van Damme J, De Boeck S, Vandekerckhove J. Protein identification based on matrix assistedd laser desorption/ionisation-post source decay-mass spectrometry. Electrophoresis 22 (2001) 1645–1651.

Gharahdaghi F, Weinberg CR, Meagher DA, Imai BS, Mische SM; Mass spectrometric identification of proteins from silver-stained ployacrylamide gel: A method for the removal of silver ions to enhance sensitivity. Electrophoresis 20 (1999) 601–605.

Gharbi S, Gaffney P, Yang A, Zvelebil MJ, Cramer R, Waterfield MD, Timms JF. Evaluation of 2D-differential gel electrophoresis for proteomic expression analysis of a model breast cancer cell system. Mol Cell Proteomics (2002) in press.

Goodlett DR, Bruce JE, Anderson GA, Rist B, Pasa-Tolic L, Fiehn O, Smith RD, Aebersold R. Protein identification with a single accurate mass of a cysteine-containing peptide and constrained database searching. Anal Chem 72 (2000) 1112–1118.

Görg A, Postel W, Günther S, Weser J. Improved horizontal two-dimensional electrophoresis with hybrid isoelectric focusing in immobilized pH gradients in the first dimension and laying-on transfer to the second dimension. Electrophoresis 6 (1985) 599–604.

Görg A, Postel W, Weser J, Günther S, Strahler JR, Hanash SM, Somerlot L. Horizontal two-dimensional electrophoresis with immobilized pH gradients in the first dimension in the presence of nonionic detergent. Electrophoresis 8 (1987a) 45–51.

Görg A, Postel W, Weser J, Günther S, Strahler JR, Hanash SM, Somerlot L. Elimination of point streaking on silver stained two-dimensional gels by addition of iodoacetamide to the equillibration buffer. Electrophoresis 8 (1987b) 122–124.

Görg A, Postel W, Weser J, Günther S, Strahler JR, Hanash SM, Somerlot L, Kuick R. Electrophoresis 9 (1988a) 37–46.

Görg A, Postel W, Günther S, Friedrich C. Horizontal two-dimensional electrophoresis with immobilized pH gradients using PhastSystem. Electrophoresis 9 (1988b) 57–59.

Görg A, Postel W, Günther S. Review. The current state of two-dimensional electrophoresis with immobilized pH gradients. Electrophoresis 9 (1988c) 531–546.

Görg A, Postel W, Friedrich C, Kuick R, Strahler JR, Hanash SM. Temperature-dependent spot positional variability in two-dimensional polypeptide patterns. Electrophoresis 12 (1991) 653–658.

Görg A. High-resolution two-dimensional electrophoresis of proteins using immobilized pH gradients. In: Celis J, Ed. Cell Biology: A Laboratory Handbook. Academic Press Inc., San Diego, CA. (1994) 231–242.

Görg A, Boguth G, Obermaier C, Posch A, Weiss W. Two-dimensional polyacrylamide gel electrophoresis with immobilized pH gradients in the first dimension (IPG-Dalt): The state of the art and the controversy of vertical *versus* horizontal systems. Electrophoresis 16 (1995) 1079–1086.

Görg A, Obermaier C, Boguth G, Csordas A, Diaz J-J, Madjar J-J: Very alkaline immobilized pH gradients for two-dimensional electrophoresis of ribosomal and nuclear proteins. Electrophoresis 18 (1997) 328–337.

Görg A, Boguth G, Obermaier C, Harder A, Weiss W: 2-D electrophoresis with immobilized pH gradients using IPGphor isoelectric focusing system. Life Science News 1 (1998) 4–6.

Görg A, Boguth G, Obermaier C, Weiss W: Two-dimensional electrophoresis of proteins in an immobilized pH 4–12 gradient. Electrophoresis 19 (1998) 1516–1519.

Görg A, Obermaier C, Boguth G, Weiss W: Recent developments in Wide pH gradients up to pH 12, longer separation distances and simplified procedures. Electrophoresis 20 (1999) 712–717.

Görg A, Obermaier C, Boguth G, Harder A, Scheibe B, Wildgruber R, Weiss W: The current state of two-dimensional electrophoresis with immobilized pH gradients. Electrophoresis 21 (2000) 1037–1053.

Görg A, Weiss W. Two-dimensional electrophoresis with immobilized pH gradients. In Rabilloud T, Ed. Proteome research: Two-dimensional gel electrophoresis and identification methods. Springer, Berlin Heidelberg New York (2000) 107–126.

Gyenes T, Gyenes E. Effect of "stacking" on the resolving power of ultrathin-layer two-dimensional gel electrophoresis. Anal Biochem 165 (1987) 155–160.

Gygi SP, Rist B, Gerber SA, Turecek F, Gelb MH, Aebersold R. Quantitative analysis of complex protein mixtures using isotope-coded affinity tags. Nature Biotech 17 (1999) 994–999.

Hanash SM, Strahler JR, Neel JV, Hailat N, Melham R, Keim D, Zhu XX, Wagner D, Gage DA, Watson JT. Highly resolving two-dimesnional gels for protein sequencing. Proc Natl Acad Sci USA 88 (1991) 5709–5713.

Hannig K. New aspects in preparative and analytical continuous free-flow cell electrophoresis. Electrophoresis 3 (1982) 235–243.

Hardy E, Santana H, Sosa, Hernandez L, Fernandez-Patron C, Castellanos-Serra L. Imidazole-sodium dodecyl sulphate-zinc (reverse stain) on sodium dodecyl sulphate gels. Anal Biochem 240 (1996) 150–152.

Harrington MG, Merrill C. Cerebrospinal fluid protein analysis in diseases of the nervous system. J Chromatogr 429 (1988) 345–358.

Harvey DJ. Quantitative aspects of the matrix-assisted laser desorption mass spectrometry of complex oligosaccharides. Rapid Commun Mass Spectrom 7 (1993) 614–619.

Hashimoto F, Horigome T, Kanbayashi M, Yoshida K, Sugano H. An improved method for separation of low-molecular weight polypeptides by electrophoresis in sodium dodecyl sulfate-polyacrylamide gel. Anal Biochem 129 (1983) 192–199.

Henzel WJ, Billeci TM, Stults JT, Wong SC, Grimley C, Watanabe C. Identifying proteins from two-dimensional gels by molecular mass searching of peptide fragments in protein sequence databases. Proc Natl Acad Sci USA 90 (1993) 5011–5015.

Herbert BR, Molloy MP, Gooley AA, Walsh BJ, Bryson WG, Williams KL Improved protein solubility in two-dimensional electrophoresis using tributyl phosphine as reducing agent. Electrophoresis 19 (1998) 845–851.

Herbert B, Righetti PG. A turning point in proteome analysis: Sample prefractionation via multicompartment electrolyzers with isoelectric membranes. Electrophoresis 21 (2000) 3639–3648.

Herbert B, Galvani M, Hamdan M, Olivieri E, MacCarthy J, Pedersen S, Righetti PG. Reduction and alkylation of proteins in preparation of two-dimensional map analysis: why, when, and how? Electrophoresis 22 (2001) 2046–2057.

Heukeshoven J, Dernick R. Simplified method for silver staining of proteins in polyacrylamide and the mechanism of silver staining. Electrophoresis 6 (1985) 103–112.

Ho Y, Gruhler A, Heilbut A, Bader GD, Moore L, Adams SL, Taylor P, Bennett K, Boutilier K, Yang L, Wloting C, Donaldson I, Schandorff S, Shewarane J, Vo M, Taggart J, Goudreault M, Muskat B, Alfarano C, Dewar D, Lin Z, Michalichova K, Wilems AR, Sassi H, Nielsen PA, Rasmussedn KJ, Anderson JR, Johansen LE, Hansen LH, Jespersen H, Podtelejnikov A, Nielsen E, Crawford J, Poulsen V, Sørensen BD, Matthiesen J, Hendrickson RC, Gleeson F, Pawson T, Moran MF, Durocher D, Mann M, Hogue CWV, Figeys D, Tyers M. Systematic identification of protein complexes in *Saccharomyces cerevisiae* by mass spectrometry. Nature 415 (2002) 180–183.

Hoess M, Robins P, Naven TJP, Pappin DJC, Sgouros T, Lindahl T. A human DNA editing enzyme homologous to the *E coli* DnaQ/MutD protein. EMBO J 18 (1999) 3868–3875.

Hoving S, Voshol H, van Oostrum J. Towards high performance two-dimensional gel electrophoresis using ultrazoom gels. Electrophoresis 21 (2000) 2617–2621.

Hoving S, Gerrits B, Voshol H, Müller D, Roberts RC, van Oostrum J. Preparative two-dimensional gel electrophoresis at alkaline pH using narrow range immobilized pH gradients. Proteomics 2 (2002) 127–134.

Hughes GJ, Frutiger S, Paquet N, Ravier F, Pasquali C, Sanchez JC, James R, Tissot JD, Bjellqvist B, Hochstrasser DF. Plasma protein map: an update by microsequencing. Electrophoresis 13 (1992) 707–714.

Hunt DF, Yates JR III, Shabanowitz J, Winston S, Hauer CR. Protein sequencing by tandem mass spectrometry. Proc Natl Acad Sci USA 83 (1986) 6233–6237.

Hurkman WJ, Tanaka CK. Solubilization of plant membrane proteins for analysis by two-dimensional gel electrophoresis. Plant Physiology 81 (1986) 802–806.

Husi H, Ward M, Choudhary JS, Blackstock WP, Grant SGN. Proteomic analysis of NMDA receptor-adhesion protein signalling complexes. Nature Neurosci 3 (2000) 661–669.

Ibel K, May RP, Kirschner K, Szadkowski H, Mascher E, Lundahl P. Protein-decorated micelle structure of sodium-dodecyl-sulfate protein complexes as determined by neutron scattering. Eur J Biochem 190 (1990) 311–318.

Ito T, Chiba T, Ozawa R, Yoshido M, Hattori M, Sakaki Y. A comprehensive two-hybrid analysis system to explore the yeast protein interactome. Proc Natl Acad Sci USA 98 (2001) 4569–4574.

Iglesias T, Cabrera-Poch N, Mitchell MP, Naven TJP, Rozengurt E, Schiavo G. Identification and cloning of Kidind220, a novel neuronal substrate of protein kinase D. J Biol Chem 275 (2000) 40048–40056.

James P, Quadroni M, Carafoli E and Gonnet G. Protein identification by mass profile fingerprinting. Biochem Biophys Res Commun 195 (1993) 58–64.

James P, Quadroni M, Carafoli E, Gonnet G. Protein identification in DNA databases by peptide mass fingerprinting. Protein Sci 3 (1994) 1347–1350.

Jensen ON, Podtelejnikov A, Mann M. Delayed extraction improves specificity in database searches by matrix-assited laser desorption/ionisation peptide maps. Rapid Commun Mass Spectrom 10 (1996) 1371–1378.

Johnson RS, Martin SA, Biemann K. Collision-induced fragmentation of $(M + H)^+$ ions of peptides. Side chain specific sequence ions Int J Mass Spectrom Ion Proc 86 (1988) 137 –154.

Johnston RF, Pickett SC, Barker DL. Autoradiography using storage phosphor technology. Electrophoresis 11 (1990) 355–360.

Jonscher KR, Yates JR3rd, The quadrupole ion trap mass spectrometer- a small solution to a big challenge. Anal Biochem 244 (1997) 1–15.

Karas M, Hillenkamp F. Laser desorption of proteins with molecular masses exceeding 10000 daltons. Anal Chem 60 (1988) 2299–2301.

Kauffman R, Chaurand P, Kirsch D, Spengler B. Post-source decay and delayed extraction in matrix-assited laser desorption/ionisation-reflectron time-of-flight mass spectrometry. Are there trade-offs? Rapid Commun Mass Spectrom 10 (1996) 1199–1208.

Kauffman R, Kirsch D, Spengler B. Sequencing of peptides in a time-of-flight mass spectrometer: evaluation of post-source decay following matrix-assisted laser desorption ionisation (MADLI). Int J Mass Spectrom Ion Proc 131 (1994) 355–385.

Kellner R, Lottspeich F, Meyer H, Eds. Microcharacterization of proteins. Second edition. WILEY-VCH Weinheim (1999).

Keough T, Youngquist RS, Lacey MP. A method for high sensitivity peptide sequencing using postsource decay matrix-assisted laser desorption ionisation mass spectrometry. Proc Natl Acad Sci USA 96 (1999) 7131–7136.

Keough T, Lacey MP, Fieno AM, Grant RA, Sun Y, Bauer MD, Begley KB. Tandem mass spectrometry methods for definitive protein identification in proteomics research. Electrophoresis 21 (2000) 2252–2265.

Keough T, Lacey MP, Youngquist RS. Derivatisation procedures to facilitate de novo sequencing of lysine-terminated tryptic peptides using post-source decay matrix-assisted laser -desorption/ionisation mass spectrometry. Rapid Commun Mass Spectrom 14 (2000a) 2348–2356.

Kondo H, Rabouille C, Newman R, Levine TP, Pappin D, Freemont P, Warren G. p47 is a cofactor for p97-mediated membrane fusion. Nature 388 (1997) 75–78.

Kleine B, Löffler G, Kaufmann H, Scheipers P, Schickle HP, Westermeier R, Bessler WG. Reduced chemical and radioactive liquid waste during electrophoresis using polymerized electrode gels. Electrophoresis 13 (1992) 73–75.

Klose J, Kobalz U. Two-dimensional electrophoresis of proteins: an updated protocol and implications for a functional analysis of the genome. Electrophoresis 16 (1995) 1034–1059.

Kussmann M, Nordhoff E, Rahbek-Nielsen H, Haebel S, Rossel-Larsen M, Jakobsen L, Gobom J, Mirgorodskaya E, Kroll Kristensen A, Palm L, Roepstorff P. Matrix-assited laser desorption/ionisation mass spectrometry sample preparation techniques for various peptide and protein analysis. J Mass Spectrom 32 (1997) 593–601.

Laiko VV, Baldwin MA, Burlingame AL. Atmospheric pressure matrix-assisted laser desorption/ionisation mass spectrometry. Anal Chem 72 (2000) 652–657.

Lämmli UK. Cleavage of structural proteins during the assembly of the head of bacteriophage T4. Nature 227 (1970) 680–685.

Lamond AI, Mann M. Cell biology and the genome projects – a concerted strategy for characterizing multi-protein complexes using mass spectrometry. Trends Cell Biol 7 (1997) 139–142.

Langen H, Röder D, Juranville J-F, Fountoulakis M. Effect of protein application mode and acrylamide concentration on the resolution of protein spots separated by two-dimensional gel electrophoresis. Electrophoresis 18 (1997) 2085–2090.

Langen H, Röder D. Separation of human embryonic kidney cells on narrow range pH strips. Life Science News 3 (1999) 6–8.

Langen H, Takács B, Evers S, Berndt P, Lahm H-W, Wipf, B, Gray C, Fountoulakis M. Two-dimensional map of the proteome of haemophilus influenzae. Electrophoresis 21 (2000) 411–429.

Li A, Sowder RC, Henderson LE, Moore SP, Garfinkel DJ and Fisher RJ; Chemical cleavage at aspartyl residues for protein identification. Anal Chem 73 (2001) 5395–5402.

Liang, P, Pardee AB. Differential display of eukaryotic messenger RNA by means of the polymerase chain reaction. Science 257 (1992) 967–971.

Liminga M, Borén M, Åström J, Carlsson U, Keough T, Maloisel JL, Palmgren R, Youngquist S. Water-stable chemistry for improved amino acid sequencing by derivatization post-source decay (dPSD) using Ettan MALDI-ToF with a quadratic field reflectron. Proceedings ASMS conference on mass spectrometry and allied topics, Chicago, US (2001).

Link AJ, Carmack E, Yates JR III. A strategy for the identification of proteins localised to subcellular spaces: application to E. coli periplasmic proteins. Int J Mass Spectrom Ion Proc 160 (1997) 303–316.

Link AJ, Robison K, Church GM. Comparing the predicted and observed properties of proteins encoded in the genome of Escherichia coli K-12. Electrophoresis 18 (1997) 1259–1313.

Link AJ. Ed. 2-D Proteome Analysis Protocols. Methods in Molecular Biology 112. Humana Press, Totowa, NJ (1999).

Link AJ, Eng J, Schieltz DM, Carmack E, Mize GJ, Morris DR, Garvik BM, Yates JR III. Direct analysis of protein complexes using mass spectrometry. Nature Biotech 17 (1999) 676–682.

Lobada A, Krutchinsky A, Bromirski M, Ens W, Standing KG. A tandem quadrupole/time-of-flight mass spectrometer with a matrix-assisted laser desorption/ionisation source: design and performance. Rapid Commun Mass Spectrom 14 (2000) 1047–1057.

Louris, JN, Amy, JW, Ridley, TY and Cooks, RG, Injection of ions into a quadrupole ion trap mass spectrometer. Int J Mass Spectrom Ion Proc 88 (1989) 97–111.

Mann M, Højrup P, Roepstorff P. Use of mass spectrometric molecular weight information to identify proteins in sequence databases. Biol Mass Spectrom 22 (1993) 338–345.

Mann M, Wilm M. Error-tolerant identification of peptides in sequence databases by peptide sequence tags. Anal Chem 66 (1994) 4390–4399.

Mann M. A shortcut to interesting human genes: peptide sequence tags, expressed-sequence tags and computers. Trends Biol Sci 21 (1996) 494–495.

Martin SE, Shabonowitz J, Hunt DF, Marto JA; Subfemtomole MS and MS/MS peptide sequencing using nano-HPLC micro-ESI fourier transform ion cyclotron mass spectrometry. Anal Chem 72 (2000) 4266–4274.

Mastro R, Hall M. Protein delipidation and precipitation by Tri-*n*-butylphosphate, acetone, and methanol treatment for isoelectric focusing and two-dimensional gel electrophoresis. Anal Biochem 273 (1999) 313–315.

Matsui NM, Smith DM, Clauser KR, Fichmann J, Andrews LE, Sullivan CM, Burlingame AL, Epstein LB. Immobilized pH gradient two-dimensional gel electrophoresis and mass spectrometric identification of cytokine-regulated proteins in ME-180 cervical carcinoma cells. Electrophoresis 18 (1997) 409–417.

McCormack AL, Schieltz DM, Goode S, Yang G, Barnes D, Drubin D, Yates JR III. Direct analysis and identification of proteins in mixtures by LC-MS/MS and database searching at the low femtomole level. Anal Chem 69 (1997) 767–776.

McFarlane RD, Torgerson DF. Californium-252 plasma desorption mass spectroscopy. Science 191 (1976) 920–925.

McGillivray J, Rickwood D. The heterogeneity of mouse-chromatin nonhistone proteins as evidenced by two-dimensional polyacrylamide-gel electrophoresis and ion-exchange chromatography. Europ J Biochem 41 (1974) 181–190.

Medzihradszky KF, Campbell JM, Baldwin MA, Falick AM, Juhasz P, Vestal ML, Burlingame AL. The characteristics of peptide collision-induced dissociation using a high performance MALDI-TOF/TOF tandem mass spectrometer. Anal Chem 72 (2000) 552–558.

Mirgorodskaya OA, Shevchenko AA, Chernushevich IV, Dodonov AF, Miroshnikov AI. Electrospray ionisation time-of-flight mass spectrometry in protein chemistry. Anal Chem 66 (1994) 99–107.

Mock KK, Davy M, Cottrell JS. The analysis of underivatised oligosaccharides by matrix-assisted laser desorption mass spectrometry. Biochem Biophys Res Commun 177 (1991) 644–651.

Molloy MP. Two-dimensional electrophoresis of membrane proteins using immobilized pH gradients. Anal Biochem 280 (2000) 1–10.

Moritz RL, Eddes JS, Reid GE, Simpson RJ. S-pyridylethylation of intact polyacrylamide gels and in situ digestion of electrophoretically separated proteins: a rapid mass spectrometric method for identifying cysteine-containing peptides. Electrophoresis 17 (1996) 907–917.

Morris HR, Paxton T, Dell A, Langhorne J, Berg M, Bordoli RS, Hoyes J, Bateman RH. High sensitivity collisionally-activated decomposition tandem mass

spectrometry on a novel quadrupole/orthogonal-acceleration time-of-flight mass spectrometer. Rapid Commun Mass Spectrom 10 (1996) 889–896.
Munchbach M, Quadroni M, James P. Quantitation and facilitated de novo sequencing of proteins by isotopic N-terminal labelling of peptides with a fragmentation directing moiety. Anal Chem 72 (2000) 4047–4057.
Neubauer G, King A, Rappsilber J, Calvio C, Watson M, Ajuh P, Sleeman J, Lamond A, Mann M. Mass spectrometry and EST database searching allows characterisation of the multi-protein spliceosome complex. Nature Genet 20 (1998) 46–50.
Neuhoff V, Arold N, Taube D and Ehrhardt W. Improved staining of proteins in polyacrylamide gels including isoelectric focusing gels with clear backgroud at nanogram sensitivity using Coomassie Brilliant Blue G-250 and R-250. Electrophoresis 9 (1988) 255–262.
O'Farrell PH. High-resolution two-dimensional electrophoresis of proteins. J Biol Chem 250 (1975) 4007–4021.
O'Farrell PZ, Goodman HM, O'Farrell PH. High resolution two-dimensional electrophoresis of basic as well as acidic proteins. Cell 12 (1977) 1133–1142.
Oda Y, Huang K, Cross FR, Cowburn D, Chait BT. Accurate quantitation of protein expression and site specific phosphorylation. Proc Ntl Acad Sci USA 96 (1999) 6591–6596.
Ohlmeier S, Scharf C, Hecker M. Alkaline proteins of *Bacillus subtilis:* first steps towards a two-dimensional alkaline master gel. Electrophoresis 21 (2000) 3701–3709.
Pandey A, Mann M. Proteomics to study genes and genomes. Nature 405 (2000) 837–846.
Pappin DJC, Højrup P, Bleasby AJ. Rapid identification of proteins by peptide mass fingerprinting. Curr Biol 3 (1993) 327–332.
Pappin DJC, Rahman D, Hansen HF, Bartlet-Jones M, Jeffery W, Bleasby AJ. Chemistry, mass spectrometry and peptide-mass databases. Evolution of methods for the rapid identification and mapping of cellular proteins. In Mass Spectrometry in the Biological Sciences. Burlingame AL, Carr SA, Eds. (1996) 135–150.
Pappin DJC, Rahman D, Hansen HF, Jeffery W, Sutton CW. Peptide mass fingerprinting as a tool for the rapid identification and mapping of cellular proteins. In Appella E, Atassi MZ, Eds. Methods in protein structure analysis. Kluwer Academic/Plenum Publishers, Dordrecht (1995) 161–173.
Pardo M, Ward M, Bains S, Molina M, Blackstock, Gil C, Nombela C. A proteomic approach for the study of Saccharomyces cerevisiae cell wall biogenesis. Electrophoresis 21 (2000) 3396–3410.
Parness J, Paul-Pletzer K. Elimination of keratin contaminant from 2-mercaptoethanol. Anal Biochem 289 (2001) 98–99.
Pasa-Tolic L. High throughput proteome-wide precision measurements of protein expression using mass spectrometry. J Am Chem Soc 121 (1999) 7949–7950.
Patestos NP, Fauth M, Radola BJ. Fast and sensitive protein staining with colloidal Acid Violet 17 following isoelectric focusing in carrier ampholyte generated and immobilized pH gradients. Electrophoresis 9 (1988) 488–496.
Patterson SD, Aebersold R. Mass spectrometric approaches for the identification of gel-separated proteins. Electrophoresis 16 (1995) 1791–1814.
Patton W. Detecting proteins in polyacrylamide gels and on electroblot membranes. In Proteomics: from protein sequence to function. Pennington SR, Dunn MJ, Eds. Bios Scientific Publishers Ltd. (2001) 65–86.

Pennington SR, Wilkins MR, Hochstrasser DR, Dunn MJ. Proteome analysis: from protein characterization to biological function. Trends Cell Biol 7 (1997) 168–173.
Perkins DN, Pappin DJC, Creasy DM, Cottrell JS. Probability-based protein identification by searching sequence databases using mass spectrometry data. Electrophoresis 20 (1999) 3551–3567.
Pieles U, Zurchner W, Schar M, Moser HE. Matrix-assisted laser desorption time-of-flight mass spectrometry: a powerful tool for the mass and sequence analysis of natural and modified oligonucleotides. Nucleic Acids Res 21 (1993) 3191–3196.
Poduslo JF. Glycoprotein molecular-weight estimetion using sodium dodecyl sulfate – pore gradient electrophoresis: comparison of Tris-glycine and Tris-borate buffer system. Anal Biochem 114 (1981) 131–139.
Posch A, van den Berg BM, Burg HCJ, Görg A. Genetic variability of carrot seed proteins analyzed by one- and two-dimensional electrophoresis with immobilized pH gradients. Electrophoresis 16 (1995) 1312–1316.
Qin J, Herring CJ, Zhang X. De novo peptide sequencing in an ion trap mass spectrometer with ^{18}O labelling. Rapid Commun Mass Spectrom 12 (1998) 209–216.
Qin J, Steenvorden RJJM, Chait BT. A practical ion trap mass spectrometer for the analysis of peptides by matrix-assisted laser desorption/ionisation. Anal Chem 68 (1996) 1784–1791.
Rabilloud T, Gianazza E, Cattò N, Righetti PG. Amidosulfobetaines, a family of detergents with improved solubilization properties: Application for isoelectric focusing under denaturing conditions. Anal Biochem 185 (1990) 94–102.
Rabilloud T. A comparison between low background silver diammine and silver nitrate protein stains. Electrophoresis 13 (1992) 429–439.
Rabilloud T, Valette C, Lawrence JJ. Sample application by in-gel rehydration improves the resolution of two-dimensional electrophoresis with immobilized pH gradients in the first dimension. Electrophoresis 15 (1994) 1552–1558.
Rabilloud T, Vuillard L, Gilly C, Lawrence JJ. Silver staining of proteins in polyacrylamide gels: a general overview. Cell Mol Biol 40 (1994) 57–75.
Rabilloud T. Use of thiourea to increase the solubility of membrane proteins in two-dimensional electrophoresis. Electrophoresis 19 (1998) 758–760.
Rabilloud T. Silver staining of 2-D electrophoresis gels. Methods Mol Biol 112 (1999) 297–305.
Rabilloud T, Blisnick T, Heller M, Luche S, Aebersold R, Lunardi J, Braun-Breton C. Analysis of membrane proteins by two-dimensional electrophoresis: Comparison of the proteins extracted from normal or *Plasmodium falciparum*-infected erythrocyte ghosts. Electrophoresis 20 (1999) 3603–3610.
Rabilloud T, Chevallet M. Solubilization of proteins in two-dimensional electrophoresis. In Rabilloud T, Ed. Proteome research: Two-dimensional gel electrophoresis and identification methods. Springer, Berlin Heidelberg New York (2000) 9–29.
Rabilloud T, Chalmont S. Detection of proteins in two-dimensional gels. In Rabilloud T, Ed. Proteome research: Two-dimensional gel electrophoresis and identification methods. Springer, Berlin Heidelberg New York (2000) 107–126.
Rabilloud T, Strub J-M, Luche S, van Dorsselaer A, Lunardi J. A comparison between Sypro Ruby and ruthenium II tris (bathophenanthroline disulfonate) as fluorescent stains for protein detection in gels. Proteomics 1 (2001) 699–704.

Ramagli LS. Quantifying protein in 2-D PAGE solubilization buffers. In Link AJ. (Ed) 2-D Proteome Analysis Protocols. Methods in Molecular Biology 112. Humana Press, Totowa, NJ (1999) 99–103.

Regula JT, Ueberle B, Boguth G, Görg A, Schnölzer M, Herrmann R, Frank R. Towards a two-dimensional proteome map of *Mycoplasma pneumoniae*. Electrophoresis 21 (2000) 3765–3780.

Righetti PG, Tudor G, Gianazza E. Effect of 2-mercaptoethanol on pH gradients in isoelectric focusing. J Biochem Biophys Methods 6 (1982) 219–227

Righetti PG. In: Work TS, Ed. Burdon RH. Isoelectric focusing: theory, methodology and applications. Elsevier Biomedical Press, Amsterdam (1983).

Righetti PG, Gelfi C. Immobilized pH gradients for isoelectric focusing. III: preparative separations in highly diluted gels. J Biochem Biophys Methods 9 (1984) 103–119.

Righetti PG. In: Burdon RH, van Knippenberg PH. Ed. Immobilized pH gradients: theory and methodology. Elsevier, Amsterdam (1990).

Rimpilainen M, Righetti PG. Membrane protein analysis by isoelectric focusing in immobilized pH gradients. Electrophoresis 6 (1985) 419–422.

Roepstorff P and Fohlman J, Proposal for a common nomenclature for sequence ions in mass spectra of peptides. Biomed Mass Spectrom 11 (1984) 601.

Rosenfield J, Capdevielle J, Guillemot JC, Ferrara P. In-gel digestion of proteins for internal sequence analysis after one- or two-dimensional electrophoresis. Anal Biochem 203 (1992) 173–179.

Rosengren A, Bjellqvist B, Gasparic V. A simple method for choosing optimum pH conditions for electrophoresis. In: Radola BJ, Graesslin D. Ed. Electrofocusing and isotachophoresis. W. de Gruyter, Berlin (1977) 165–171.

Roth KDW, Huang ZH, Sadagopan N, Throck Watson J. Charge derivatisation of peptides for analysis by mass spectrometry. Mass Spectrom Reviews 17 (1998) 255–274.

Sabounchi-Schütt F, Aström J, Olsson I, Eklund A, Grunewald J, Bjellqvist B. An Immobiline DryStrip application method enabling high-capacity two-dimensional gel electrophoresis. Electrophoresis 21 (2000) 3649–3656.

Sanchez J-C, Appel RD, Golaz O, Pasquali C, Ravier F, Bairoch A, Hochstrasser DF. Inside SWISS-2DPAGE database. Electrophoresis 16 (1995) 1131–1151.

Sanchez J-C, Rouge V, Pisteur M, Ravier F, Tonella L, Moosmayer M, Wilkins MR, Hochstrasser DF. Improved and simplified in-gel sample application using reswelling of dry immobilized pH gradients. Electrophoresis 18 (1997) 324–327.

Santoni V, Molloy M, Rabilloud T. Membrane proteins and proteomics: Un amour impossible? Electrophoresis 21 (2000) 1054–1070.

Schägger H, von Jagow G. Tricine-sodium dodecyl sulfate-polyacrylamide gel electrophoresis for the separation of proteins in the range from 1 to 100 kDa. Anal Biochem 166 (1987) 368–379.

Scheele GA. Two-dimensional gel analysis of soluble proteins. Characterization of guinea pig exocrine pancreatric proteins. J Biol Chem 250 (1975) 5375–5385.

Schlosser A, Pipkorn R, Bossemeyer D and Lehman WD; Analysis of protein phosphorylation by a combination of elastase digestion and neutral loss mass spectrometry. Anal Chem 73 (2001) 170–176.

Schnolzer M, Jedrzejewski P, Lehman WD. Protease-catalysed incorporation of ^{18}O into peptide fragments and its application for protein sequencing by electrospray and matrix-assisted laser desorption/ionisation mass spectrometery. Electrophoresis 17 (1996) 945–953.

Sechi S and Chait BT. Modification of cysteine residues by alkylation. A tool in peptide mapping and protein identification. Anal Chem 70 (1998) 5150–5158.
Shevchenko A, Wilm M, Vorm O, Mann M. Mass spectrometric sequencing of proteins silver-stained polyacrylamide gels. Anal Chem 68 (1996a) 850–858.
Shevchenko A, Jensen ON, Podtelejnikov A, Sagliocco F, Wilm M, Vorm O, Mortensen P, Shevchenko A, Boucherie H, Mann M. Linking genome and proteome by mass spectrometry: large scale identification of yeast proteins from two dimensional gels. Proc Natl Acad Sci USA 93 (1996b) 1440–1445.
Shevchenko A, Chemushevich IV, Ens W, Standing KG, Thomson B, Wilm M, Mann M. Rapid 'de novo' peptide sequencing by a combination of nanoelectrospray, isotopic labelling and a quadrupole/time-of-flight mass spectrometer Rapid Commun Mass Spectrom 11 (1997) 1015–1024.
Shevchenko A, Loboda A, Shevchenko A, Ens W, Standing KG. MALDI quadrupole time-of-flight mass spectrometry: A powerful tool for proteomic research. Anal Chem 72 (2000) 2132–2141.
Shi SDH, Hemling ME, Carr SA, Horn DM, Lindh I, McLafferty FW; Phosphopeptide/phosphoprotein mapping by electron capture dissociation mass spectrometry. Anal Chem 73 (2001) 19–22.
Silles E, Mazon MJ, Gevaert K, Goethals M, Vanderkerckhove, Lebr R, Sandoval IV. Targeting of aminopeptidase I to the yeast vacuole is mediated by Ssa1p, a cytosolic member of the 70 kDa stress protein family. J Biol Chem 275 (2000) 34054–34059.
Sinha P, Köttgen E, Westermeier R, Righetti PG. Immobilized pH 2.5–11 gradients for two-dimensional electrophoresis Electrophoresis 13 (1992) 210–214.
Sinha P, Poland J, Schnolzer M, Rabilloud T. A new silver staining apparatus and procedure for matrix-assisted laser desorption/ionisation-time of flight analysis of proteins after two-dimensional electrophoresis. Proteomics 1 (2001) 835–840.
Skilling J, Cottrell J, Green B, Hoyes J, Kapp E, Landgridge J, Bordoli B; Automated de novo peptide sequencing using Q-Tof ESI-MS/MS and software interpretation of the data. Proc. 47th ASMS conference Mass spectrometry and allied topics, Dallas (1999).
Smolka M, Zhou H, Aebersold R. Quantitative protein profiling using two-dimensional gel electrophoresis, isotope coded affinity tag labeling and mass spectrometry. Mol Cell Proteomics 1 (2002) 19–29.
Spengler B, Luetzenkirchen F, Metzger S, Chaurand P, Kaufmann R, Jeffrey W, Bartlet-Jones M, Pappin DJC. Peptide sequencing of charged derivatives by postsource decay MALDI mass spectrometry. Int J Mass Spectrom Ion Proc 169/170 (1997) 127–140.
Spengler B; Post-source decay analysis in matrix-assisted laser desorption/ionization mass spectrometry of biomolecules. J Mass Spectrom 32 (1997) 1019–1036.
Stafford GC, Kelley PE, Syka JEP, Reynolds WE, Todd, JFJ. Int J Mass Spectrom Ion Proc 60 (1984) 85–98.
Stahl B, Steup M, Karas M, Hillenkamp F. Analysis of neutral oligosaccharides by matrix-assisted laser desorption/ionisation mass spectrometry. Anal Chem 63 (1991) 1463–1466.
Stasyk T, Hellman U, Souchelnytskyi S. Optimizing sample preparation for 2-D electrophoresis. Life Science News 9 (2001) 9–12.
Staudenmann W, Dainese-Hatt P, Hoving S, Lehman A, Kertesz M, James P. Sample handling for proteome analysis. Electrophoresis 19 (1998) 901–908.
Steen H, Küster B, M Fernandez, Pandy A, Mann M. Characterisation of STAM2 using a new method for tyrosine phosphorylation analysis by tandem mass

spectrometry. Proceedings of the 49th ASMS conference on mass spectrometry and allied topics, Chicago (2001).

Stegemann H. Proteinfraktionierungen in Polyacrylamid und die Anwendung auf die genetische Analyse bei Pflanzen. Angew Chem 82 (1970) 640.

Steinberg TH, Hangland RP, Singer VI. Applications of SYPRO Orange and SYPRO Red protein gel stains. Anal Biochem 239 (1996) 238–245.

Stensballe A and Jensen ON; Simplified sample preparation method for protein identification by matrix-assisted laser desorption/ionisation mass spectrometry: In-gel digestion on the probe surface. Proteomics 1 (2001) 955–966.

Strahler JR, Hanash SM, Somerlot L, Weser J, Postel W, Görg A. High-resolution two-dimensional polyacrylamide gel electrophoresis of basic myeloid polypeptides: Use of immobilized pH gradients in the first dimension. Electrophoresis 8 (1987) 165–173.

Strupat, K., Karas, M., and Hillenkamp, F. 2,5-Dihydroxybenzoic acid: a new matrix for laser desorption-ionisation mass spectrometry. Int J Mass Spectrom Ion Proc 111 (1991) 89–102.

Sutton CW, Pemberton KS, Cottrell JS, Corbett JM, Wheeler CH, Dunn MJ, Pappin DJ. Identification of myocardial proteins from two-dimensional gels by peptide mass fingerprinting. Electrophoresis 16 (1995) 308–316.

Svensson H. Isoelectric fractionation , analysis and characterization of ampholytes in natural pH gradients. The differential equation of solute concentrations as a steady state and its solutin for simple cases. Acta Chem Scand 15 (1961) 325–341.

Taylor G. Disintegration of water drops in an electric field. Proc R Soc Lond A280 (1964) 383–397.

Thiellement H, Bahrman N, Damerval C, Plomion C, Rossignol M, Santoni V, de Vienne D, Zivy M. Proteomics for genetic and physiological studies in plants. Electrophoresis 20 (1999) 2013–2026.

Tugal T, Zou-Yang XH, Gavin K, Pappin D, Canas B, Kobayashi R, Hunt T, Stillman B. The Orc4p and Orc5p Subunits of the Xenopus and human origin recognition complex are related to Orc1p and Cdc6p. J Biol Chem 273 (1998) 32421–32429.

Uetz P, Giot L, Cagney G, Mansfield TA, Judson RS, Knight JR, Lockson D, Narayan V, Srinivasan M, Pochart P, Qureshi-Emili A, Li Y, Godwin B, Conover D, Kalbfleisch T, Vijayadamodar G, Yang M, Johnston M, Fields S, Rothberg JM. A comprehensive analysis of protein-protein interactions in *Saccharomyces cerevisiae.* Nature 403 (2000) 623–627.

Ûnlü M, Morgan EM, Minden JS. Difference gel electrophoresis: a single gel method for detecting changes in protein extracts. Electrophoresis 19 (1997) 2071–2077.

Urwin V, Jackson P. Two-dimensional polyacrylamide gel electrophoresis of proteins labeled with the fluorophore monobromobimane prior to first-dimensional isoelectric focusing: imaging of the fluorescent protein spot patterns using a cooled charge-coupled device. Anal. Biochem 209 (1993) 57–62.

Uttenweiler-Joseph S, Neubauer G, Christoforidis S, Zerial M, Wilm W. Automated *de novo* sequencing of proteins using the differential scanning technique. Proteomics 1 (2001) 668–682.

Vandahl BB, Birkelund S, Demol H, Hoorelbeke B, Christiansen G, Vandekerckhove J, Gevaert K. Proteome analysis of the Chlamydia pneumoniae elementary body. Electrophoresis 22 (2001) 1204–1223.

Vandekerckhove J, Bauw G, Puype M, Van Damme J, Van Montegu M. Protein blotting on polybrene coated glass-fiber sheets: a basis for acid hydrolysis and gas phase sequencing of picomole quantities of protein previously separated

on sodium dódecyl sulfate polyacrylamide gel. Eur J Biochem 152 (1985) 9–19.
Verentchikov AN, Ens W, Standing KG. Reflecting time-of-flight mass spectrometer with an electrospray ion source and orthogonal extraction. Anal Chem 66 (1994) 126–133.
Vestal ML, Juhasz P, Martin SA. Delayed extraction matrix-assisted laser desorption time-of-flight mass spectrometry. Rapid Commun Mass Spectrom 9 (1995) 1044–1050.
Vesterberg, O. Synthesis and isoelectric fractionation of carrier ampholytes. Acta Chem. Scand 23 (1969) 2653–2666.
von Haller PD, Donohoe S, Goodlett DR, Aebersold R, Watts JD. Mass spectrometric characterisation of proteins extracted from Jurkat T cell detergent-resistant membrane domains. Proteomics 1 (2001) 1010–1021.
Vorm O, Roepstorff P, Mann M. Improved resolution and very high sensitivity in MALDI-TOF of matrix surfaces made by fast evaporation. Anal Chem 66 (1994) 3281–3287.
Vuillard L, Marret N, Rabilloud T. Enhancing protein solubilization with nondetergent sulfobetains. Electrophoresis 16 (1995) 295–297.
Wagner H, Kuhn R, Hofstetter S. Praxis der präperativen Free-Flow-Elektrophorese. In: Wagner H, Blasius E. Ed. Praxis der elektrophoretischen Trennmethoden. Springer-Verlag, Heidelberg (1989) 223–261.
Washburn MP, Wolters D, Yates JR III. Large scale analysis of the yeast proteome by multidimensional protein identification technology. Nature Biotechnol 19 (2001) 242–247.
Wasinger VC, Cordwell SJ, Cerpa-Poljak A, Yan JX, Gooley AA, Wilkins MR, Duncan MW, Harris R, Williams KL, Humphery-Smith I. Progress with gene-product mapping of the mollicutes: *Mycoplasma genitalium*. Electrophoresis 16 (1995) 1090–1094.
Weber K, Osborn M. The reliability of molecular weight determinations by dodecyl sulphate-polyacrylamide gel electrophoresis. J Biol Chem 244 (1968) 4406–4412.
Weiss W, Vogelmeier C, Görg A. Electrophoretic characterization of wheat grain allergens from different cultivars involved in baker's asthma. Electrophoresis 14 (1993) 805–816.
Wenger P, de Zuanni M, Javet P, Righetti PG. Amphoteric, isoelectric Immobiline membranes for preparative isoelectric focusing. J Biochem Biophys Methods 14 (1987) 29–43.
Wessel D, Flügge UI. A method for the quantitative recovery of protein in dilute solution in the presence of detergents and lipids. Anal Biochem 138 (1984) 141–143.
Westermeier R. Electrophoresis in Practice. WILEY-VCH, Weinheim (2001).
Westermeier R, Postel W, Weser J, Görg A. High-resolution two-dimensional electrophoresis with isoelectric focusing in immobilized pH gradients. J Biochem Biophys Methods 8 (1983) 321–330.
Whittal RM, Li L. High resolution matrix-assisted laser desorption ionisation in a linear time-of-flight mass spectrometer. Anal Chem 67 (1995) 1950–1954.
Wildgruber R, Harder A, Obermaier C, Boguth G, Weiss W, Fey SJ, Larsen PM, Görg A. Towards higher resolution: Two-dimensional electrophoresis of Saccharamyces cerevisiae proteins using overlapping narrow immobilized pH gradients. Electrophoresis 21 (2000) 2610–2616.
Wiley WC and McLaren IH. Time-of-flight mass spectrometer with improved resolution. Rev Sci Instrum 26 (1955) 1150–1157.

Williams RCJ, Shah C, Sackett D. Separation of tubulin isoforms by isoelectric focusing on immobilized pH gradient gels. Anal Biochem 275 (1999) 265–267.

Wilm M, Mann M. Analytical properties of the nanoelectrospray source. Anal Chem 68 (1996) 1–8.

Wilm M, Neubauer G, Mann M. Parent ions scans of unseparated peptide mixtures. Anal Chem 68 (1996) 527–533.

Wilm M, Shevchenko A, Houthaeve T, Breit S, Schweigerer L, Fotsis T, Mann M. Femtomole sequencing of proteins from polyacrylamide gels by nano-electrospray mass spectrometry. Nature 379 (1996) 466–469.

Yan JX, Harry RA, Spibey C, Dunn MJ. Postelectrophoretic staining of proteins separated by two-dimensional gel electrophoresis using SYPRO dyes. Electrophoresis 21 (2000) 3657–3665.

Yan JX, Wait R, Berkelman T, Harry RA, Westbrook JA, Wheeler CH, Dunn MJ. A modified silver staining protocol for visualization of proteins compatible with matrix-assisted laser desorption/ionization and electrospray ionization-mass spectrometry. Electrophoresis 21 (2000) 3666–3672.

Yates JR III, Speicher S, Griffin PR, Hunkapiller T; Peptide mass maps: a highly informative approach to protein identification. Anal Biochem 214 (1993) 397–408.

Yates JR III, Eng JK, McCormack AL, Schieltz D; Method to correlate tandem mass spectra of modified peptides to amino acid sequences in the protein database. Anal Chem 67 (1995) 1426–1436.

Yun M, Mu W, Hood L, Harrington MG. Human cerebrospinal fluid protein database: edition 1992. Electrophoresis 13 (1992) 1002 –1013.

Zerr I, Bodemer M, Otto M, Poser S, Windl O, Kretzschmar HA, Gefeller O, Weber T. Diagnosis of Creutzfeldt-Jacob disease by two-dimensional gel electrophoresis of cerebrospinal fluid. Lancet 348 (1996) 846–849.

Zhou G, Li H, DeCamp D, Chen S, Shu H, Gong Y, Flag M, Gillespie J, Hu N, Taylor P, Buck ME, Liotta LA, Petricoin III EC, Zhao Y. 2-D differential in-Gel electrophoresis for the identification of human esophageal squamous cell cancer specific protein markers. Mol Cell Proteomics (2002) in press.

Index

THE PREMIER INTERNATIONAL SOURCE FOR INFORMATION IN THE FIELD OF PROTEOMICS

PROTEOMICS

Integrates

- the methodological developments in protein separation and characterization
- the advances in bioinformatics
- the novel applications of proteomics

in all areas of the life sciences and industry.

NEW JOURNAL

First Issue in January 2001

■SOMETHING OLD –
■SOMETHING NEW

■SOMETHING RED –
■SOMETHING BLUE

EDITOR-IN-CHIEF:
Michael J. Dunn
Imperial College, London, UK

Available as a separate journal and also as a supplement of *Electrophoresis*.

WILEY-VCH ■ Journals Department ■ P.O. Box 10 11 61 ■ 69451 Weinheim ■ Germany ■ Fax: +49 (0) 6201 606 117 ■ e-mail: subservice@wiley-vch.de ■ http://www.wiley-vch.de

Order Details at:
www.interscience.wiley.com/Proteomics

WILEY-VCH

71131122_bu